PALGRAVE STUDIES IN THE HISTORY OF
SCIENCE AND TECHNOLOGY

James Rodger Fleming (Colby College) and Roger D. Launius (National Air and Space Museum), Series Editors

This series presents original, high-quality, and accessible works at the cutting edge of scholarship within the history of science and technology. Books in the series aim to disseminate new knowledge and new perspectives about the history of science and technology, enhance and extend education, foster public understanding, and enrich cultural life. Collectively, these books will break down conventional lines of demarcation by incorporating historical perspectives into issues of current and ongoing concern, offering international and global perspectives on a variety of issues, and bridging the gap between historians and practicing scientists. In this way they advance scholarly conversation within and across traditional disciplines but also to help define new areas of intellectual endeavor.

Published by Palgrave Macmillan:

Continental Defense in the Eisenhower Era: Nuclear Antiaircraft Arms and the Cold War
By Christopher J. Bright

Confronting the Climate: British Airs and the Making of Environmental Medicine
By Vladimir Jankovic´

Globalizing Polar Science: Reconsidering the International Polar and Geophysical Years
Edited by Roger D. Launius, James Rodger Fleming, and David H. DeVorkin

Eugenics and the Nature-Nurture Debate in the Twentieth Century
By Aaron Gillette

John F. Kennedy and the Race to the Moon
By John M. Logsdon

A Vision of Modern Science: John Tyndall and the Role of the Scientist in Victorian Culture
By Ursula DeYoung

Searching for Sasquatch: Crackpots, Eggheads, and Cryptozoology
By Brian Regal

Inventing the American Astronaut
By Matthew H. Hersch

The Nuclear Age in Popular Media: A Transnational History
Edited by Dick van Lente

Exploring the Solar System: The History and Science of Planetary Exploration
Edited by Roger D. Launius

The Sociable Sciences: Darwin and His Contemporaries in Chile
By Patience A. Schell

The First Atomic Age: Scientists, Radiations, and the American Public, 1895–1945
By Matthew Lavine

NASA in the World: Fifty Years of International Collaboration in Space
By John Krige, Angelina Long Callahan, and Ashok Maharaj

NASA in the World

Fifty Years of International Collaboration in Space

John Krige,
Angelina Long Callahan,
and
Ashok Maharaj

First published in 2013 by
PALGRAVE MACMILLAN®
in the United States—a division of St. Martin's Press LLC,
175 Fifth Avenue, New York, NY 10010.

Where this book is distributed in the UK, Europe and the rest of the world,
this is by Palgrave Macmillan, a division of Macmillan Publishers Limited,
registered in England, company number 785998, of Houndmills,
Basingstoke, Hampshire RG21 6XS.

Palgrave Macmillan is the global academic imprint of the above companies
and has companies and representatives throughout the world.

Palgrave® and Macmillan® are registered trademarks in the United States,
the United Kingdom, Europe and other countries.

ISBN: 978–1–137–34091–7 (hc)
ISBN: 978–1–137–34092–4 (pb)

Library of Congress Cataloging-in-Publication Data

Krige, John.
    NASA in the world : fifty years of international collaboration in space / John
Krige, Angelina Long Callahan, and Ashok Maharaj.
        pages cm.—(Palgrave studies in the history of science and technology)
    "This work was supported by NASA Contract No. NNH07CC56C."
    ISBN 978–1–137–34091–7 (alk. paper)—
    ISBN 978–1–137–34092–4 (alk. paper)
        1. United States. National Aeronautics and Space Administration—History.
    2. Astronautics—International cooperation. 3. Astronautics—United States—
    History. 4. Outer space—Exploration—United States—History. I. Callahan,
    Angelina Long, 1981– II. Maharaj, Ashok, 1975– III. Title.

TL521.312.N3725 2013
629.40973—dc23   2013014745

A catalogue record of the book is available from the British Library.

Design by Newgen Knowledge Works (P) Ltd., Chennai, India.

First edition: August 2013

10 9 8 7 6 5 4 3 2 1

Contents

Illustrations

Tables

Figures

Acknowledgments

Our first debt of gratitude goes to Steve Dick and Steve Garber in the NASA History Office. In selecting this team for this project they not only placed their faith in experience. They sought also to extend and consolidate the community of space historians, and gave two young scholars the opportunity to grow personally and intellectually. Bill Barry provided extensive and invaluable feedback on an earlier version of this text when he was appointed NASA chief historian, and did all he could to facilitate its publication with an academic press.

The advice and generous help of the archivists in the NASA History Office made the product possible. Jane Odom was not only our guide and counselor. She also had the task of frequently dealing with the formalities needed for two of us to enter the NASA Headquarters Building. Nadine Andreassen provided invaluable additional support. Collin Fries, John Hargenrader, and Elizabeth Suckow pulled file after file from storage cabinets, tracked down obscure references, and lived with four huge boxes of our material on the floor of the archive for months on end. This study has benefited extensively from their knowledge of the primary source material, their imagination in suggesting new leads, and their generous availability whenever we called on them. Thanks too to Patrick Reust of the World Meteorological Organization Library, Geneva, and Betsey Stout and Elizabeth Rogers of the Marriott Library Special Collections at the University of Utah for their advice and help.

Primary sources were supplemented with about two dozen formal interviews, now transcribed and available in the NASA Historical Reference Collection, and with many informal discussions with eminent members of the space community. Those interviewed included many dedicated NASA staff who gave generously of their time. All interviewees are acknowledged individually in the "List of Interviews": their insights added important texture to the documentary record. John Hall and Paula Geisz deserve a special word of thanks for their careful reading, and rereading, of the chapter on export control compliance.

The Georgia Institute of Technology was an immensely supportive environment for doing this work. We want to thank Chris Fehrenbach, above all, for administering this contact with consummate skill and good humor. Janis Goddard, in the Office of Sponsored Programs, was our interface with NASA itself. Katharine Calhoun, Gladys Toppert, and Bruce Henson in the library made untiring efforts to get books, articles, and dissertations in record time. The faculty and the graduate community in the School of History, Technology and Society were always engaged and willing to give advice and feedback. We particularly want to thank Ron Bayor, Laura Bier, Doug Flamming, Prakash Kumar, Ken Knoespel, Nathan Moon, Jenny Smith, Tim Stoneman, and Steve Usselman whose insight and personal encouragement was invaluable.

Preliminary versions of our findings have been presented at many conferences, workshops, and university seminars. Engaged audiences were constructive critics. We owe a particular debt of gratitude to Jacques Blamont, Roger Bonnet, Jason Callahan, Jean Pierre Causse, Indira Choudhary, Martin Collins, Mark Finlay, James Flemming, Kristine Harper, Tom Lassman, Pam Mack, Ian Pryke, Yasushi Sato, Asif Siddiqi, Raman Srinivasan, Roger Turner, and Odd Arne Westad for thoughtful comments and criticisms on various incarnations of this research.

In transiting from a NASA publication to an academic press we owe an immense debt of gratitude to Roger Launius and Jim Fleming, who immediately agreed to include it in their series. Chris Chappell and Sarah Whalen at Palgrave Macmillan provided outstanding editorial support.

This book would not have been possible without the unstinting support, professional and personal, that many colleagues, friends and family have provided. Special friends too numerous to mention have stood by us through thick and thin. And then there are those who have intimately shared our anxieties and our excitement—Lydie, and David, Peter, Simon, and Sara; Adrienne, Cyndi, Frances, and Jason; Dolly and Nate, and Appa, Amma, Kannan, Rajesh, and Anand. We could not have done this without knowing that you were there.

Archives Consulted

Federal Records Center, Suitland, Washington, DC
The Historical Archives of the European Union, Florence, Italy
Jimmy Carter Presidential Library, Atlanta, GA
Library of Congress Manuscripts Division, Washington, DC
Lyndon Baines Johnson Presidential Library, Austin, TX
NASA Historical Reference Collection, NASA Headquarters, Washington, DC
National Archives and Records Administration, College Park, MD
The National Archives, London, England
Nehru Memorial Museum and Library, New Delhi, India
North Carolina State University—Special Collections Research Center, Raleigh, NC
Space Application Center Library, Ahmedabad, India
Vikram Sarabhai Archives, Ahmedabad, India

Interviews

The following interviews were conducted in the framework of this project and have been transcribed and lodged in the NASA Historical Reference Collection, NASA HQ, Washington, DC.

Richard Barnes, with John Krige, Washington, DC, March 27, 2007

Richard Barnes, with Ashok Maharaj, Washington, DC, April 28, 2008

Karin Barbance, with John Krige, Paris, June 22, 2007

Jacques Blamont, with John Krige, CNES, Paris, July 4, 2007

John Casani, with John Krige, Pasadena, May 18, 2009

Lynn Cline, with John Krige, NASA HQ, Washington, DC, March 30, 2009

Peter Creola, with John Krige, Berne, May 25, 2007

Charles Elachi, with John Krige, Pasadena, June 10, 2009

Margaret Finarelli, with John Krige, Washington, DC, April 20, 2010

Arnold Frutkin with John Krige, accompanied by Angelina Long and Ashok Maharaj, Virginia, August 19, 2007

Roy Gibson, with John Krige, Montpelier, June 15, 2007

Andre Lebeau, with John Krige, Paris, June 4, 2007

Reinhardt Loosch, with John Krige, Bonn, June 29, 2007

Fernando de Mendonça, with Ashok Maharaj, Washington, DC, April 28, 2008

Robert Mitchell, with John Krige, Pasadena, June 19, 2009

Michael O'Brien, with John Krige, NASA HQ, Washington, DC, March 30, 2009

Karl Reuter, with John Krige, Munich, June 17, 2007

John Schumacher, with John Krige, Washington, DC, April 19, 2010

David Southwood, with John Krige, ESA, Paris, July 16, 2007

Substantial informal exchanges were also had with the following persons:

With John Krige

Wolfgang Finke, Bonn

Doug Millar, NASA HQ, Washington, DC

With Ashok Maharaj

Binod C. Agrawal, Anthropologist, ISRO, SITE Project

Prof. Asnani, Senior Scientist, ISRO, Remote Sensing, based in Pune, near Bombay

Dr. Bhatia, Retired Scientist, ISRO, based in Ahmedabad

Prof. Bhavsar, Retired Scientist, ISRO, based in Ahmedabad

Jacques Blamont, CNES, Paris

Chandrasekhar, Former Scientist, ISRO, based in Bangalore

E. V. Chitnis, Former Director of Space Application Center, ISRO, based in Pune

Padmanabh Joshi, Retired Scientist, ISRO, based in Ahmedabad

Abdul Kalam, Former Director of ISRO, Former President of India, based in Delhi

Pramod Kale, Former Director of Space Application Center, ISRO, based in Pune

K. Kasturirangan, Former Director of ISRO, currently Member of the Parliament (MP, upper house), based in New Delhi

Yash Pal, Retired Scientist, ISRO, based in Delhi

N. Pant, Retired Scientist, ISRO

Radhakrishnan, Retired Scientist/Payload Specialist, ISRO

Gopal Raj, Senior Science Correspondent, *The Hindu*, Trivandrum, Kerala

Manoranjan Rao, Retired Scientist, ISRO, Official Historian of ISRO

U. R. Rao, Former Chairman of ISRO

Prof. Vasagam, Project Leader for Apple Satellite, ISRO

Abbreviations

ACDA	Arms Control and Disarmament Agency
ACRV	Assured Crew Return Vehicle
AO	Announcement of Opportunity
ARC	Ames Research Center
ASRM	Advanced Solid Rocket Motor
ASTP	Apollo-Soyuz Test Project
ATS	Application Technology Satellite
CCL	Commerce Control List
CEPT	European Conference of Postal and Telecommunications Administrations
CERN	European Organization for Nuclear Research, *now* European Laboratory for Particle Physics
CETS	European Conference on Satellite Communications
CIA	Central Intelligence Agency
CIS	Commonwealth of Independent States
CNES	Centre nationale d'études spatiales
CNET	Centre nationale d'études de télécommunications
COPUOS	Committee on the Peaceful Uses of Outer Space
COSPAR	Committee on Space Research
CRA	Centro Ricerche Aerospaziali
CRAF	Comet Rendezvous Asteroid Flyby
DAE	Department of Atomic Energy, India
DOD	US Department of Defense
DOS	Department of Space, India
DRS	Direct Reception System
DVL	Deutsche Versuchsanstalt für Luftfahrt
EAR	Export Administration Regulations
EBU	European Broadcasting Union
EEC	European Economic Community
ELDO	European (Space Vehicle) Launcher Development Organization
ESA	European Space Agency
ESC	European Space Conference
ESRO	European Space Research Organization
ESTEC	European Space Research and Technology Centre
ETF	Environmental Task Force

EVA	Extravehicular Activity
FAA	Federal Aviation Authority
FCC	Federal Communications Commission
FGB	Functional Cargo Blok
FOC	Faint Object Camera
GARP	Global Atmospheric Research Program
GCC	Gore-Chernomyrdin Commission
GHOST	Global Horizontal Sounding Technique
GSFC	Goddard Space Flight Center
IAEA	International Atomic Energy Agency
IAU	International Astronomical Union
ICSC	Interim Communications Satellite Committee
IGA	Intergovernmental Agreement
IGY	International Geophysical Year
IIOE	International Indian Ocean Experiment
IIS	Institute for Industrial Science, Japan
IISC	Indian Institute of Science
IKI	Moscow Space Research Institute
INA	Iran Nonproliferation Act
INAS	Indian National Academy of Science
INCOSPAR	Indian National Committee for Space Research
INMARSAT	International Maritime Satellite Organization
INSAT	Indian National Satellite
Intelsat	International Telecommunications Satellite Organization
IQSY	International Quiet Sun Year
ISAS	Institute of Space and Aeronautical Science, Japan
ISPM	International Solar-Polar Mission
ISRO	Indian Space Research Organization
ISS	International Space Station
ITAR	International Traffic in Arms Regulations
JAXA	Japanese Aerospace Exploration Agency
JPL	Jet Propulsion Laboratory, Pasadena
JWG	Joint Working Group
LBJL	Lyndon B. Johnson Presidential Archives, University of Austin, Austin, TX
LTV	Ling-Temco-Vought
MIBP	Moscow Institute of Biomedical Problems
MIT	Massachusetts Institute of Technology
MOM	(Russian) Ministry of General Machine Building
MOU	Memorandum of Understanding
MTCR	Missile Technology Control Regime
MTPE	Mission to Planet Earth
NARA	National Archives and Records Administration, College Park, MD
NASA	National Aeronautics and Space Administration
NASDA	National Aeronautics and Space Development Agency, Japan
NATO	North Atlantic Treaty Organization
NHRC	NASA Historical Records Collection, NASA HQ, Washington DC

NIS	Newly Independent States
NOAA	National Oceanic and Atmospheric Administration
NPT	Nuclear Nonproliferation Treaty
NPO	(Russian) Scientific Production Association
NRSA	National Remote Sensing Agency, India
NSAM	National Security Action Memorandum
NSDC	National Space and Development Center, Japan
NSDM	National Security Decision Memorandum
NSSM	National Security Study Memorandum
OMB	Office of Management and Budget
OMC	Office of Munitions Control
OTA	Office of Technology Assessment
PNE	Peaceful Nuclear Explosion
PRC	People's Republic of China
PRL	Physical Research Laboratory, Ahmedabad
RAM	Research Applications Model
RAS	Russian Academy of Sciences
RKA	Roskosmos
RMS	Remote Manipulator System (Canadarm)
RSA	Russian Space Agency
SAC	Space Application Center Library, Ahmedabad, India
SDI	Strategic Defense Initiative
SEATO	Southeast Asia Treaty Organization
SITE	Satellite Instructional Television Experiment
SLV	Satellite Launch Vehicle
SPOT	Système pour l'observation de la terre
SSF	Space Station Freedom
STA	Science and Technology Agency, Japan
STG	Space Task Group
STL	Space Technology Laboratories of TRW
STS	Space Transportation System
TAG	Technology Advisory Group
TERLS	Thumba Equatorial Rocket Launching Station
TIFR	Tata Institute of Fundamental Research, India
TOMS	Total Ozone Mapping Spectrometer
TPS	Thermal Protection System
TRW	Thompson Ramo Wooldridge
TsKB	Research and Production Rocket Space Center
UN	United Nations
US	United States of America
USAF	United States Air Force
USML	United States Munitions List
USSR	Union of Soviet Socialist Republics
WNRC	Washington National Records Center, Suitland, MD
WWP	World Weather Program
WWW	World Weather Watch

Part I

Fifty Years of NASA and the World

John Krige

Chapter 1

Introduction and Historical Overview: NASA's International Relations in Space

"That's one small step for man, one giant leap for mankind." These "eternally famous words," as James Hansen calls them in his biography of Neil Armstrong, expressed both a NASA and an American triumph.[1] They also reached out to the millions watching the spectacle on television screens all over the world, allowing them to make it their own. About 30 minutes into the mission, and shortly after having been joined by Buzz Aldrin, Armstrong read the words on a plaque attached to one of the ladder legs of the lunar module. The *Eagle*—a name deliberately chosen by the astronauts as the symbol of America—had no territorial ambitions: as Armstrong said, "We came in peace for all mankind."[2] "For one priceless moment in the history of man," President Nixon told the astronauts as they explored the lunar surface, "all the people on this earth are truly one ... "[3]

The spectacles of the moon landing and the moonwalk are suffused with quintessentially American tropes: white, athletic males burst the grip of gravity to conquer a new frontier.[4] All the same, we should not be overwhelmed by the political and ideological staging of Apollo 11 as an American-led achievement in the context of Cold War competition. For the mission also had genuine international components. Beginning with Apollo 11, NASA astronauts collected over 840 pounds of moon rock, and distributed hundreds of samples for public viewing and scientific research all over the world.[5] The first video images of Armstrong's and Aldrin's steps on the moon were picked up, not in the United States, but by antennae at Honeysuckle Creek and the Parkes Observatory near Canberra in Australia, a tribute to the vast global data and tracking network that supports NASA's missions.[6] And one of the few scientific experiments conducted on the lunar surface during Armstrong and Aldrin's 160-odd minutes of surface activity on the night of July 20, 1969, had a foreign principal investigator.

During their brief sojourn on the moon the astronauts engaged in six scientific experiments, all chosen by a NASA scientific panel for their interest and excellence. Five of these were part of the Early Apollo Scientific Experiment Package. They included a passive seismometer to analyze lunar structure and detect moonquakes, and a device to measure precisely the distance between the moon and the earth. The sixth was an independent Solar Wind Composition Experiment submitted from abroad. To perform this experiment the astronauts had to unroll a banner of thin aluminum metal foil about 12 inches wide by 55 inches long, and orient one side of it toward the sun. The foil trapped the ions of rare gases emitted from the fireball. It was brought back to earth in a teflon bag, sent to Europe, cleaned ultrasonically, and melted in an ultra-high vacuum, releasing the gases that were analyzed in a mass spectrometer.[7] The results provided insights into the dynamics of the solar wind, the origin of the solar system, and the history of planetary atmospheres.

Johannes Geiss, a leading Swiss scientist, was responsible for this experiment. The payload was manufactured at Geiss's University of Bern and was paid for by the Swiss National Science Foundation.[8] What is more, apart from Armstrong's contingency collection of lunar samples immediately on emerging from the lunar module, this was the first experiment deployed by the astronauts. Indeed, to ensure that the foil was exposed to the sun for as long as possible, it was even deployed *before* Armstrong and Aldrin planted the American flag in the lunar surface and spoke to the president. Scientific need trumped political and ideological statement. NASA's commitment to international cooperation could not be expressed by having the flags of many countries, or perhaps just the flag of the United Nations, left on the moon. Congress decided that this was an American project and that the astronauts would plant the US flag.[9] Instead NASA's international agenda fused seamlessly with the "universalism" of science to create a niche for flying an experiment built by a university group in a small, neutral European country.

It is striking that even though the Solar Wind Experiment is routinely mentioned in writings on the Apollo 11 mission, the European source of the experiment is not.[10] This is partly because of the iron grip human space flight has on the imagination, a mindset constructed by enthusiasts whose shrill voices and skillful marketing have capitalized on the frontier myth that is deeply ingrained in America's sense of itself and its destiny, so playing down alternative, less glamorous visions of spaceflight using benign technologies.[11] It is the challenges faced by the astronauts as they conquer new domains, not the scientific content of the Apollo missions, that resonate culturally, that entertain and inspire, that showcase American technological success and project American power abroad.

The foreign contribution to Apollo 11 is also ignored because so much space history in the United States—as in all space-faring nations—is nationalistic and celebratory, a symptom of the high value placed on technological achievement as a marker of national prowess. Today historians are increasingly aware of the need to situate national narratives in transnational or global frameworks, in recognition of the interdependence and interconnectivity of the modern state. This focus is all the more important in the case of NASA since the Space Act of 1958 both mandated the new agency to secure American space leadership and to pursue an active program of international cooperation. An emphasis on purely national narratives occludes one of the agency's core activities.

There have been many scholarly studies of various aspects of NASA's international relations. They have two dominant features. First, they concentrate on a single project or program (Germany's Helios probe to the sun,[12] the Satellite Instructional Television Experiment SITE developed with India, the Apollo-Soyuz Test Project, the International Space Station[13]), like so much of space history itself. Second, they mostly treat the political and diplomatic context in which NASA engages in international collaboration as a taken-for-granted backdrop. NASA's international activities are seen as subsidiary to its prime mission of building US space leadership. Its history is defined as a history of the agency's ability to secure resources for that mission from Congress and the American people, and to bring its scientific and technological ambitions to fruition (or not, as the case may be).

This book takes a different approach. It covers 50 years of NASA's international relations, and although it is necessarily mission-oriented—for it is around missions that NASA organizes collaboration—it selects from the vast panorama of these missions those that reveal the different scientific and technological *but also political, industrial, and ideological* rationales for embarking on particular space ventures with foreign partners (including the Soviet Union). This book treats NASA as an organization dedicated to the exploration of space that acts in a complex foreign policy context whose definition is itself fluid and contested both at home and abroad. The authors are not only interested in NASA as a national space agency, then, but in NASA as an actor in the world, in NASA as the bearer and defender of American interests on the world stage. They explore the articulation between the pursuit of scientific, technological, and industrial preeminence in space and the consolidation of American global leadership, the intersection between space science and technology and international relations.

One dominant thread runs through the analysis, and shapes some of the key questions we address. Simply put it is this: how did NASA reconcile America's conquest of space with its collaborative activities? How did it harmonize the pursuit of space leadership, premised on scientific and technological leadership, with the increasingly insistent demands of foreign partners to have meaningful access to American scientific, technological, and industrial know-how? Almost since its inception, the exploration and exploitation of space has not been a level playing field: the United States, despite some spectacular Soviet firsts, has always been the leading spacefaring nation on the globe. This means that NASA has had to devise policies to protect US industrial competitiveness and national security while, at the same time, engaging in suitably advanced levels of scientific, technological, and industrial cooperation to satisfy its partners. It had to strengthen the programs of the free world, and sustain civil relationships with its communist rivals, without seriously undermining its position as the world's leading space agency.

Harmonizing leadership with collaboration was an ongoing process: though certain general principles were quickly laid down by NASA to shape the engagement, their implementation in practice varied depending on the nature of the mission (science, applications, technology, especially launcher technology), the space strengths of the other (a threat but also a resource to draw on to enhance US capabilities), and the political and ideological stakes involved. American global leadership in any domain is not a given. It requires ongoing work, and an ability to adjust to the changing balance of power between the United States and its partners in all of the domains in which NASA was engaged.

Knowledge is the key site around which international collaboration is organized in much of this study. Knowledge, for our purposes, is not restricted to propositional knowledge, of course, but is also embedded in multiple material substrates, including technology, and is embodied in diverse human skills, including project management. International collaboration involves the management of such flows across the interface between US entities and their partners. It also called for the transfer of knowledge embedded in environmental, capitalistic, and trade regimes that were deployed to restructure the ex-Soviet space system in the 1990s. The policies that NASA put in place to manage cross-border flows of knowledge of all kinds define the dynamic equilibrium between scientific and technological denial, on the one hand, and controlled assistance and collaboration on the other. They constitute the sinews of international collaboration in a domain as tightly bound up with national competitiveness and national security as is space, and they often provide the main leitmotif for the case studies explored in this book.

The intellectual orientation provided in this chapter extends beyond the framework of analysis just sketched to provide a quick survey of 50 years of NASA's international activities in space. This overview gives one some idea of the extensive scope of NASA's international activities, and of how its dynamic has changed over time. It is also a pocket guide to what follows in the rest of the book: the chapter introduces readers briefly to the collaborative missions and countries or regions that are described in more detail in the body of the work, and provides a rationale for focusing on them. Since 1960 NASA has embarked on something like four thousand international projects. It is extraordinary that so few people realize this or understand its place in the panorama of NASA's, and the US government's, activities. The authors hope that this book will fill a yawning gap in the understanding of NASA, and transform it from being seen as a purely national agency into a global actor that embodies the highest ideals, and the internal contradictions, of American foreign policy at the "new frontier" that is space.

International Collaboration in the 1958 Space Act

The National Aeronautics and Space Act of 1958 was signed into law by President Eisenhower on July 29, 1958. [14] It distinguished between the civilian and defense-oriented aspects of aeronautical and space activities, and called for the establishment of a new agency to provide for the former in parallel to the Department of Defense (and—although this was not specified in the Act—to the Central Intelligence Agency and later to a highly secret covert agency, the National Reconnaissance Office, established in September 1961[15]). The primary mission of the resulting National Aeronautics and Space Administration (NASA), which formally came into being on October 1, 1958, reflected the dynamics of superpower rivalry with the Soviet Union in the wake of the Sputnik shocks the year before. The Space Act called on the new agency to ensure "the role of the United States as a leader in aeronautical and space science and technology and in the application thereof to the conduct of peaceful activities within and outside the atmosphere (Sec. 2 (c) 5)."

Other countries, above all from the free world, were to be enrolled in this endeavor. To this end the Space Act also included among NASA's missions "Cooperation by the United States with other nations and groups of nations [...]" (Sec. 2 (c) 7)." This objective was developed in a short, separate section headed "International Cooperation." Here it was specified that "[t]he Administration, under the foreign policy guidance of the President, may engage in a program of international cooperation in work done pursuant to the Act, and in the peaceful application of the results thereof, pursuant to agreements made by the President with the advice and consent of the Senate (Sec. 205)." International collaboration thus went hand in hand with foreign policy: NASA was to be an arm of American diplomacy.

Eisenhower stressed from the outset that this clause was not intended to engage presidential authority for all bilateral or multilateral programs undertaken by NASA. Its aim, rather, was to allow for the rare occasions when cooperation engaged such important questions of foreign policy that it had to be underpinned by international treaties. The *Final Report of the Senate Special Committee on Space and Aeronautics*, dated March 11, 1959, confirmed this interpretation.[16] As a result, as Arnold Frutkin has put it, the pace of the cooperative program "was to be faster and its procedures far simpler than would have otherwise been the case." In particular, "NASA's international program was thus immediately distinguished from that of the Atomic Energy Commission which, under its legislation, was required to obtain approval of its international efforts from the Congress."[17] The Space Act thus gave NASA considerable latitude to engage in international collaboration as its officers saw fit, and to handle the diplomatic dimensions of its policies and practices through interagency consultation, above all with the State Department.

A commitment to the "peaceful use" of outer space was essential to the successful exploitation of space for civilian scientific and applications programs on both a national and international collaborative level. As Eilene Galloway, who was involved in drafting the Space Act, has put it, the emphasis on peaceful use was intended to preserve space "as a dependable orderly place for beneficial pursuits."[18] To that end the United States moved rapidly to set up an international regime forbidding the militarization of space. In the face of considerable Soviet hostility and suspicion the United States took the lead in establishing an ad hoc Committee on the Peaceful Uses of Outer Space (COPUOS), which became a regular committee of the UN General Assembly in December 1959.[19]

No clear definition of "peaceful use" was laid down by COPUOS, nor has one been established since. This is because of the immense importance of military space programs, and above all of the role that intelligence and reconnaissance satellites have played since the dawn of the space age. As one scholar puts it, from the get go "[t]he term 'peaceful' in relation to outer space activities was interpreted by the United States to mean 'non-aggressive' rather than 'non-military.'" In international law this entails that all military uses are permitted and lawful as long as they do not engage the threat or the use of force. This interpretation has been essential to the preservation of both international stability and the national security of the space powers.[20] It is now a central plank of the military's expanding reliance on space.

The Locus and Scope of International Collaboration

NASA's collaborative effort was originally located institutionally in the Office of International Programs. The first director, Henry E. Billingsley, was quickly replaced by Arnold W. Frutkin in September 1959. Frutkin joined NASA from the National Academy of Sciences. There he had been the deputy director of the US National Committee for the International Geophysical Year and had also served as an adviser to the academy's delegate to the first and second meetings of COSPAR.

Frutkin had a long and distinguished career at NASA. In 1978 NASA administrator Robert Frosch appointed him deputy associate administrator, then associate administrator for external relation. The post was not to his liking, and Frutkin left government service shortly thereafter, in June 1979.[21] Some have suggested that his resistance to collaborating with Japan, which was emerging as major global power (see chapters 9 and 10), led to his relocation and his eventual decision to resign. His activities were taken over by Norman Terrell for a couple of years, before Kenneth Pedersen joined the agency as director of the International Affairs Division of the Office of External Relations. Pedersen had been an assistant professor of political science at San Diego State University from 1968 to 1971, before taking on various policy analysis activities in the federal government.

Frutkin laid down the basic principles that guided NASA's international collaborative projects for two decades in which the United States was the leading space power in the free world. Pedersen frequently remarked that he was dealing with a different geopolitical situation in which the United States' historical rival for space superiority, the Soviet Union, was showing a greater willingness to open out to international partners and in which the space programs in other regions and countries, notably Western Europe and Japan, had matured significantly. In September 1985 Pedersen was named deputy associate administrator for external relations and was replaced by Richard Barnes, who was Frutkin's right-hand man during the 1960s and 1970s.[22] In 1991 Pedersen returned to academia and was replaced as associate administrator for external relations by Margaret (Peggy) Finarelli.[23] Finarelli joined NASA in 1981 after serving in various government agencies. She was NASA's chief negotiator for the international agreements with Canada, Europe, and Japan regarding cooperation in the Space Station Freedom program.[24]

Over the past two decades the management of NASA's external relations has been reorganized several times reflecting the increasing scope and complexity of the agency's international activities. In 2010 they were handled by the Office of International and Interagency Relations. Associate Administrator Michael O'Brien and his deputy Al Condes watched over a variety of activities that include, for example, distinct divisions for "international efforts to pioneer approaches in aeronautics research and the exploration of the Moon and Mars and beyond," for "international and interagency policy issues" for science, and for the administration of NASA's export control program. NASA had field offices, not only in Europe, but in Japan and Russia too.[25]

The scope of NASA's international collaboration is truly vast. In 1970, when many countries only had embryonic programs of their own, Arnold Frutkin reported that NASA had already collaborated with scientists in 70 different countries, and had established 225 interagency or executive agreements with 35 countries.[26] Addressing a congressional subcommittee in 1981, Ken Pederson remarked that NASA's international activities had grown to over 1,000 agreements with 100 countries, and that these programs had resulted in more than $2 billion of economic benefits for the country. Michael O'Brien has recently counted over 4,000 international agreements of all kinds.[27]

Looking only at scientific collaboration with Europe, we find that this has increased rapidly in recent times. John Logsdon counted just 33 projects between 1958 and 1983.[28] Roger Launius later reported that there were 139 cooperative agreements with European nations between 1962 and 1997, that is, about 100 agreements were signed between 1984 and 1997.[29]

Numbers alone cannot capture this immense enterprise. Table 1.1 surveys the range of international activities that NASA was engaged in for the first 26 years of its existence. These include infrastructural components like tracking and data acquisition, and launch provision. They cover collaboration in science using bal-

Table 1.1 Cumulative statistical summary of NASA's international programs through January 1, 1984

Type of Arrangement	A	B	Type of Arrangement	A	B
Cooperative arrangements			*Reimbursable launchings*		
Cooperative spacecraft projects	8	38	• Launching of non-US spacecraft	15	95
Experiments on NASA missions			• Foreign launchings of NASA spacecraft	1	4
• Experiments with foreign principal investigators	14	73	*Tracking and data acquisition*		
• US experiments with foreign coinvestigators or team members	11	56	NASA overseas tracking stations/facilities	20	48
• US experiments on foreign spacecraft	3	14	NASA-funded SAO optical and laser tracking facilities	16	21
Cooperative sounding rocket projects	22	1774[a]	Reimbursable tracking arrangements		
Joint development projects	5	9	• Support provided by NASA	5	48
Cooperative ground-based projects			• Support received by NASA	3	12
• Remote sensing	53	163	*Personnel exchanges*		
• Communication satellite	51[b]	19	Resident research associateships	43	1417
• Meteorological satellite	44[c]	11	International fellowships		358
• Geodynamics	43	20	Technical training	5	985
• Space plasma	38	10	Foreign visitors	131	85,177
• Atmospheric study	14	11			
• Support of manned space flights	21	2			
• Solar system exploration	8	10			
• Solar terrestrial and astrophysics	25	11			

continued

Table 1.1 Continued

Type of Arrangement	A	B	Type of Arrangement	A	B
Cooperative balloons and airborne projects					
• Balloon flights	9	14			
• Airborne observations	12	17			
International solar energy projects	24	9			
Cooperative aeronautical projects	5	40			
US/USSR coordinated space projects	1	9			
US/China space projects	1	5			
Scientific and technical information exchanges	70	3			

Notes: A: Number of countries/international organizations
B: Number of projects/investigations/actions completed or in progress as of January 1, 1984
[a] Number of actual launches
[b] AID-sponsored international applications demonstration
[c] Automatic picture transmission stations.
Source: Anon., 26 Years of NASA International Programs (Washington, DC: NASA, n.d.). Thanks to Dick Barnes for providing a copy of this booklet.

loons, sounding rockets and satellites and applications in areas like remote sensing, communications and meteorology.

In addition NASA has sponsored an education and training program through fellowships, research associateships, and by hosting foreign visitors. There is no doubt that the agency has played a fundamental role in encouraging and strengthening the exploration and exploitation of space throughout the world, or at least among friendly nations. NASA has helped many countries kick-start their space programs and has enriched them once they had found their own feet. More than that, it has helped give thousands of people in over one hundred nations some stake in space, some sense of contributing, albeit in perhaps a small way, to the challenges and opportunities, the excitement and the dangers that the conquest of space inspires.

Frutkin's Guidelines for International Collaboration

The original stimuli to international collaboration were two; both of them were referred to in the episode described at the start of this book and are illustrated in table 1.1. First, there was the wish, inspired by major international initiatives such as the International Geophysical Year (IGY), and coherent with an abiding thread in American foreign policy, to engage other countries, especially friendly and neutral countries, in an exciting new scientific and technological adventure where they could benefit from American leadership and largesse.[30] Second, there was the need for global coverage in some applications and for a worldwide tracking and data handling network to support NASA's multiple space missions from planetary probes to human exploration. Sunny Tsiao has recently covered the latter dimension in depth.[31] This book will concentrate on the scientific and technological aspects of international collaboration in scientific and applications satellites and in human spaceflight from the creation of NASA into the twenty-first century.

In 1965 Arnold Frutkin published an important book in which he identified a number of criteria for a successful international collaborative project.[32] Twenty years later they were presented more or less unchanged as the basic guidelines for NASA's relationship with its partners.[33] In this summary form they read:

- Designation by each participating government of a government agency for the negotiation and supervision of joint efforts;
- conduct of projects and activities having scientific validity and mutual interest;
- agreement upon specific projects rather than generalized programs;
- acceptance of financial responsibility by each participating agency for its own contributions to joint projects;
- provision for the widest and most practicable dissemination of the results of cooperative projects.

This list requires some elaboration.

The first requirement was that NASA have just one interlocutor to deal with in the partner country, and an interlocutor that had official authority to engage the resources, human, financial, and industrial in the collaborative project. Frutkin was aware that, at the dawn of the space age, many individuals, pressure groups, and government departments would be jockeying for control of the civilian space program, as they had in the United States. He wanted to avoid NASA becoming enrolled in these domestic conflicts or, indeed, unwittingly being used to promote the interests of one party over the other. Hence his reluctance to negotiate with anyone but an official representative. This policy, coupled with NASA's offer to fly foreign payloads in March 1959 (see chapter 2), not only stimulated the creation of space programs in foreign countries, but also encouraged the national authorities to designate one body as responsible for international collaboration, and in some cases led to the rapid establishment of a national or regional space agency. Whereas Frutkin originally left the door open for collaborating with "a central, civilian, and government sponsored, if not governmental authority," by 1986 space agencies were so widespread internationally that NASA could simply designate them as its preferred partners.[34]

The second criterion was obviously meant to make scientific exploration, not political exploitation, the core of any collaborative space program. Frutkin was emphatic that each country "poll its scientific community for relevant ideas" and, in consultation with NASA, "develop full-fledged proposals for cooperative experiments having a character of their own."[35] This would also deflect charges that the United States was using its superior space capabilities to "dominate" its partners.

This concern also informed the criterion that all agreements should be on a project-by-project basis. An open-ended engagement to collaborate could lead to NASA committing itself to costly projects that were of no interest to US investigators. By evaluating each proposal on a case-by-case basis, it could be assessed for its novelty and compatibility with the general thrust of the American space effort, so contributing to the knowledge base of both partners. For that reason too, both would be willing to invest resources in their part of the project without seeking help from the other. This clause, summarized by the slogan "no exchange of funds," was a cornerstone of NASA policy, and a touchstone for the willingness of its partners to take space collaboration seriously and to invest their (often scarce) resources in a project.

The demand for full disclosure in the fifth and last criterion listed above flows from this. It was also meant to ensure that the joint program did not touch directly on matters of national security at home or in the foreign country. Frutkin was well aware of the tight interconnection between the civil and the military in space matters. The requirement that the results of any joint effort be disseminated as widely as possible was at once a gesture to this commingling and an attempt to carve out a space for civil, peaceful activities that could be conducted internationally alongside military, and so predominantly national programs.

Frutkin's principle of "clean technological and managerial interfaces" was an ingenious solution to resolving NASA's two, potentially conflicting, missions as mandated by the Space Act: to collaborate without jeopardizing leadership. Leadership depended on the capacity to define the frontier of space science and technology. Scientific and technological collaboration, unless carefully managed, could undermine that leadership. By maintaining "clean" technological interfaces, and by regulating knowledge flows across them, NASA was able to protect its cutting-edge science and technology to secure American preeminence while sharing knowledge and skills that foreign partners still valued.

It is not surprising that of 38 international cooperative spacecraft projects undertaken or agreed on between 1958 and 1983, 33 were with Western Europe,

Table 1.2   Benefits of NASA's international programs in Western Europe

Scientific/ Technical Benefits

Attracts brainpower to work on challenging research problems

Shapes foreign programs to be compatible with US effort by encouraging others to "do it our way"

Limits foreign funds for space activities that are competitive or less compatible with US space interests

Obtains outstanding experiments from non-US investigators

Obtains coordinated or simultaneous observations from multiple investigators

Opens doors for US scientists to participate in foreign programs

Economic Benefits

Has contributed over $2 billion in cost savings and contributions to NASA's space effort

Improves the balance of trade by creating new markets for US aerospace products

Political Benefits

Creates a positive image of the United States among scientific, technical, and official elites

Encourages European unity by working with multinational institutions

Reinforces the image of US openness in contrast to the secrecy of the Soviet space program

Uses space technology as a tool of diplomacy to serve broader foreign policy objectives

Source: Adapted from John Logsdon, "US-European Cooperation in Space Science: A 25-Year Perspective," *Science* 223:4631 (January 6, 1984): 11–16.

given its relative wealth and industrial capacity. Of a total of 73 experiments with foreign principal investigators, 52 were with this region. Canada, Japan, and the Soviet Union, along with several developing countries made up the balance.[36] This was quite unlike a program like Atoms for Peace that proliferated research and some power reactors throughout the developed and developing world in the late 1950s driven by foreign policy and commercial concerns that had little regard for indigenous capability. This difference was deliberate: Frutkin was emphatic that space collaboration should never become a form of foreign aid, so effectively restricting the scope of NASA's activities to industrialized or rapidly industrializing countries with a strong science and engineering base.

This also explains the insistence that collaborative experiments should be of "mutual interest" (second criterion above). How could a foreign experiment that had "a character of its own" be of some value to NASA and to American investigators? For Frutkin, it had to dovetail with the broad interests of the American program, if only to justify the expenditure of US dollars. Thus, each cooperative project had to be "a constructive element of the total space program of the United States space agency, approved by the appropriate program officials and justifying the expenditure of funds for the US portion of the joint undertaking."[37]

John Logsdon has put together some of the "constructive" contributions that international collaboration, notably with Western Europe, made between 1958 and 1983, not only to the US space effort as such, but also to the American economy and to the pursuit of American foreign policy. His findings are summarized in table 1.2. This table not only shows the concrete ways in which foreign experiments were to be of "mutual interest" scientifically, but also draws attention to the economic and political benefits of space collaboration, including channeling foreign resources down avenues that would not undermine American scientific and technological leadership, creating markets, projecting a positive image of the United States abroad, and promoting foreign policy agendas, including the postwar integration of Europe.

These putative benefits were not always welcomed by those actually engaged in the practicalities of international collaboration. American scientists and engineers, flush with the enormous success of their own program, feared that their partners were less capable than they were, and might not fulfill their commitments. They balked at the additional layers of managerial complexity, and the assumed added cost of international projects. As resources for NASA's space science program shrunk in the 1970s they sometimes resented the presence of foreign payloads on NASA satellites, suspecting that they had been chosen less on the basis of merit than because they were free to the agency. And they noted that by encouraging foreign powers to develop space capabilities NASA was undermining American leadership in high-technology industry: it was producing its own competitors.[38] International collaboration was not uncontested at home, particularly as NASA's partners gained in maturity, and were competitors as much as collaborators.

The weight of the several factors (scientific and technical/economic/political) that were brought into play in the first two decades of international collaboration varied depending on circumstances. A scientific experiment built with a foreign principal investigator and paid for by a national research council—like Geiss's solar wind experiment on Apollo 11—raised few if any broader economic or political issues. Complex and expensive projects calling for major technological developments and managerial inputs were at the other end of the spectrum.

The 1975 Apollo-Soyuz Test Project (ASTP) is an example of this (see chapter 7). Often reduced to simply a "handshake in space," it involved docking an American Apollo and a Soviet Soyuz spacecraft with each other in orbit 120 miles above the earth. During the two days in which the hatch between Apollo and Soyuz was open, three American astronauts and two Soviet cosmonauts exchange pleasantries and gifts, and conducted a few scientific experiments together. This was above all a political statement, a concrete manifestation of the new climate of détente with the Soviet Union being pursued by President Nixon and his national security adviser and secretary of state Henry Kissinger.[39]

Political concerns also provided a trigger for two other major projects in the 1960s and 1970s. One was Helios, the $100-million venture to send two probes built in West Germany, and weighing over 200 kilograms each, to within 45 million kilometers of the sun (see chapter 2). Helios was the most ambitious joint project agreed to in the 1960s between NASA and a foreign partner. It was the result of an invitation for space collaboration made by President Lyndon Johnson to Chancellor Ludwig Erhard during a state banquet at the White House in December 1965. For Erhard a major civil space project was one way of reducing German obligations to buy military equipment from the United States as required by the offset agreements between the two countries. For Johnson it was a gesture of support for America's most faithful ally in Europe at a time when the Vietnam War was increasingly unpopular, and the French were increasingly hostile to NATO. Of course, once the official offer had been made these political concerns receded into the background. Scientific and technical success, however, should not be decoupled from the political will that created the essential window of opportunity for scientists, engineers, and industry to embark on such an ambitious project so early in Germany's postwar space history with NASA's help.

The same can be said of the Satellite Instructional Television Experiment (SITE), another impressive international project that was agreed on with the Indian authorities in 1970 (see chapter 12). Here an advanced application satellite (ATS-6) broadcast television programs to village receivers directly, or via relay stations provided by the Indian authorities. For India the satellite was an ingenious way of bringing educational television, produced locally and dealing with local needs such as family planning, into otherwise inaccessible rural areas, while giving an important popular boost to the indigenous space program. For the United States it served a variety of political and economic needs. It promoted the modernization of India as an alternative model to China for developing countries. It was part of broader strategy to channel Indian resources down the path of civilian technologies. And, by withdrawing the satellite from service after a year, NASA successfully encouraged the Indian government to buy additional models from US business. SITE, while being of undoubted benefit to various constituencies in India, also served multiple geopolitical needs for the United States in the region.

In all three of the cases just described, while political (and economic) motives were part of the broader context inspiring the collaboration in question, they were essentially left behind or bracketed during the scientific and technical definition of the projects and their implementation. Once the programs got under way the fundamental maxims of clean interfaces and no exchanges of funds dominated development.

There was a notable exception to this: the major initiative, inspired by NASA administrator Tom Paine, to engage Europe at the technological core of the post-Apollo program between 1969 and 1973 (see chapters 4–6). In a nutshell, with NASA's budget shrinking dramatically after the "golden years" of the Apollo lunar missions, Paine hoped to get Europe to contribute as much 10 percent (or $1 billion) of an ambitious program that initially included a space station and a shuttle to service it. Foreign participation would also help win the support of a reluctant Congress and president for NASA's plans. And it would undermine those who insisted that Europe needed independent access to space—Europeans were told that they were wasting valuable resources by developing their own expendable launcher to compete with a reusable shuttle that, it was claimed, would reduce the cost per kilogram into orbit by as much as a factor of ten. For several years joint working groups invested hundreds of hours discussing a variety of projects. Some, like having European industry build parts of the orbiter wing, threw clean interfaces to the winds. Others, like the suggestion that Europe build a space tug to transfer payloads from the shuttle's low-earth orbit to a geosynchronous orbit, a project of interest to the Air Force, touched directly on matters of national security. The entire process was reconfigured soon after President Nixon authorized the development of the shuttle in January 1972. Clean interfaces and no exchange of funds imposed their logic on the discussion (and were reinforced by anxieties about European capabilities to fulfill commitments and by fears that NASA was becoming entangled in unwieldy and costly joint management schemes). The European "contribution" was reevaluated, many existing projects were cancelled, and Germany decided to take the lead in building Spacelab, a shirt-sleeve scientific laboratory that fitted into the shuttle's cargo bay and that satisfied all the standard criteria of international collaboration. So too did Canada's construction of the Remote Manipulator System (RMS), a robotic arm that grabbed satellites in space, or lifted them from the shuttle's payload bay prior to deployment. Once built both Spacelab and the RMS were handed over entirely to NASA to operate.

The willingness to share technology in the post-Apollo program (and also in support of the European Launcher Development Organization in the mid-1960s—see chapter 3) was part of a general sentiment in Washington that something had to be done to close the technological gap that had opened up between the two sides of the Atlantic at the time. Space technology was seen as a crucial sector for closing this gap.[40] Technological sharing would undermine European criticisms of American dominance in high-tech areas, while helping to build a European aerospace industry that could eventually serve as a reliable partner sharing costs in civil and military areas: Europe would assume some of the burden for its own defense.

The Changing Context in the 1980s

The context of international cooperation changed importantly in the 1980s. In essence the technological gap between NASA and its traditional partners began to close in a variety of space sectors. At the same time the Soviet Union began to be more open to international collaboration. NASA had to find ways to retain leadership while collaborating with partners who were also competitors in many space sectors.

Launchers were at the cutting edge of this transformation. On Christmas Eve 1979 the European Space Agency (ESA) successfully tested its first Ariane rocket. After overcoming the normal teething troubles Ariane soon proved to be a spectacular success. Arianespace (the company that commercialized Ariane) had acquired about 50 percent of the commercial market for satellites by the end 1985, helped on by the lower-than-expected launch rate of the US space shuttle. A second major new player entered the field of rocketry in the late 1980s. Japan developed its H-series to replace the N-series built under American tutelage (see chapter 10). China's Long March 3 placed a satellite in geostationary orbit in April 1984; the authorities immediately announced that they too were keen to find clients abroad. Finally the Soviet Union was showing a greater willingness to offer its previously closed and secretive launcher system for commercial use, and was even seeking a contract to launch a satellite for the International Maritime Satellite Organization (INMARSAT), something that was simply inconceivable several years before (see chapter 8).

Launch technology was not the only area where American leadership was being challenged. Advanced communications satellites and remote sensing satellites with technologies more sophisticated than those available in the civil sector in the United States were being built in Europe, Japan, and Canada. The French had taken the lead in commercializing images from SPOT, an earth remote sensing satellite that technologically outstripped the earlier NASA Landsat system, then bogged down in negotiations over privatization. Australia as well as a number of rapidly industrializing countries—Brazil, China, India—had constructed solid national space programs, and many Third World countries, along with the Soviet Union (in a reversal of its historic policy), were clamoring for a greater say in international bodies such as Intelsat, which governed the global satellite telecommunications system. Summing up the situation, a special task force of the NASA Advisory Council reported in November 1987 "that there is in process an accelerating equalization of competence in launching capability, satellite manu-facturing and management for communications, remote sensing and scientific activity, and in the prospective use of space for commercial purposes."[41] For Pedersen writing in 1986 this meant that for NASA now "'power' is much more likely to mean the power to persuade than the power to prescribe."[42]

The 1990s and Beyond

The end of the Cold War forced yet another reassessment of NASA's role. The rigid-ity that had marked 40 years of US-Soviet rivalry and the framework for collabora-tion that it had defined had now collapsed. The space program "lost an enemy." The political and military rationales for collaboration with Western allies—and the subordination of economic considerations to geostrategic concerns during the Cold War—would come back to haunt the United States: the technological gap was no more and erstwhile allies were now economic competitors. As the Soviet Empire crumbled "the Bush administration, in a sharp reversal of prior practice, [. . ,] announced that it [would] henceforth review license applications to export dual-use technology to the CIS (Commonwealth of Independent States) countries with a 'presumption of approval.'"[43] The hallowed principles of no exchange of funds and clean interfaces to restrict technology transfer were being overturned. Efforts were made to retain the infrastructure and institutional memory of the major Soviet space programs in Russia and later the Ukraine, though technology transfer was restricted by the Missile Technology Control Regime. As a report for the Office of Technology Assessment pointed out in 1995 Russian industrialists involved in the International Space Station would be obliged to abide by Western nonproliferation rules, for example, by not selling sensitive booster technology to unreliable partners.[44] Scientists and engineers were given strong incentives to ally themselves with US- and Western-style reforms in an attempt to stem "the flow of indigenous high-risk technologies and expertise from those locations [the CIS states] to outside destinations, principally Third World Nations."[45]

This change in context had palpable effects on the evolution of the plans for the Space Station (see chapter 13). NASA had already shown a new flexibility in defining this huge technological venture with representatives of ESA, Canada, and Japan even before the president authorized the scheme in 1984; in recognition of the technological maturity of its partners, and the absolute necessity to have them share the cost, NASA's "coordination in the early planning phases indicated a con-sideration of foreign partner interests and objectives unprecedented in space coop-eration hitherto."[46] With the inclusion of Russia in the venture beginning in 1993 there was an increased move to multilateralization and interdependence. NASA and American industry could benefit directly by collaborating closely with a partner that had extensive experience in human space flight. It was reported in 1995 that US firms and their counterparts in Canada, Europe, and Japan had entered into space-station-related contracts and other agreements worth over $200 million. NASA had procured about $650 million of material from Russian suppliers over four years.[47] Russia became functionally integrated into the station in 1998, pro-viding critical path infrastructure elements on what became a US-Russian core.

In 1984 NASA administrator James Beggs had warned his senior staff involved in the Space Station program that they were to be careful to avoid "adverse technology transfer" in international programs, notably where the Soviet Union was involved, and expressed concern about "careless and unnecessary revelation of sensitive technology to our free world competitors—sometimes to the serious detriment of this nation's vital commercial competitive position" (see chapter 15).[48]

Economic concerns were complemented by new military demands. As satellite technology became more sophisticated, the military began to make increasing use of space-based hardware as a "force-multiplier," that is, they exploited its capacity to enhance traditional military operations. Satellites began to be used to improve the effectiveness of battlefield surveillance, tactical targeting, and communications.[49] These advantages were dramatically demonstrated in Operation Desert Storm, the UN-sanctioned, US-led assault on Iraqi forces that had occupied Kuwait in 1991. The *Final Report to the President on the U.S. Space Program* of 1993 stressed this dimension of the conflict. "Control of space was essential to our ability to prosecute the war quickly, successfully, and with a minimum loss of American lives." Communications, navigation, weather reporting, reconnaissance, surveillance, remote sensing, and early warning—all these were mentioned in the report as essential to US victory.[50] The defense space budget climbed in line with demand. NASA's budget remained roughly unchanged in constant dollars between 1975 and 1984 (hovering between $8 and $9 billion 1986 dollars). The defense space budget came from behind to equal NASA's around 1981. By 2000 they were approximately the same at $12.5–$13 billion current dollars. The terrorist attacks on American soil on September 11, 2001, accelerated demands for the protection of space as a key asset in America's defensive arsenal.[51] It was recently reported that for FY2005 Congress allocated $19.8 billion for space to the Department of Defense, and $16.2 billion to NASA.[52]

Already in the 1980s there were major concerns that the Soviet Union had taken advantage of the liberalized trade agreements that were part of the policy of détente to acquire, by every means possible, knowledge and training in superior American high technology to build their industrial and military strength. Beginning in the mid-1990s, and with increasing emphasis today, it is the determination of the People's Republic of China to reap the fruits of America's scientific and technological research system to enhance its global standing, either by exploiting openness or by espionage. The International Traffic in Arms Regulations (ITAR), which have always impacted the circulation of satellite and launcher technology, have been tightened up, and heavy fines imposed on those who break them. NASA has responded to this situation by centralizing its export control activities in a special division and by engaging with the State Department in ongoing discussions on ways to improve the implementation of ITAR (chapter 14). It is significant to note, however, that it succeeded in making the International Space Station an "ITAR-free" project (chapter 13).

The Organization of This Book

Much of this book concentrates on the first 30–40 years of NASA's international collaboration, concluding with brief overviews of the International Space Station and of ITAR today. The aim has been to throw light on the general policies that informed the agency's actions in select regions of the globe, and to illustrate their application in practice in specific and important cases. The vast scope of NASA's international activities demanded that choices be made. Those choices, in turn, were constrained by the usual factors: the availability of sources, the capacities and interests of the authors, and the time foreseen to complete the work, which competed with many other responsibilities.

The book is divided into three major sections, and a conclusion. It deals successively with NASA's relations with Western Europe, with the Soviet Union and Russia, and with two countries in Asia—Japan and India.

John Krige's section on Western Europe distinguishes collaboration in space science (chapter 2) from technological collaboration (chapters 3–6). This distinction reflects the very different issues that arise in the two domains. Clean interfaces and no exchange of funds are relatively easy to respect in space science; the management of knowledge and dollar flows is far more contentious in advanced technological collaboration. Here working with foreign partners engages multiple arms of the administration in NASA's programmatic affairs—the State Department, the Department of Defense, the Department of Commerce, most obviously. It can also raise eyebrows in Congress. For a leading space power like the United States, the desire of some to promote international collaboration by sharing technology is in a state of dynamic equilibrium with the determination of others to deny technology in the interests of national security, economic competitiveness, and American global leadership.

Chapter 2 complements national studies of four major Western European countries, Britain, France, Italy, and West Germany, with two more detailed case studies from the ESA period: the International Solar Polar Mission (ISPM) and the Cassini-Huygens planetary probe. The core theme of the national studies, readily and warmly acknowledged by many actors of the day, is NASA's generosity in catalyzing programs in countries in which the space community was small, industrial capacity was minimal, and political will was diffuse or nonexistent. The ISPM project, in which NASA withdrew its satellite from a joint venture with ESA, is important if only because of the shock that the agency's move caused in Europe. Cassini-Huygens, by contrast, was not only a superb scientific achievement: its survival in Goldin's NASA, which was committed to "faster, better, cheaper" projects, was a tribute to the ability of international collaboration to protect a mission.

The exchange of sensitive technology is at the core of the next four chapters. The first treats NASA and the State Department's attempts to save the European Launcher Development Organization (ELDO) from collapse in the mid-1960s, describing their efforts to cope with restrictions on collaboration in this sensitive domain that had been imposed by various national security directives from the White House. The next three chapters are a detailed examination of the lengthy negotiations, initiated by NASA administrator Tom Paine in 1969, to engage Western Europe (and others, notably, Australia and Japan) in the post-Apollo program. This event is important for the light it throws on the perennial conflict between technological sharing and technological denial. It explores the crucial role that the parallel negotiations over the definitive Intelsat agreements played in the discussions between the United States and Europe over launcher availability, and it studies the effects of the eventual decision to limit technological exchange to the barest minimum.

The two chapters by Angelina Long Callahan on the Soviet Union, its collapse, and the emergence of Russia as a major space player span a collaborative regime that might appear to have moved from one extreme to another. Collaboration in the 1960s was, in some respects, arm's length; the 1990s did indeed bring about increasing dependence on Russia for construction, design, and access to the ISS. However, the notion of a wholesale shift between two extremes of criticality is to some extent artificial, an artifact of Cold War thinking. Historiography has often eclipsed parallel efforts made to maintain dialog and some measure of trust. What we learn from this section is that, embedded in the upheavals that have marked the history of the Soviet empire in the past 50 years, and the tension that has so marked its relationships with the rest of the free world, there lies a slow, cumulative attempt to build durable relationships in space. NASA policymakers often initiated these efforts; the Soviet Union typically made critical technological contributions. Low-key initiatives—meteorology, the Bion satellites, atmospheric sciences—and the odd space spectacular like the ASTP (chapter 7) were a forerunner of the full-blown effort in technological collaboration in the 1990s when Russia's experience in human spaceflight in the Shuttle-Mir and ISS program led to its integration into the core of the space station (chapter 8). It is emphasized that that process was one element in the sweeping measures embarked on by the US administration to rebuild and to reintegrate a transformed group of nations into the capitalist world economy, dismantling their military infrastructure and sealing their allegiance to international agreements concerning the proliferation of military technology.

The third section of the book, written by Ashok Maharaj, looks at NASA's relationships with Japan and India. A longitudinal overview of each (chapters 9 and 11) is supplemented with case studies of particularly important joint ventures. The first is the controversial decision, spearheaded by the State Department, that the United States should share Thor-Delta launcher technology with Japan (chapter 10). The second case study is the marvelous SITE project that was so near and dear to Frutkin's heart: the use of an ATS-6 communications technology satellite to beam educational and other programs to about 2,200 villages in rural India (chapter 12).

As I have said, there is an inherent contradiction in NASA's twin missions to maintain leadership *and* to foster international collaboration. By helping others acquire space capabilities NASA at once enhances the capacity, visibility, and reach of the US space program, and contributes to the efforts of those who may eventually compete with it. The dilemma is particularly acute when collaboration involves managing dual-use technologies that are of both commercial and military significance. Chapters 13 and 14 discuss the contradictory forces now at work on international collaboration in the civilian aerospace sector. The new interdependence between partners embodied in the International Space Station (ISS) is contrasted with the stricter implementation of the International Traffic in Arms Regulations (ITAR) in the rest of the civilian space sector. It is impossible to know which of the two models—closer interdependence or retreat behind high technological walls—will prevail. The recent tectonic shifts in the global world order, the emergence of new space powers, many of them perceived as a threat to the United States, the increasing pressure for the commercialization of space, and the growing importance of space technologies as force-multipliers in war define the contours of a transition whose future it is difficult to predict.

Part II

NASA and Western Europe

John Krige

Chapter 2

NASA, Space Science, and Western Europe

Although the Space Act of 1958 specifically mandated NASA to promote international collaboration, it was up to officials in the new organization to define the terms and conditions under which they would work with partners, most of whom had little or no experience in the domain. Chapter 1 described the parameters that Frutkin believed were essential if international collaboration was to be a success.[1] He was emphatic that those who wished to work with NASA had to come to the table with their own scientific ideas, their money to fund them, an industrial capacity able to produce the hardware, and with official support for their program. In the late 1950s very few countries were in any position to meet these demands. In the context of the widespread interest in space at the time, this meant that scientists, industries, and national administrations had to come up with experiments, payloads, adequate funding, and an institutional home for space before they could exploit the opportunities that NASA offered. In short, NASA played a major role in kick-starting and orienting incipient space programs in many friendly countries.

A significant step toward international collaboration in space science was taken about six months after the Space Act was signed into law. It was announced at the second meeting of the Committee on Space Research (COSPAR). COSPAR was set up by the International Committee of Scientific Unions to maintain the momentum of the IGY.[2] At the meeting in The Hague in March 1959 the American delegate to the committee, Richard Porter, made a formal offer of international cooperation on behalf of the United States. NASA's offer, conveyed by Porter, was that the United States would support the work of COSPAR by launching "worthy experiments proposed by scientists of other countries," either as "single experiments as part of a larger payload, or groups of experiments comprising complete payloads."[3] In the former case the proposer would be "invited to work in a United States laboratory on the construction, calibration, and installation of the necessary equipment in a U.S. research vehicle." If that was not possible, a US scientist could help the originator bring the payload to fruition or, less desirably, the investigator abroad could simply provide the payload as a "black box" for installation. Entire payloads, which could weigh anywhere from 100 to 300 pounds and be placed in orbits ranging from 200 to 2,000 miles, would be accepted if recommended by COSPAR. In such cases the United States was willing to "advise on the feasibility of proposed experiments, the design and construction of the payload package, and the necessary pre-flight environmental testing." NASA also offered substantial funding for resident research associate positions in both experimental and theoretical space research.

This offer had an electrifying impact on those present. As Frutkin writes, "[T]he future of international cooperation in space exploration was raised at a stroke from the token to the real."[4] COSPAR emerged as an essential forum bringing together "precisely those individuals and national agencies best situated to motivate a positive response from their governments," so facilitating bilateral and multilateral programs. NASA's prestige and desirability as an international partner of choice was also confirmed at The Hague.

European space scientists were quick to respond to NASA's offer, and remain unanimous in their praise for the help that NASA provided them, particularly in the 1960s. Reimar Lüst, who was a leading figure in the European space effort for many years, spoke for many when he gave the Fulbright Fortieth Anniversary lecture in Washington in April 1987. "That the Europe of today can be seen as an autonomous, real and reliable partner of the United States in various fields of science and technology," said Lüst, "is thanks to the immensely unselfish help given to it by the United States."[5] This chapter fleshes out that remark by describing the crucial role that the agency and the US administration played in kick-starting the space programs in the "big four" European countries: Britain, France, Italy, and West Germany. This country-based approach is followed by a brief description of Europe's contribution to the Hubble Space Telescope and a survey of two major projects undertaken between NASA and the European Space Agency (ESA). The first, the International Solar Polar Mission, is interesting for the light it throws on the misunderstandings that can arise in international collaboration when partners are constrained by different funding mechanisms for space projects. The second, the Cassini-Huygens planetary mission to Saturn and Titan, is a prime example of a successful joint venture of immense cost, complexity, and scientific interest.

The tapestry that weaves together NASA and Western Europe in space science has hundreds of threads and tens of knots, and it grows ever richer. This chapter can do no more than highlight some salient features. Some readers may have preferred an account that chose to be broader rather than to probe deeper. They are referred to a National Academies publication that surveyed about a dozen collaborative projects between the United States and Europe in selected domains of space science, and that drew lessons for future collaboration from them.[6]

The United Kingdom

In October 1957 Harrie Massey, the leader of the British space science community, was at a reception at the Soviet Embassy in Washington, DC, when the launch of Sputnik was announced.[7] He was taken completely by surprise. In fact 18 months earlier the British IGY Committee had concluded that American and Soviet plans to launch satellites were not likely to succeed, and that even if they did, they were not likely to be of much scientific interest. This is not to say that space research was entirely neglected in the country. On the contrary, the Royal

Society and the Ministry of Supply—responsible for the country's guided missile program—supported the development of a sounding rocket program for upper atmosphere research. After several test flights of their own Skylark, launched from the Anglo-Australian rocket range in Woomera, an experimental program got under way just after Sputnik II orbited the earth.[8] The excitement of exploiting the new device, the conviction that any useful scientific data beyond the atmosphere would be freely available to all, and of course the cost led British scientists and policymakers to remain aloof at first from embarking on a satellite-based research program.

Two main factors inspired a change of thinking. One was internal pressure from leading figures such as Bernard Lovell, the director at Jodrell Bank, whose giant radio-telescope had tracked the trajectory of Sputnik over British soil.[9] Lovell and some sections of the government, notably the Foreign Office, argued that the United Kingdom would be "classed as an underdeveloped country" if she did not launch her own satellite.[10] Then there was the wish to improve relations with the United States. The launch of the Sputniks transformed the parameters of scientific and technological collaboration between the two nations. For a decade the British had resented the constraints imposed on scientific and technological exchange by Washington with one of its most faithful allies, particularly in sensitive areas.[11] A few days after Sputnik II was launched the minister of defense told Parliament that the Soviet satellites had "helped precipitate closer collaboration with the United States" and remarked that this new impetus "offers new prospects which we dared not hope for a few months ago." The country's crash program to develop a hydrogen bomb was given a major boost.[12] Cooperation in space science also moved center stage. In September 1958, six months before NASA's offer at COSPAR, American officials offered to launch a British payload on an American satellite more-or-less free of charge. In October the administration released a report praising British achievements and extolling the importance of international collaboration in space. The State Department hoped that Britain could quickly launch a payload into space, so becoming the first nation after the superpowers to do so, denying the communist bloc another space first.

The British community was in no position to meet a tight deadline: the use of the Skylark sounding rocket still defined the limits of its space ambitions. However, the new opportunities for UK-US cooperation led the Royal Society and the Ministry of Supply to reconsider a satellite program. Massey masterminded two major policy statements in October 1958 in which he made a strong plea for an autonomous British program; he was even opposed to launching a British satellite with an American rocket, as this would involve "an obvious loss of prestige." This national approach was quickly undermined, both for lack of domestic support and by NASA's offer at COSPAR in March 1959. Six weeks after Porter made his statement at The Hague the British National Committee on Space Research, chaired by Massey, had established working groups to define experiments that could be launched by NASA.

In June 1959 Massey led a cohort of British scientists armed with 11 experiment proposals on a trip to NASA.[13] They were warmly received. A provisional agreement was reached for using a Scout rocket to launch three satellites at roughly annual intervals.[14] There would be no exchange of funds and clean interfaces. The scientific instruments would have to be tested using sounding rockets, preferably British but if need be also American. The first launches would probably be from Wallops Island and if the launch failed every experiment would be launched a second time. NASA would provide the body of the first satellite and auxiliary services such as power supplies and telemetry: Britain would gradually assume responsibility for the entire satellite. The agency's tracking and telemetry network, possibly supplemented by British facilities, would be available. For the first satellite NASA also offered to receive the telemetry tapes from the ground stations, catalog and edit them, and compress them into digital form before sending them to the United Kingdom for further analysis. NASA also provided equipment to have a quick look at the first data received from the satellite so that experimenters could gauge if their instruments were performing as hoped. The entire process of data reception and analysis would be handed over to the British teams after this first "tutorial."

The design of the satellite required close collaboration between teams on both sides of the Atlantic. The performance characteristics of the Scout rocket had still not been fixed early in 1960: payload capacity and orbit capability were in flux, and temperature and vibration conditions at launch were only vaguely known. As Massey and Robins explain, the scope for mishap was immense. NASA had to design "the satellite structure, power supplies, data storage devices and encoders, and telemetry transmitters." Groups in Britain, for their part, "would design the instrument sensors, which would interlock and interact with practically every aspect of the NASA design activities."[15] They had chosen their seven experiments by December 1959 by drawing as much as possible on the experience gained with Skylark. This favored instruments that studied the high altitude ionosphere.

To manage the project a joint US/UK working group, which met every three months, was established. By the end of 1960 a number of major difficulties still loomed on the horizon. Data encoding and transmission was proving more demanding than anticipated. Mechanically complex external paddles were needed to support the solar power cells. For safety reasons the maximum launch inclination of the Scout from Wallops Island was 52°, somewhat less than the British scientists had planned for.

The restraint on launch inclination was lifted during the course of 1961, when it became clear that the Scout would not be available early in 1962, as planned. NASA accepted the additional cost of launching with the already developed and reliable Thor-Delta rocket from Cape Canaveral. The British were more than willing to oblige, and took advantage of the enhanced telemetry coverage by enrolling radio stations in Singapore, Port Stanley in the Falkland islands, and on a Royal Navy ship off the coast of Tristan da Cunha.

The first launch of the satellite, on April 10, 1962, was aborted due to a fault in the fuel system of the rocket. A few weeks later, on April 26, 1962, Britain's first scientific payload was successfully lofted into space on board a satellite baptized Ariel 1. Its capacity was degraded unexpectedly three months later by a high-altitude test of a hydrogen bomb that temporarily perturbed the operation of its instruments and permanently damaged the solar cells providing electric power. Ariel 2 (launched from Wallops Island in March 1964) and Ariel 3 (launched from the Western Test Range in May 1967) followed, with the United

Kingdom taking increasing responsibility for the engineering design, construction, and testing of the satellite itself, as well as for data handling and analysis.

As was mentioned earlier, NASA originally hoped that, in the context of Cold War rivalry, a British satellite might secure a space first for the free world. This was not to be. On September 29, 1962, Canada's Alouette 1 was successfully orbited by a Thor-Agena rocket.[16] NASA provided the launch vehicle, launch facilities, and a worldwide network of ground stations. The satellite was designed and built by the Defense Research Telecommunications Establishment (DTRE) in Ottawa. The project was intended to help Canada gain a space capability, to acquire new data for the engineering of high-frequency communication links, and to enhance the DTRE's considerable in-house expertise on the effects of the ionosphere on the scattering and deflection of radar beams. Alouette 1 was the first successfully launched satellite to be developed indigenously by a country other than the United States or the Soviet Union.

France

I have always claimed with gratitude that CNES is the child of NASA, and I would add, the loving child of NASA. There has always been a great friendship and mutual understanding between the two agencies...

Jacques Blamont[17]

Nazi missiles raining down on their country stimulated the French military's interest in rocketry.[18] About 100 V-1s fell between June and September 1944; almost 80 V-2s struck in four weeks from September to October that year. Henri Moreau, the director of a Parisian laboratory, was so impressed with the weapons that he made several trips to Germany to study them more closely, including a visit to the infamous production facility at Nordhausen. Moreau brought back nine wagon loads of missile parts and signed an agreement with the American authorities to receive ten complete V-2s. These were never delivered, presumably because of the presence of communist ministers in the postwar French government and in important scientific organizations.

A ballistic missile research laboratory was established at Vernon in May 1946 to exploit the spoils of war, a test range was built at Colomb-Béchar in the Sahara Desert, and 123 German engineers and technicians who had been involved in Von Braun's program at Peenemünde were employed under contract to work on missiles for the French military. One of them, Karl-Heinz Bringer, was to stay in France and play a crucial role in developing the propulsion systems for the French sounding rocket Véronique as well as its first missile-derived satellite launcher, Diamant, and the immensely successful European rocket, Ariane.[19]

France was ill-prepared for the opportunities provided by the IGY. Contrary to Britain, it had no space policy, no institutions to promote it, no technological or industrial capability in the space sector, and no space science community. This was partly because of the weakness of science in France after the war, and its inability to organize groups having a critical mass, partly due to interservice rivalry between the technical branches of the three arms of the military, and partly due to the huge investment, undertaken in 1956, to test a French atomic bomb within four years. In summer 1958 the Ministry of Foreign Affairs lamented the country's marginal influence on the international scene. The dispersal of already limited resources between different administrative organs made it impossible for France to speak with one voice. The essentially military character of its rocket program excluded it from playing a role in COSPAR.

The arrival of President General de Gaulle to power in June 1958 was transformative. De Gaulle was determined to strengthen the country's scientific and technological capability, believing that it was essential to reestablishing "la grandeur de la France" and to its strategic independence. A major missile program was established to provide an independent nuclear deterrent. A new civil Committee for Space Research was set up in January 1959 at the request of the minister of foreign affairs. Its brief was to take stock of the resources already at hand, to draw up a plan for the future, and to advise the prime minister on national and international space policy.

With space assuming a new significance, considerable resources were released for a campaign using an enhanced Véronique-IGY sounding rocket. The first launches that got under way in March 1959 were a spectacular success. The payloads were provided by a newly minted PhD, Jacques Blamont, who had worked at the University of Wisconsin in 1957. Blamont visited the Air Force Cambridge Research Laboratories near Boston on his way home, where he was given the blueprint of the mechanism for ejecting sodium vapor into the atmosphere that was being used with the American Aerobee sounding rocket. It was perfectly adapted to the limitations of the French situation at the time: cheap, solid, simple, of proven success, and it did not require any electronic equipment. Three German engineers prepared the rockets for launch at Colomb Béchar. Though the first launch did not attain the expected height the next two achieved their objectives. The ejector released a huge orange sodium cloud over Algeria between 90 and 130 kilometers, and then between 90 and 180 kilometers.

On Blamont's telling, in addition to its scientific achievements, this campaign had two major consequences. First, there was renewed interest in having a French space program. The rocket-borne sodium clouds that could be seen hundreds of kilometers away for over an hour were given wide media coverage. The public was so enthralled that hundreds of newborn girls were named Veronique.[20] Second, it brought him together with Robert Aubinière, "a brilliant army colonel whose ambitions were inspired by technology and the future."[21] Strong bonds were quickly established between the two men and with Aubinière's support previously unimaginable resources were made available for Blamont and for French space science. What is more, the authorities were persuaded that France now had the means to move beyond sounding rockets to ballistic missiles and satellite launchers. In March 1962 the French national space agency, CNES (Centre national d'études spatiales) came into being to replace the Committee for Space Research. Over the years the agency developed launchers, built a national satellite industry, a tracking network, and a dedicated equatorial launch pad in Kourou, French Guyana, as well as being responsible for international affairs.

Relationships with the United States were an important source of training and of legitimacy for the young community of French space scientists and engineers. Bell labs helped engineers from the national center for telecommunications research (CNET—Centre nationale d'études de télécommunications) to build a ground station at Pleumeur-Boudou to receive signals from Telstar 1.[22] Blamont's sodium vapor experiment was followed by an invitation to the Goddard Space Flight Center in October 1959. NASA encouraged Blamont to extend the range of his investigations to higher altitudes and in 1960 and 1961 he launched his payload with Javelin sounding rockets from Wallops Island, reaching an altitude of 600 kilometers (compared to 200 meters for Veronique). In March 1961 a formal agreement was signed in Washington for launching French payloads on American rockets and for hosting French engineers in NASA centers in the framework of the COSPAR offer. A French group took over a major balloon project that had lost support in the United States, and which they baptized Eole. In 1963 CNES and NASA signed a protocol defining a two-phase FR1 program: sounding-rocket studies of the upper atmosphere between 75 and 100 kilometers followed by the launch of a scientific satellite using a Scout.

The origins of Eole can be traced back to a project called GHOST (Global Horizontal Sounding Technique) promoted by Vincent Lally at the Air Force Cambridge Research Laboratories. Lally suggested floating 2,000 mylar balloons in low earth orbit along with a system of satellites that would localize them and relay meteorological measurements made at different heights back to earth.[23] This corresponded with a surge of interest in mathematical models of the atmosphere that needed an input of fresh data points at least once a day. Blamont realized that a project of this kind was one that was both prestigious and politically visible and NASA agreed that France pursue it. Eole was led by Pierre Morel using mylar balloons imported from the United States. About 500 balloons were launched from stations constructed in Argentina for the project. The lifetime of each was about 103 days, and each took some 8 days to go around the world. The project was haunted by the fear of a collision with high-flying aircraft and was gradually wound down. Morel's conclusion is uncompromising. Eole, he says, was a courageous and risky choice but it was not a scientific success. His team launched less balloons than they had hoped. The project was premature given the state of knowledge at the time, and it was undertaken in a hemisphere about which the French scientists knew very little.

NASA's help was unstinting in the FR1 program. Arnold Frutkin and Jack Townsend arranged for 12 young, enthusiastic French engineers to spend six months at the Goddard Space Flight Center (GSFC). Each worked in a separate technical domain and was instructed to establish bonds of mutual respect and friendship with their American colleagues. Whenever possible, contracts for the hardware were placed with French firms; otherwise NASA helped arrange for orders to be given to American companies that were visited regularly by CNES engineers to improve their own skills. To facilitate communications with NASA's tracking network the French used the already crowded VHF bands that NASA used, 136 MHz for telemetry, 148 MHz for tele-command. Relationships were warm, and with the help of NASA the French were able to proceed far more rapidly, and with a reduced risk, than if they had worked alone. Sam Stevens, the project leader at NASA was particularly effective. Jean Pierre Causse, the first director of the satellite division at CNES, affectionately remembers him as a kind of elder brother who freely gave of his advice without ever imposing his solutions. In fact this support meant so much to him that at a recent conference Causse exclaimed, "Thank you Sam! Bravo NASA and the United States!"[24]

The construction of FR1 also established close ties between Thompson Ramo Wooldridge (TRW) and Matra.[25] TRW sought international partners to strengthen its bid for communications satellites being built by Comsat on behalf of Intelsat (see chapter 5), while the French firm sought an American partner to build its credibility as a prime contractor for projects being developed by CNES and by ESRO (the European Space Research Organization). In 1965 a "Technical Assistance and License Agreement" was signed between Matra and TRW's Space Technology Laboratories division (STL) that allowed Matra to have access to the patents and know-how of STL through visits and internships of French engineers and technicians at its headquarters in California. The inter-penetration of practices between the two firms was so great that one senior ESA official reputedly remarked that "[w]hen one spoke with people from Matra one had the impression that one was speaking to American industrialists."[26]

In 1964 NASA established an office in Paris that gave the agency a permanent representative in Europe. The first to arrive was Gilbert Ousley, who left GSFC in 1964 to take up his new post. He has described his role at the time as primar-ily being "to find cooperative programs which would benefit NASA and which in our judgment could be done with a partner that would live up to their side of the agreement." The training offered at GSFC was not simply intended to bring young French scientists and engineers up to speed, however. It was also intended to export NASA's way of running projects abroad. As Ousley puts it, it

was a great excuse for us to really share technology and training but we also had a selfish purpose. It was to get young engineers that were experienced to participate in our program and later come back to France speaking the same terminology that NASA uses, that understood our review process and did not feel insulted by peers looking at what was being done and making constructive criticism.[27]

Jean Pierre Causse amplified this by stressing how important the NASA man-agement principles of "no exchange of funds," memoranda of understanding, a single project manager, design reviews, systematic testing by engineers in the project and in industry, and so on were to the success of the French teams sent to Goddard.[28] This flow of management practices across the Atlantic from TRW and from GSFC was a characteristic feature of NASA's relationship with European projects in the 1960s and 1970s, as Stephen Johnson has shown, and played a major role in helping Europeans acquire the skills needed to bring com-plex space projects to fruition.[29]

Close collaboration with France also had an important political and ideologi-cal role. Many French scientists were left-wing. Working with NASA sharpened their perception of the differences between the two world systems. Roger Bonnet, for example, who grew up in a communist family was first attracted to space by Soviet achievements. And even if he would have liked to work closely with Soviet colleagues, he found that, by adopting an "open policy of information which we could not always get from the Russians," NASA "could attract and involve the best foreign scientists in their programs, directly or indirectly [...] So, ultimately there was a greater appeal to cooperate with the Americans."[30]

From NASA's point of view, collaboration with France did not simply kick-start the national space program, and build a community that adopted NASA's management practices, so facilitating the day-to-day technical cooperation between people on both sides of the Atlantic. It was also an instrument of "soft power" that provided a counterweight to the attraction that some French scientists felt for working with the highly successful Soviet program.[31]

Italy

The San Marco project, named after the patron saint of seamen, was a major cooperative effort to build an Italian satellite and to launch it using a Scout rocket.[32] In October 1962 NASA deputy administrator Hugh Dryden described it as "the biggest and most important international program in which NASA was presently participating."[33] Its novelty lay in the use of a sea-borne platform to launch a payload that measured the atmospheric density and the character of the ionosphere in the equatorial region.

The driving force behind the project was Luigi Broglio, a professor at the University of Rome, a lieutenant colonel in the Italian Air Force, and the recognized Italian authority in the field of aeronautics. Broglio discussed the San Marco project tentatively with NASA officials at the COSPAR meeting in Florence in April 1961. US interest in the scheme led him to coauthor a proposal to Prime Minister Fanfani. He was attracted by the idea: it was suitably ambitious to capitalize on the Italian public's fascination with space flight, it harnessed science and technology to industrial development and national pride, and it would provide government support for the aerospace industry. In May 1962, just a year after the preliminary contacts were made at COSPAR in Florence, Broglio and Dryden signed a memorandum of understanding (MoU) between the Italian Space Commission and NASA for the realization of the San Marco project. It entered into force in September 1962.

As defined in the MoU the project had three phases. In the first a satellite would be designed and built by the Italians and its instruments would be tested on sounding rockets launched from Wallops Island. A prototype of the satellite would then be launched by a Scout rocket from the same base. Finally, in phase three the satellite would be launched by a Scout from an Italian platform located in equatorial waters.

NASA offered to help the Italian scientists and engineers at all stages of the project, in the spirit of Porter's proposal at COSPAR in March 1959. It would provide sounding rockets and two Scout launchers. It would provide technical support and training for the design, fabrication, and testing of the payloads, and in vehicle assembly, launch, and range safety. NASA would also provide tracking and data-acquisition facilities for the sounding rockets and the first Scout launch from Maryland. The Italians would take over this function when they launched the second San Marco satellite from their floating platform in the Indian Ocean.

During 1963 and 1964 over 70 engineers from Broglio's group were trained in the United States. They learnt about spacecraft at the GSFC. At the Langley Research Center they were trained to use NASA's Shotput sounding rockets, a two-stage unguided vehicle stabilized by aerodynamic fins and developed at Langley by combining standard solid-propellant motors.[34] They learnt range procedures and safety practices on Wallops Island. The prime contractor for NASA's Scout rockets, Ling-Temco-Vought (LTV) instructed them on the assembly and checkout of the vehicle. All of these exchanges seem to have gone smoothly until Broglio asked if he could buy all the components of the Scout in the United States and assemble it in Italy to save costs and to acquire significant technical information. This was refused point blank, in line with a general policy of not proliferating sensitive rocket/missile technology even with one's closest allies. A compromise was struck in which the cost of the launcher was reduced by $150,000 (to $495,000). Broglio and LTV also signed an agreement in which three of the contractor's senior engineers would assemble the Scout in Rome along with people from the CRA (Centro Ricerche Aerospaziali—Aerospace Research Center) and from Italian industry.

The Italian spacecraft was tested using several Shotput launches in the first six months of 1963. In parallel Broglio began setting up the floating launch platform off the African coast. An oil-rig platform was purchased and towed to Formosa Bay off the coast of Kenya. The site chosen was near-equatorial at about latitude 3°S and longitude 40°E. The Santa Rita platform, as it was called, was validated using three Nike-Apache rockets in the spring of 1964. In December that year, with extensive help from NASA and LTV, an all-Italian CRA crew successfully orbited the San Marco 1 satellite from Wallops Island with a Scout rocket.

The floating platform was the centerpiece of the final phase of the joint project. The Italian authorities decided to use a new platform for launching purposes, and to commission the Santa Rita platform as a control center. The new San Marco platform, acquired from the US Army, was a rectangular steel barge 90 feet wide, 300 feet long, and 13 feet deep. It was towed to Kenya via the NATO Mediterranean base in La Spezia, just south of Genoa. Once embedded in the ocean floor it supported the launcher and its transporter, as well as the electrical and mechanical ground support system for servicing and testing the rocket. Santa Rita, anchored about 1,800 feet away in the bay, housed the range control, blockhouse and telemetry gear, and living quarters for about 80 people. A small tower attached to the platform supported the generators that provided the electrical power for the launch complex. On April 26, 1967, the San Marco 2 spacecraft was successfully launched into an equatorial elliptic orbit by a Scout Mark II rocket. It remained in orbit for almost six months, providing valuable new scientific data on the structure of the ionosphere and on local variations in its electronic density.[35]

The San Marco project was an essential component of the early Italian space program. NASA and the Scout's prime contractor LTV did not simply provide invaluable technical training and support in all aspects of satellite construction and integration, launcher use, range management, and tracking and data analysis. They also provided Broglio with the arguments and the additional credibility that he needed to persuade his authorities to invest in a major space effort, and to release funds to support the people, the institutions, and the industries that would become the backbone of an autonomous space program. For the State Department the venture provided an opportunity to express US solidarity with an administration that was a faithful American and NATO ally, and that was under constant domestic left-wing and communist pressure in the 1960s. For NASA the project was coherent with its mission to promote international cooperation. It produced valuable scientific data on the ionosphere in the not-easily accessible equatorial region.

To conclude it is worth quoting Frutkin's account of his visit to the San Marco complex shortly before the launch took place. It provides an entertaining antidote to the dry account one gains from official records, which really cannot do justice to the spirit of adventure and personal satisfaction derived from these early, sometimes artisanal collaborative space research efforts:

> We had the agreement for the Italian San Marco project for a launch from their platform launch site. It's a marvelous, marvelous program, and the greatest fun in the world. [...] You see what happened was our project people within NASA who were pursuing, monitoring the Italian effort to get prepared for a Scout launch from this platform came in and said, "We're not going to be able to do this...We've been out there to that platform, it's a mess. It's a god-awful mess." Well, that was the first occasion when I was threatened with a cancellation. So I got hold of one of my buddies, somebody in the program, a very able, capable guy, Jack Townsend, who was then number three man at Goddard, and we went out to Africa together and climbed up onto that platform and looked around. There was water on the deck and there were wires snaking all around in the water and every-thing else, and it did look a bit of a mess. After a careful look-around Townsend said, "No problem, it'll work." [...]
> We went out to see the first launch, went up to the top of this Texas tower they were using and when it came time for lunch, they said, "Let's go up to the *terrazzo*." We went up to an upper deck under a striped awning where a great tribal warrior with scars, ritual scars, on his face made the pasta. [...] The Italians are more fun than anybody.
>
> —Arnold W. Frutkin, in conversation with the author.[36]

West Germany

The trajectory of West Germany's entry into the space age was marked by her history. The horrors of the Nazi regime, its promotion of advanced technolo-gies like the lethal V-2 missile developed by Wernher Von Braun and his team at Pennemünde, and widespread fears of a resurgence of German nationalism and militarism led the allies to impose severe constraints on the country's scientific and technological development after the war.[37] In the mid-1950s the division of Germany became accepted as a (temporary) fait accompli in the context of Cold War rivalry. The Federal Republic was given its sovereignty and entered NATO. A major effort was also made to integrate West Germany into the embryonic supranational nuclear power organization, Euratom, and into the European Common Market. The State Department actively promoted these initiatives. Its policy was guided by what diplomatic historians call double containment—restraining both Soviet expansion and German nationalism by building a strong, integrated Western Europe under American leadership.[38]

In October 1954 Chancellor Konrad Adenauer solemnly pledged that the country would never develop nuclear weapons on its soil. With this path to superpower status denied them, an alternative path to international signifi-cance was actively promoted by Franz Josef Strauss. Strauss was the minister of atomic affairs for one year beginning in October 1955, after which he was nominated the federal minister of defense. He was convinced that "the indus-trial competitiveness of a country as well as its international political weight was going to become increasingly dependent upon the national ability to master new technologies."[39] This national agenda was translated into his local political ambitions. As Niklas Reinke puts it, Strauss was a crafty strategist who, "notwithstanding his undoubted devotion to his homeland...acted with an eye to his political power base in Bavaria."[40] He adopted "a state-supported industrial policy aiming at creating innovative high technologies [...]."[41] When he was minister of atomic energy he actively promoted nuclear energy at Garching near Munich, and lobbied for the interests of German firms that wanted to develop civilian nuclear power. As minister of defense he ensured that the Deutsche Versuchsanstalt für Luftfahrt (DVL, German Aeronautical Test Establishment) was also established in Bavaria. It was again Strauss who in 1961 enabled Ludwig Bölkow, who had done sophisticated design work for Messerschmitt in the Third Reich, to create a big complex of industrial research laboratories for the aerospace industry next to his military production facilities in the south of Munich.[42]

Germany's pool of skilled scientists and engineers was seriously depleted by the emigration—sometimes forced—of thousands to the allied powers after the war. Those who remained gradually built up small communities of space scientists and engineers in the early 1950s. Helmut Trischler tells us that these groups served two important functions. First, they helped reinterpret spaceflight in the political and popular imagination as a peaceful activity, dedicated to scientific exploration and technological advance. Second, they built a network of space enthusiasts dedicated to rocketry and the space sciences. This network established international linkages, including with the United States, successfully lobbied for the foundation of a university chair, established officially sanctioned research institutes, and built ties with German industry.

Increasing industrial capacity, along with growing scientific interest, notably after the IGY, were not sufficient to galvanize the German government into action. Nor did the launch of Sputnik, which was seen as just another factor in an arms race between the superpowers in which Germany was not a participant.[43] By the end of the 1950s "space activities enjoyed a degree of political support from various ministries, but this did not as yet amount to space politics or a space policy." In addition the minister for economic affairs, Ludwig Erhard, did not approve of Strauss's views on state-interventionism in the economy, and did not see space activities as being significant drivers of economic development. It took the initiative in June 1960 by two of the founding fathers of CERN (European Organization for Nuclear Research) to build a collaborative European space effort to "arouse the authorities from their research policy torpor."[44] Eighteen months later Chancellor Adenauer put an end to interdepartmental rivalry, and gave the Atomic Affairs Ministry overall responsibility for space science and space transport research and development. This department was transformed into the Federal Ministry of Scientific Research in May 1963.

The scientific community, with industry's support, made three main criticisms of the Federal Republic's space policy in the 1960s.[45] They opposed Erhard's free-market philosophy, insisting that the federal government should take responsibility for space research and development that was far from the market, and promote it as a core national asset. Second, they insisted that it was imperative to provide sufficient funds to develop a strong national capability and to participate internationally, both with Europe and the United States. Third, they emphasized that working with other space programs was meaningless without significant investments in a domestic effort. As one document submitted to the now-chancellor Erhard put it in July 1965, "All experience in science and technology shows that unless national funding is at least two to three times greater than contributions to international programs, much of the money contributed to those programmes must be considered as subsidies on which there is no return. In those circumstances, we are simply supporting space research in other countries."[46]

This financial aspect of this plea was not heeded. The political imperative of being engaged in the European program, including in the development of a European launcher, skewed space expenditure away from the national.[47] As the German authorities struggled to find the right balance between a national program and a European collaborative effort, bilateral programs emerged as a means to lever limited resources to kick-start space activities: Trischler remarks that this was not indicative of a clear political strategy; it was dictated by pragmatism. The preferred collaborators were the United States, where Germany would necessarily be a junior partner, and France, where the asymmetry between the nations was less marked.[48]

The first links with NASA were established by the minister for atomic affairs, Siegfried Balke, in February 1961, who visited the United States again in May 1962. Balke's successor, Federal Research Minister Hans Lenz, crossed the Atlantic with his counselor Max Mayer in June 1963. During these visits German officials became painfully aware of the limits of US support. Rocket technology would not be shared on a bilateral basis. The amount of funding Germany intended to allocate to space research produced what Mayer called "sympathetic smiles."[49] A visit to von Braun at the Marshall Space Flight Center in Huntsville was also disappointing. Mayer asked if some of the German-born members of von Braun's team who were "experienced policy and contract planners" could be released for a limited period of time to work with Lenz. They would be on the government's payroll and would help the minister "put the show on the road." Von Braun refused.[50] It became clear that if West Germany was going to build a bilateral program with NASA it would have to bring something to the table. A small research satellite footed that bill.

In April 1964 a 60- to 80-kilogram scientific satellite labeled Project 625A emerged as the centerpiece of the Federal Republic's first national satellite project.[51] The concept was presented to NASA as a suitable candidate for a bilateral program six months later. Its mission was to explore the interaction between cosmic rays and the magnetosphere, notably in the region of the inner Van Allen radiation belt and of the Northern Lights, as well as during temporary changes in solar wind during eruptions on the sun. On July 17, 1965, a groundbreaking memorandum of understanding was signed between the Ministry and NASA.

Project 625A meshed with Germany's wish to build its national scientific and industrial strength. Early in 1965 the Federal Research Ministry had received over 100 experiment proposals from academic institutions, independent research establishments, and industry. Seven of these were selected. As for the satellite itself, the prime contractor was Bölkow, and was responsible for payload integration, testing, and launch support. Many other firms were involved, including AEG, Dornier, and Siemens. These firms improved their technical capability by testing new processes and techniques involving components developed in the United States. As in the French case, with the help of TRW they also gained insight into NASA's management methods to better cope with an enormous technological challenge for which, as Bölkow put it, they were "scarcely prepared."[52]

The arrangement with NASA respected Frutkin's criteria for international cooperation. It was concluded with a government ministry. There were clean technological interfaces. There was no exchange funds. West Germany's contribution was some 80 million DM. NASA provided for preliminary testing of experiment payloads on sounding rockets. It also provided a Scout rocket for the launch, and initial tracking and data services for the satellite. These were later taken over by a newly created German Space Operations Center at Oberpfaffenhofen whose personnel had been trained by NASA. The only condition imposed by NASA was that the German project should not duplicate work already done in the United States and that all of the new data obtained should be made freely available to the entire scientific community.

The 71-kilogram satellite was launched on November 8, 1969, when it was baptized Azur. The tape recorder failed five weeks after its launch, after which data could only be received in real-time. For reasons that are still not clear, all contact was lost with the satellite late in June 1970, over a year before its expected demise. All the same, as Reinke puts it, "the political hopes vested in the venture were not disappointed: the involvement of many firms in the Azur mission expanded the expertise of German industry and the German science community in the space sector and prepared them for many tasks."[53] Azur was not only Germany's own spacecraft. Twenty-five years after "the end of the calamitous Peenemünde project, German science and industry had successfully demonstrated its capacity and its determination to peacefully re-enter space."[54]

Another important step toward NASA-West German collaboration was taken a few months after the memorandum of understanding that led to Azur was signed.[55] The plan was publicly announced at a state banquet in honor of Chancellor Ludwig Erhard a few days before Christmas 1965. In a brief toast to his guest President Johnson remarked that the time had come for the two countries and other European partners, "to do together what we cannot do so well alone." He identified a probe to the sun and a probe to Jupiter as appropriate ventures that were both "very demanding" and "quite complex." Both would contribute "vastly to our mutual knowledge and our mutual skills." Johnson did not fail to couple this proposal with broader foreign policy considerations, thanking Erhard for "the support which your Government has given to the common cause in Viet Nam, and which you may give in the days ahead."[56] The president's high-profile offer to collaborate in space was also a public act of gratitude to a faithful ally.

In February 1966 Arnold Frutkin and Homer Newell (responsible for space science) visited several European capitals to sound out their interest in the president's proposal, which NASA had suggested to him under the label of the Advanced Cooperation Project.[57] The two NASA officials began their trip in West Germany, and also visited Britain, France, Italy, and The Netherlands.[58] The project was also presented to ESRO, which was NASA's preferred partner. The American delegation emphasized that the Jupiter probe—though only illustrative of what might be done—was technologically and managerially challenging, and would significantly advance European industry. The solar probe would be used to investigate magnetic fields and the interplanetary environment near the sun. "The reaction," writes Newell, "was surprising. [...] Only West Germany was interested in an expanded program with the United States."[59]

Newell has given several reasons for European skepticism. They doubted that either project would advance European technology. With resources for space research limited, they wondered whether it would not be preferable to devote their available funds to developing applications satellites. They also suspected that NASA was less interested in promoting European capabilities than in having Europe contribute money to large projects that Congress was reluctant to support. Some critics went further. They were suspicious that "America was dangling the Jupiter probe in front of Europe to divert attention toward science and away from more practical projects like communications satellites."[60]

West Germany's "interest" was of course sparked by the presidential initiative during Erhard's visit in December 1965. The German chancellor was far from enthusiastic about the idea, however.[61] Erhard's retained his skepticism about space projects as candidates for federal funding. Indeed just before he left to meet Johnson the Research Ministry was complaining bitterly about the tight-fisted approach of the administration. It had only managed to secure long-term financial support for Azur because a memorandum of understanding had been signed with NASA. By contrast, "funding for the development of a second scientific satellite and the conduct of further experiments, already planned under a specific program and agreed in preliminary talks with NASA [presumably the project officially announced at the State Banquet], has so far been refused" by the Finance Ministry.[62]

This financial prudence also reflected Erhard's concerns about West Germany's budget. Among a wide range of issues that were raised during his visit to Washington in December 1965 one of the most pressing concerns was the question of Germany's offset payments to the United States. West Germany was required to "offset" with military purchases the approximate costs to the American government of retaining US forces in its territory. From the US point of view this arrangement both provided a market for US weapon's systems and improved the balance of payments. From Germany's point of view, it secured an American commitment to hold the front line against Soviet expansion in the Cold War.

That said, the scheme was not popular in the Federal Republic. The flow of dollars abroad was significant For example, Erhard was supposed to place $1.35 billion of weapons orders in the United States by December 31, 1966, and to make a further $1.4 billion of offset payments by June 1967.[63] In addition, offset payments were associated in the public's mind with a series of crashes of the F-104G Starfighter jets—ten in the first half of 1966 alone—giving the impression that the United States was selling unreliable and unnecessary military equipment to its ally.[64] To add to Erhard's woes the Federal parliament had just imposed a 10 percent budget cut on the chancellor.

The offset issue was raised when Erhard met Johnson in December 1965.[65] On that occasion Johnson told him that

the Viet-Nam conflict is beginning to put a strain on our budget which will have to expand to accommodate the necessary expenditures...The President said he expected the FRG to make another payment under the offset agreement this month so as not to upset the quarterly balance of US finances and not to weaken the international confidence in the dollar.

Erhard remarked that he had taken some extreme measures to meet the budget cut imposed on him at home. He assured the president that he wanted to respect his commitments, but suggested that he was looking for greater flexibility in the US approach: "The FRG would be willing to talk about this matter but at present it had considerable difficulties," said Erhard.

Erhard returned to Washington for two days in September 1966. The American ambassador in Germany, George McGhee, advised Johnson ahead of the trip that this meeting would be "the most critical one you have yet held with the German leader." The offset agreements were now "the greatest single source of friction" between Washington and Bonn.[66] McGhee insisted that Johnson had to be flexible: Erhard's political future depended on it. The ambassador (and others) made a number of suggestions for how the burden on Germany could be reduced, including "limited purchases in the field of space and foreign aid," which would probably not exceed about $20–50 million annually.[67]

Johnson propelled space collaboration into prominence by accompanying Erhard down to Cape Kennedy during this very brief visit. In an official address in the still incomplete Vehicle Assembly Building the president personally thanked all those who had come to the United States from Germany, including von Braun, for the "great efforts" they had made to the American space program. He also enumerated the many projects that NASA had engaged in with European partners, and reiterated his desire to "vigorously pursue" international cooperation in space science, and to provide launchers for space efforts of mutual interest.[68] On the flight back to Washington NASA administrator James Webb took the opportunity to talk at length with the German chancellor. As he wrote to Secretary of State Dean Rusk, he assured the chancellor

> that the President was, in fact, offering him more than friendship and more than dollars. In fact he was offering a partnership in the development of technology that could permit Germany to increase its own capability, gain a better understanding of its own needs and opportunities for multilateral and bilateral cooperation, establish a basis for leadership in the direction it felt its leadership could be effective in Western Europe, and could set a pattern of university/industry/government cooperation suited to the needs of Germany, benefiting throughout from our own experience.[69]

Webb left his guest with the impression that "Erhard had a different attitude when we left the Cape than when we arrived. In fact," Webb wrote Rusk, "he did say that it was impossible to learn from pictures, television, and documents the true scope and magnitude of what was being done and that he had a much better appreciation of its importance."[70]

Did the trip to the Cape also signify the president's willingness to allow the purchase of civilian space technology to offset the German debt? Reinhard Loosch, who was engaged in these early discussions and who later had an important administrative role in the Federal Republic's space program, says that it did. Loosch stressed that the possibility of doing a joint satellite project with the United States not only "gave us at least the feeling that we would then be at the forefront of technology," but was also a response to the question "what can we do, mutually agreeable, in order to help in offsetting the foreign exchange expenditures of the United States government." Loosch emphasized that the FRG authorities did not object to the principle of offset. It was the implementation that was straining the alliance:

It was clear for us from the very beginning—I should say from 1955 on, when we finally came back into the international political scene—that we would have to pay for that. This was taken for granted. But then, let's do something more than just *pay,* help pay for the costs, but something where we could get something out of it. And in this respect, I think, the [collaboration with NASA] was quite, quite good.[71]

The memorandum of understanding for the cooperative satellite called Helios was signed in June 1969, more than three years after the first official contacts were made with Germany.[72] In December 1974 and in January 1976 two German spacecraft weighing about 205 kilograms each, Helios 1 and Helios 2, were launched by Titan rockets from Cape Canaveral into elliptical orbits about the sun. They were designed to fly closer to the sun than any previous spacecraft (approaching to within 45 million kilometers) and to provide novel scientific information about solar processes and solar-terrestrial relationships. The probes were designed, manufactured, and integrated by Messerchmitt-Bölkow-Blohm, who worked closely with the Federal Ministry for Research and Technology and the German Aeronautical and Space Research Test Center.[73] Each carried ten experiments, the majority of which were German (though there were also contributions from the United States, Australia, and Italy). The spacecraft, which cost Germany about $100 million, were operated and controlled from a national facility. NASA provided the deep-space tracking network to support the mission, and participated in the Joint Working Group responsible for technical implementation.[74]

Helios was the most ambitious bilateral scientific project that NASA had undertaken to date. The Helios spacecraft not only imposed advanced technical requirements on German industry, particularly for the development of the on-board power system, on-board data-processing system, and thermal controls that had to survive high levels of solar radiation, it also introduced German engineers and project managers to the way space projects were implemented in the United States.[75] Admittedly quite a bit of the equipment in these early projects was not of German origin. However, the "conscientious imitation" of successful technologies and management methods were fundamental to building an independent national effort.[76]

NASA-ESA Relations in the 1970s and 1980s: The Hubble Space Telescope and the International Solar Polar Mission

The European collaborative space program went through a number of crises in the early 1970s that were resolved by making some important institutional and programmatic changes. ESRO's mission was broadened to include both applications and scientific satellites.[77] The European Launcher Development Organization (ELDO) was dissolved, France took the lead in providing Europe with an autonomous launch capability, and a new body, the European Space Agency (ESA), was formed. ESA, like NASA, was now responsible for all aspects of the collaborative European space program (though countries could still pursue bi- and multilateral programs in parallel). To satisfy the diverse and sometimes conflicting needs of its member states, however, no single country was obliged to participate in a program if it did not want to. The exception to this was the science program that was mandatory: no government could opt out of it.[78]

This section discusses two programs to illustrate NASA-ESA relations in this period, the contribution to the Hubble Space Telescope and the International Solar Polar Mission (ISPM), which was renamed Ulysses. While the former can be counted a success, the latter, in which NASA withdrew its spacecraft from a two-satellite mission, has been regarded by Europeans as a cooperative failure never to be repeated. As Roger Bonnet, who became director of ESA's Scientific Programme in 1983, puts it, "No one can deny that the ISPM crisis had a profound and lasting effect on the attitude of ESA toward NASA and on international cooperation in general."[79] For this reason alone it commands more space than the Hubble.

The Hubble Space Telescope was, as its historian Robert Smith tells us, "designed to be the most powerful optical telescope ever constructed."[80] Its centerpiece was a 2.4-meter primary mirror, whose collected light was reflected back through a hole in the mirror to be analyzed by five instruments and the telescope's fine guidance system, which served as the sixth instrument. The main scientific justification for the Hubble was "the large increase in capability promised by the instrument's resolving power, rather than its ability to tackle any particular scientific questions"[81] Free from the interference of atmospheric absorption, the giant telescope made observations at wavelengths ranging from 120 nanometers to 1 millimeter, covering the ultraviolet, infrared, submillimeter, and optical regions of the spectrum. The telescope was approved by Congress and the White House in 1974, construction began in 1977, and it was launched by the Space Shuttle in 1990.

During the planning stages of the Hubble (the famous astronomer's name was actually only added in 1983) NASA discussed a possible contribution from Europe. A Faint Object Spectrograph was one interesting candidate for European participation that NASA quickly ruled out: it was seen as one of the most important instruments on the telescope and the major partner was obviously not going to hand it over to a junior participant. The alternative that emerged was a Faint Object Camera (FOC) that made use of a technique called the Imaging Photon Counting System developed by University College, London. The FOC's task was to examine exceptionally faint objects that could only be "seen" by collecting light during many orbits of observation time. NASA agreed that this instrument could be one of those included in the system. What is more it was willing to accept Europe's demand that this contribution need not be subject to open competition with other instrument proposals coming from the space science community. A place on board was guaranteed—on condition that NASA was satisfied that the Europeans had the technological capability required to build such a sophisticated piece of equipment.

A "tiger team" of US engineers and astronomers visited the laboratories and industrial plants engaged in the project to see for themselves. They concluded that the technology, the facilities, and the expertise required to build an FOC existed in Europe. But they were unhappy with the design being proposed. They felt that the inclusion on the camera of two possible light paths and a spectrograph complicated the device's mechanism unnecessarily and might cause catastrophic failure in orbit. It was an FOC "with bells and whistles attached."[82]

The negotiations over the space telescope were not without conflict. Some European scientists felt that it was unnecessary to use scarce resources for an expensive, dedicated instrument when NASA was soliciting proposals for experiments in open competition. Others resented the implication in the United States that European industry was not up to building a device as complex as the FOC. In any event it took an "unaccountable number of meetings" to find a suitable agreement.[83] Nancy Roman, who was responsible for astronomy at NASA, was a central figure in these negotiations and is fondly remembered by the Europeans for her generous hospitality. The final arrangement gave ESA 15 percent of the observing time on all instruments in return for contributing one of them. This has been more than respected: in fact Europe's share of observing time has been closer to 20 percent after proposal selection through NASA's competitive peer review system (and thanks to their additional contribution of solar arrays, later replaced by the United States).[84]

The ISPM "was born to be the paradigm of ideal cooperation between NASA and ESA."[85] Its aim was to send two spacecraft, symmetrical with respect to the plane of the ecliptic, to simultaneously fly above the opposite poles of the sun. Each agency would develop its own satellite, and scientific instruments from both sides of the Atlantic would be accommodated on each in open competition after peer review. There would be no exchange of funds, and both were to be launched together on the Shuttle in February 1983.

In November 1977 the ESA space science community selected its satellite, along with participation in the Hubble, rejecting four other proposals. One of the reasons ISPM won out was that "the dual mission, to which ESA with its spacecraft would make a major contribution, offers the basis for a clean interface and fruitful cooperation with NASA."[86] The experiments were jointly chosen in February 1978, offering a place on the payloads to more than 200 scientists from 65 universities and research institutes in 13 countries.

The funding procedures were very different. Funding was secured on ESA's side by the policy of ensuring cost-to-completion for projects once they were accepted by the member states. Budgetary control was exercised by demanding that the cost for the development of the satellite, its launch, and its operation did not exceed 20 percent of the envelope estimated at Phase B (project definition phase). In short, once ISPM was accepted it was extremely likely that Europe would maintain funding to completion. On the US side Congress gave the go-ahead for the ISPM by including the satellite in the FY1979 budget. At this stage of development this was, of course no more than a statement of intent, not a commitment to complete. NASA's appropriation is renegotiated annually in what is sometimes a bruising battle with the White House, the Senate, and the House. The agency is obviously never granted all the funds that it applies for, and sometimes has to make hard choices that can seriously impact the viability of some missions. In the United States, in other words, there is no guarantee that a project will be funded to completion when start-up funds are allocated to NASA. Budgetary control takes place annually.

The Europeans were aware of this. They hoped, though, that the international MoU detailing their respective obligations in the ISPM mission, while not having the force of a treaty, would bind NASA and the US administration tightly into the collaboration, and protect ISPM from the annual vagaries of the budget allocation process in Washington. This despite the clause in the MoU stipulating that the execution of the project was "subject to the availability of funds" by both partners. "Unfortunately," as Bonnet and Manno put it, "the events which followed shattered this quiet conviction and initiated a new era in the relations between ESA and NASA."[87]

NASA's difficulties with this mission were created by the need to complete the Shuttle and by increasingly deep cuts to its space science budgets by successive administrations. The warnings were there when NASA was instructed by the Carter administration to slash its budget for FY1981 in advance of the elections. One measure that it took was to postpone the launch of ISPM by two years to 1985. This decision was discussed with the Europeans, who reluctantly accepted it. Once President Reagan entered office in 1982 the downward pressure on NASA's budget increased further. David Stockman, the new director of the Office of Management and Budget (OMB), was determined to reign in federal spending. NASA responded to his cuts by reducing its budget for space science by 30 percent. In doing so it eliminated its ISPM satellite without consulting ESA.

ESA's director general and the director of the science program objected strongly. The cancellation of NASA's spacecraft degraded the scientific objectives of the dual mission and eliminated about 80 European and American investigators at a stroke. If ESA followed suit and cancelled its mission it stood to lose about $100 million. NASA stood firm. It would continue to provide the launch, a radioisotopic thermal generator that was on the payload, and the retrieval and dissemination of data from ESA's satellite.

Faced with this situation, ESA officials came up with a new idea: that Dornier, the prime contractor on the European spacecraft, should produce a second unit for NASA at little additional cost to the agency. It would not be as sophisticated as the original American satellite but most of the scientific mission would be salvaged. In a desperate attempt to save the dual mission, the ESA executive visited Congress, the State Department, the Office of Science and Technology Policy, and the Office of Management and Budget. Congress was sympathetic, and NASA was willing to reinstate the ISPM, but only if it was granted additional money by the OMB. It had other international obligations—to Galileo (with West Germany) and to the Hubble Space Telescope (with ESA). It was not prepared to jeopardize either to save ISPM. NASA administrator James Beggs delivered the coup de grace in September 1982: he informed ESA director general Erik Quistgaard that the agency would not include any request for a second ISPM spacecraft in its new budget request. ESA decided to go it alone with one spacecraft, renamed Ulysses, whose launch was further delayed by four years by the Challenger accident in January 1986.

In their account of this unfortunate affair, Roger Bonnet and Vittorio Manno are uncompromising in their critique of the way NASA and the US administration handled matters, notably the failure to consult.[88] NASA official Lynn Cline understands the frustration but noted that the European view doesn't capture the full picture of what NASA faced on its side. As she put it,

> We were going through our budget review between NASA and Office of Management and Budget. NASA was directed to take a large cut in its budget, and we were told that we weren't allowed to take the cut in certain areas. So that left us with some science programs as the particular area that was under debate. So the question was, did we take a budget cut in Hubble Space Telescope, [that] was one of the options. That happened to be a cooperative project with the European Space Agency, and obviously, for our science community, a very prestigious, high-priority project. The second option was to take a cut in the Galileo mission, and I happened to be the German desk officer, so that was the one I was working on, [and I saw] all the reasons why we shouldn't do that. And then the third option was to take the budget cut from the International Solar Polar Mission. All three of those were international missions, two with ESA, one with Germany.[89]

Why then was the ISPM cut? Cline explains:

> One reason was that you could cut out one spacecraft and not terminate the entire mission. Secondly, NASA would still be able to provide the launch and all of the tracking capabilities, as well as its science instruments, for the one remaining spacecraft. So while we were losing a portion of the mission by eliminating one of the spacecraft and losing some of the flight opportunities for science instruments, that was less severe than lose a Hubble mission or lose a Galileo mission, and so that was the lesser of the evils, if you will.

Why did NASA not discuss this decision with the Europeans before it was made public? Why the failure to consult? Cline points out that this procedure was not of the agency's choosing:

> NASA went to the Office of Management and Budget and asked for permission to talk with Europe about this, and we were told by the administration that the budget was embargoed and we were not allowed to consult with the European Space Agency on this. So the first time we were able to directly address it with Europe was when it was broadly [...] public and a fait accompli, which obviously was not well received, and we went through all of the protests from Europe about not consulting and weren't there other options and can we restore this, and a whole series of activities.

The legacy of the ISPM affair lives on in relationships between the United States and Europe, notwithstanding the fact that Ulysses mission was carried out, and, more generally, that this was an isolated, if unfortunate case, and did not in any way signify a retreat from a commitment to international cooperation. As Lynn Cline put it,

> Now fast-forward to years later when I was lead negotiator for the Solar Terrestrial Physics Program, which was a NASA-ESA collaboration, at virtually every negotiating session I was treated to a lecture from the Europeans on how horrible we were as a partner, and new language they needed in the agreements to guarantee

some greater protection for them on consultations and follow-through from the U.S., as a result of that experience. I heard about it again when I did Cassini, and I heard about it again when I did the International Space Station negotiations.

It has to be admitted that the United States handled the ISPM situation badly through lack of consultation, though even here NASA had its hands tied by the administration. More fundamentally, though, this kind of situation is always possible because the budget of the US space program is subject to annual review and cuts. The central lesson of the ISPM affair is that this procedure cannot be overridden by legal instruments like a memorandum of understanding, even at the international level. Hard choices are imposed by the funding regime under which NASA is obliged to operate and—as the ISPM affair illustrates—no international partner can count on their collaborative project being immune to unexpected budget cuts, or even to cancellation.

Cassini-Huygens

It is clear that Europe cannot allow itself to be reduced to a subordinate or subsidiary role in space ventures if it is to maintain its current hard-won position [...] The need for international collaboration on major space undertakings is not disputed, but Europe wishes to enter such undertakings on an "equal partnership" basis, this concept applying at all levels, including operational control.

—Reimar Lüst, 1987[90]

The new determination by ESA to be taken seriously as an international partner by the United States required a change of approach in Europe. More resources were needed for space science along with a coherent plan that could be used to win the broad support of the multidisciplinary space science community and of potentially reluctant member states. Existing procedures for selecting experiments were creating some resentment in the United States and also had to be revised. The Cassini-Huygens spacecraft, whose launch in October 1997 by a Titan IV-Centaur vehicle was "regarded as a miracle by some people involved in the mission," not only promised to be of immense scientific importance, it also benefited from these institutional reforms, along with the possibility of levering the Clinton administration's commitment to international space collaboration in the mid-1990s.[91]

The Cassini mission extended the avenues opened up by the data from the Voyager 1 and Voyager 2 flybys of Saturn in 1980 and 1981. The mother craft that would approach Saturn was provided by NASA, and the probe that would land on Saturn's moon, Titan, was provided by ESA. The Italian Space Agency provided telecommunications and microwave systems. US-European cost-sharing on Cassini-Huygens was about 70 : 30. Eighteen instruments would "conduct orbital remote sensing of Saturn's atmosphere, icy satellites and rings; in situ orbital measurements of charged particles, dust particles, and magnetic fields; and detailed measurements [would be made] with six instruments on the Huygens probe during descent though Titan's dense, nitrogen atmosphere to the surface."[92] The probe would also make surface science measurements if it survived impact. The range of questions addressed by the mission was such as to attract broad-based support in the planetary science community.

The intricacies of the decision-making processes and funding battles that accompanied the acceptance and development of the Cassini mission on both sides of the Atlantic have been adequately described elsewhere.[93] This brief account will focus on those features of the mission—be they scientific, institutional, or political—that provide insight into the international aspect of the cooperation.[94]

When Roger Bonnet was nominated ESA's director of the scientific program in 1983 he was determined to place it on a stable base. He believed that ESA required a long-term science plan that was ambitious enough to demand regional collaboration, and broad enough to satisfy the diverse needs of the European space science community. It also had to be challenging enough to attract international cooperation with leading space powers, notably NASA, without being vulnerable to the kinds of setback that had bedeviled ISPM. Bonnet's solution was Horizon 2000.[95]

Horizon 2000 emerged after intensive and extensive consultation with the European space science community. It comprised four costly, long-term "cornerstones." Two were in the field of solar system exploration (solar-terrestrial physics and cometary science), and two were in the field of astronomy/astrophysics (X-ray spectroscopy and a far-infrared telescope). These cornerstones were to be under ESA's leadership and to be consistent with Europe's own technical and financial means "in order for ESA to be master of its own future and not to be dependent upon decisions taken outside its own control."[96] The cornerstones were complemented by small- and medium-sized satellites with no a priori exclusion of disciplines, and were to be selected one by one. This introduced the flexibility needed to respond to changing scientific demand and to take advantage of opportunities for international cooperation.

The broad scientific support for Horizon 2000, the lucidity of its logic, and the scope that it gave national administrations to plan their financial appropriations in advance had an immediate effect. Meeting in Rome in January 1985, the ESA member states agreed to increase the science budget by 5 percent annually in real terms (i.e., after adjustment for inflation) for ten years. This was the first time that the science budget had been increased for fifteen years, and it made it possible not only to rationalize coordination between ESA and national science programs, but also to coordinate the agency's initiatives more effectively with its international partners. With more money available for space science, and with a protective wall around the major ESA-led cornerstones, Horizon 2000 enabled the European community to engage with NASA from a position of strength that combined competition with cooperation.

Another important source of friction between ESA and NASA was removed in 1983.[97] In line with the announcement at COSPAR in March 1959, NASA had a policy of allowing any interested party to respond to an Announcement of Opportunity (AO) on its space science satellites. This caused little difficulty when other programs were in their infancy. But as they matured, and more and better foreign proposals were received, some American scientists began to feel that the agency preferred payloads submitted from abroad because they were free of charge to the US, as opposed to US entities having to pay the cost of their experiments. This frustration was heightened by ESA's restriction of its AO to proposals from member state scientists, as required by its charter. Nor did ESA feel that it should be called upon to reciprocate each individual agreement that a member state had negotiated bilaterally with NASA without involving the European agency.

Bonnet was called upon to resolve this thorny issue as soon as he took up his new position at ESA in 1983.[98] The matter was resolved after a spirited discussion thanks to the previous progress made by a committee of "wise men" that ESA had set up to tackle the problem and make recommendations. To defuse the obvious ill-will that the European policy was causing it was agreed that ESA, like NASA, would open flight opportunities to foreign investigators.[99]

In November 1988 ESA's Science Program Committee selected the Titan probe as the first medium-sized mission in the new Horizon 2000 paradigm, and baptized it Huygens to emphasize its European provenance. A year later the US Congress approved start-up funds for the Cassini and CRAF (Comet Rendezvous Asteroid Flyby) missions, the latter a joint venture with Germany. ESA and NASA issued separate but coordinated AOs for their respective contributions to Cassini. Sixteen European countries and the United States provided 18 instruments distributed over both mother craft and probe, with two–ten countries providing parts of each instrument. The overall management of the program was based at NASA Headquarters. Project managers for the Cassini mother craft and the Huygens probe established offices at the Jet Propulsion Laboratory (JPL) and at ESTEC (the European Space Research and Technology Centre), respectively. They were advised by Project Science Groups that gathered together all principal investigators, scientists, and team leaders that had instruments on the parts of the spacecraft that they managed. These groups served as a valuable forum "to optimize scientific return and to resolve the usual conflicts between the engineering and science sides of the mission."[100]

In fall 1991 the trajectory of the joint project hit a bump that threatened to sour the good relationships that had been established between the partners. A House-Senate committee cut the budget allocation to Cassini/CRAF for 1992 by $117 million, which NASA absorbed by deciding to delay the launch of Cassini from 1995 to 1997. The chairman of ESA's Space Science Advisory Committee, David Southwood, immediately contacted Berrien Moore, the chairman of NASA's Space Science and Applications Advisory Committee. Southwood emphasized that the increase in the cost of the Huygens probe caused by the delay would create an "intolerable stress" on ESA's program. It had not been easy to get the member states to agree on funding for the probe and for instrumentation for Huygens and Cassini. Their delegates had been "dragooned, cajoled and otherwise persuaded" to do so, "by emphasizing the importance of not delaying the NASA timetable." A launch delay imposed by NASA "within a year of the selection" would increase costs by about 15 percent, and could seriously undermine the "climate of cooperation."[101] Southwood's letter was quickly followed by one from ESA director general Jean Marie Luton to NASA administrator Richard H. Truly stressing that any delay in the launch date was "unacceptable" and would cost ESA a further $30 million.

In 1992 NASA and Germany agreed to cancel CRAF altogether. Responding to European objections, engineers at JPL in consultation with their European colleagues simplified the orbiter design to meet the domestic budget cut without delaying the launch. Instruments that were mounted on movable platforms that could be continuously pointed at their targets were bolted down so that the entire spacecraft had to be turned toward the target to take measurements. A separately steerable antenna intended to provide a communications link to the Huygens probe was removed, and just one antenna was used for the Cassini-Huygens link and for the Cassini-earth link. This meant that scientific data had to be stored in a buffer system until data-taking was suspended, whereupon the antenna could be turned toward the earth to transmit the stored information to ground stations. To absorb the increased operational costs of the program it was also decided to drop plans for the acquisition of scientific data in the journey through space to Saturn and its moon. While the scientific community was distressed by the limitations imposed by these changes, they also realized that some "descoping" was imperative if there was going to be any mission at all.

Cassini almost suffered the axe again in preparation for the president's budget request to Congress in January 1994, and the Congressional deliberations in the summer of that year. The threat-level was increased by the approach taken to satellite projects by a new NASA administrator collectively known as "faster, better, cheaper." In 1992 the National Space Council, reestablished by President George H. W. Bush, engineered the removal of Richard Truly who they felt was too committed to NASA's tradition of large and costly activities.[102] He was replaced in April by Dan Goldin, then an executive of TRW who had the reputation of favoring small, inexpensive spacecraft. In his confirmation hearings Goldin did not suggest that "faster, better, cheaper" was necessarily the best way for NASA to operate, and he did not mention it at all in his first address to NASA employees. It was only when he got down to preparing the agency's budget request for FY1994 that he felt that NASA had "unrealistic" expectations. Goldin decided to "re-invent NASA" by miniaturizing technology and by streamlining project management. Let's see, he said, how many satellites "we can build that weigh hundreds not thousands of pounds; that use cutting edge technology, not ten-year old technology that plays it safe; that cost tens and hundreds of millions, not billions; and take months and years, not decades, to build and arrive at their destination." From henceforth larger spacecraft were to be the exception, not the rule for NASA projects.

Faced with pressure from the Senate to reduce the budget, Goldin cast a skeptical eye over Cassini-Huygens in 1994. Indeed it was a prime example of the kind of mission that he wanted NASA to avoid. Howard McCurdy's calculations of the cost and weight of the satellite and its probe from various NASA sources give one an idea of why Goldin was so concerned:

> *Cost Cassini.* Launched 1997. Development, $1,422m; launch support, $422m; mission operations and data analysis, $755m; tracking and data support, $54m; foreign contribution, $660m; total, $3,313m (real-year dollars)
> *Weight Cassini.* Orbiter, 4,685lbs; Huygens probe, 705lbs; launch vehicle adapter, 298lbs; propellant, 6,905lbs; total, 12,593lbs.[103]

Contrast the $3 billion plus for this mission that matured for fifteen years, and that needed another eight years after launch to begin taking data, with the cost and time of the satellites built by NASA respecting Goldin's mantra. Beginning in 1992, the first sixteen missions flown under the new philosophy together cost less (in inflation-adjusted dollars) than did Cassini-Huygens alone. Nine of the first ten of these ventures were a success—though the initiative floundered in 1999 when four of the next five "faster, better, cheaper" missions failed.[104]

Given the inhospitable climate at NASA to a mission of this scale it is hardly surprising that a major effort was made on both sides of the Atlantic to save Cassini-Huygens from further damage. This time the scientific communities were united by their dependence on each other, as Roger Bonnet explained: "The Europeans wanted to put their probe on Cassini because they could not do the mission without it [...] For the Americans, the provision of the probe was a unique opportunity to do outstanding novel science." In Bonnet's view the Europeans also brought more, though: project stability. He remembers "Carl Sagan calling me on the phone from California asking for help because NASA was trying to stop the mission." European ambassadors to Washington were asked to impress upon the State Department "that they could not stop Cassini, with such a big involvement of Europe, both on the payload of Cassini and with the Huygens probe."[105]

European pressure over the satellite was given added leverage because the Clinton administration needed to make amends for its poor handling of the geopolitics of the International Space Station (ISS) that Canada, Europe, and Japan had joined in the 1980s. This is discussed in chapter 8 of this book and the thread is taken up again in chapter 13. For the present, suffice it to say that meeting in Vancouver in April 1993 the American and Russian presidents established the Gore-Chernomydin Commission comprising a number of working groups, including one on space, to advance bilateral cooperation. A year later, beginning around April 1994 stakeholders on both sides began to explore ways to integrate Russia into the ISS. This was formalized at a meeting in June 1994. NASA and the Russian Space Agency signed an interim agreement covering initial Russian participation in the ISS program. This included a $400-million contract with the new partner, 75 percent of the American money being for Russian space hardware, services, and data in support of the "Shuttle-Mir" project (a joint flight program leading to the development of the ISS). In doing so the agency not only suspended the principle of "no exchange of funds" that had been required of its traditional allies, but NASA also rode roughshod over their sentiments. As NASA official Lynn Cline put it to me, "This was another case where I don't think we adequately consulted with our partners. People in charge at the time told Dan Goldin that we needed to consult with our partners. He didn't want to hear it."[106]

This attitude may well explain why the ESA director general, Jean Marie Luton, bypassed Goldin and wrote directly to the vice president ten days before the June 1994 meeting of the Gore-Chernomydin Commission to plead the case for Cassini. Luton upped the stakes by stressing that a negative decision on Cassini could have implications far beyond this one case. As he put it, Europe "views any prospect of a unilateral withdrawal on the part of the United States as totally unacceptable. Such an action would call into question the reliability of the US as a partner in any future major scientific and technological collaboration."[107] A month later, in July 1994 President Clinton intervened to enable NASA to proceed with both the space station and its science program. All are agreed that in this case "the international aspect of the Cassini mission was an extremely important factor in reversing almost certain cancellation of the mission."[108] It must not be forgotten, though, that that "international aspect" coupled a satellite of predominantly scientific importance with a space station of immense technological and geopolitical significance. This strong coupling is probably what saved Cassini.

Goldin did not give up his reservations about the program even after the dramatic crisis of 1994 was resolved. In 1995, much to the distress of the European participants, the NASA administrator demanded that the entire project, including the foreign contributions be subjected to an external review. This not only struck a blow to the fine cooperative spirit that had prevailed at the scientific level, it was doubly infuriating because technical findings of the review panel that were deemed to touch on matters of national defense could not be conveyed to partners abroad. In the event the mission overcame this hurdle, but was then confronted with another: the "Stop Cassini" campaign by the Florida Coalition for Peace and Justice. The coalition objected to the use of plutonium dioxide in three radio-isotopic thermoelectric generators and on heater units. This was a technological option that the designers of the spacecraft had invoked since solar power was not feasible for a deep-space mission. Rallies and demonstrations were held on both sides of the Atlantic, letters were sent to the US president for and against the mission, and protestors threatened a sit-in on the launch pad in Cape Canaveral to force a launch abort. Their objections were overruled by a safety evaluation made by the Department of Energy and the Interagency Nuclear Safety Review Panel.

Cassini-Huygens finally lifted into space on October 15, 1997. Its long journey was punctuated by difficulties that emerged in the radio relay link between the European probe and the American spacecraft. These were overcome by having Cassini fly by Titan at a far greater distance than foreseen, so that the Huygens probe had to travel 65,000 kilometers instead of just 1,200 kilometers to enter Titan's atmosphere. Cassini went into orbit around Saturn on July 1, 2004. The probe was separated from the mother craft six months later on Christmas Day, reaching Titan's outer atmosphere on January 14, 2005. The descent of Huygens was slowed when its parachutes were deployed about 150 kilometers above the surface. It survived the impact and it continued to transmit data for over three hours. The first results were relayed via NASA's Deep Space Tracking Network to the European Space Operations Center in Darmstadt, Germany, where "scientists waiting anxiously for the data to arrive [...] hugged each other when the first signals arrived during the morning, showing that the mission, 20 years in the planning and execution, was functioning."[109]

The joint development of the Cassini mission was a fine example of international collaboration. That success only makes sense, though, if placed in historical context. The scientific importance of the trip to Saturn and Titan was as crucial as the historically maturing institutional and political factors: the new cohesion of the European space science community provided by Roger Bonnet's Horizon-2000 long-term plan, the "institutional learning" that structured the joint management of the project, the determination by scientists on both sides of the Atlantic not to let a repeat of the ISPM experience sour their cooperation, and the political backbone provided by the opportunity for ESA and its member states to escalate a threat to Cassini into a threat to US-European collaboration in any future major scientific and technological project.

Chapter 3

Technology Transfer with Western Europe: NASA-ELDO Relations in the 1960s

The previous chapter described the initiatives taken by NASA to promote scientific collaboration through bilateral agreements with friendly states in Europe, and with ESA. It was stressed that this form of collaboration, while not without its tensions, was not bedeviled by the dilemmas that accompany technology transfer. This chapter explores those dilemmas in some detail, discussing the early attempts made by NASA, in consultation with other agencies in the administration, to define and implement a policy for technology transfer. Satellite-launching technology, be that with expendable or reusable systems, was the key issue around which these debates took place both within the administration, and between NASA and Western Europe.

The issue of technology transfer with Western Europe was not on NASA's agenda until the early 1960s. A survey written by Arnold Frutkin in October 1960 projecting the scope of NASA's international activities over the next decade focused exclusively on space science and the supporting infrastructure (such as the construction of tracking stations).[1]

The terms of the debate began to change when the possibilities of using space for commercial purposes began to emerge—and missiles became standard delivery systems for nuclear warheads. On the one hand the Europeans, prodded by the British, began to think about building together a multistage satellite launcher funded and developed through a new supranational organization called ELDO (European Launcher Development Organization). The intergovernmental agreement that was signed in 1962 and ratified by national governments in 1964 provided for the shared development of a three-stage heavy launcher for civilian purposes.[2]

Telecommunications satellites provided the key rationale for developing this European rocket. As early as fall 1960 the British approached NASA to learn of its plans regarding an "active" communications satellite program.[3] A collaborative venture was quickly formalized in which the British, the French, and other friendly countries (e.g., Brazil) agreed to build ground terminals on their soil so as to participate in the testing of NASA's Relay, Telstar, and Echo II satellites.[4] These experiments were followed by the spectacular success of Early Bird launched into geostationary orbit in April 1965. Early Bird, which began commercial service on June 1, 1965, had 240 voice channels—all existing transatlantic telephone cables had just 317. And it was far cheaper: the most up-to-date underwater telephone cable cost about ten times as much.[5]

West European governments and their telecommunication operators had an immense stake in these issues. They agreed to invest heavily in space, above all in the development of an independent launch capability, because they looked to a future in which telecommunications and other applications (meteorology, navigation, etc.) were an integral part of their national and international technological strategies. They saw the 1960s as the period in which they would develop their industrial capabilities so as to position themselves internationally in the 1970s and beyond. They were not driven by Cold War rivalry with the Soviet Union, and they did not seek to establish a human presence in space—this would be left to the superpowers. What they sought was (eventually) to reap the practical benefits of space (along with the possibilities for new scientific discoveries that it offered).[6] In their eyes, the meaning of (civilian) space was transformed from a domain of esoteric scientific investigation (with multiple military implications) into a sector of immense commercial and political importance. Communications satellites, in particular, not only created new opportunities for the transmission of radio, television, and telephone signals. They also promised to be an important platform for promoting and projecting images of national culture and of national prowess to the remotest regions of the globe. In other words by the mid-1960s the Europeans were seeking to become less technologically dependent on the United States and to expand their activity in space to include both science and applications, along with an "autonomous" launch capability.

NASA and the Department of State welcomed these developments. NASA's objective was to promote the peaceful use of space. The State Department strongly favored European integration and the creation of an Atlantic community: only a united Europe, under American leadership, could contain the threat of Soviet expansion on the front lines of the Cold War. Support for an organization like ELDO, which was supranational and civilian, was compatible with these goals. To quote an early position paper on the issue, technological assistance to ELDO was coherent with "our objective of an economically and politically integrated European Community with increasingly close ties to this country within an Atlantic community." In addition, by working with a multinational organization rather than making bilateral arrangements with separate states, one could hope to discourage the proliferation "of independent national medium- and long-range nuclear delivery systems."[7] Technological collaboration, unlike scientific cooperation, was thus firmly embedded in the broader strategic and foreign policy concerns of the US administration in the European theater.

American willingness to assist Europe develop its aerospace technology was also linked to concerns about a supposed "technological gap" that had opened up between the two sides of the Atlantic. These concerns were widely aired in the media and were given an important impetus with the publication of Jean-Jacques Servan-Schreiber's *Le défi americain* (The American Challenge) in 1967.[8] Some American commentators placed the blame for Europe's relative "backwardness" squarely on the continent's own shoulders (as indeed did Servan-Schreiber).[9] Others, including NASA and the State Department, took a broader view and saw the "technological gap" as a threat to the stability of the free world. For them, European scientific and technological strength was essential if capitalism was to compete successfully with the Soviet system, and if America's partners across the Atlantic were to share the burden of the defense of the West. Space was particularly important in this regard, not because of the content and goals of the space program, but because such programs were seen to be key drivers of scientific and technological innovation.

Frutkin forcefully made this point at a meeting of the American Academy of Political and Social Science in Philadelphia in April 1966. The American space program, he said, pushed established scientific and technical disciplines to probe new frontiers, be it in fields such as physics, astronomy, and geodesy, or in materials, structures, and fuels. "In fact," he insisted, "we may with increasing confidence say that the peculiar quality of space science and technology is its forcing function, its acceleration of joint progress in a wide range of disciplines."[10] Frutkin claimed that space research and development had contributed "significantly to the fundamental strength and viability of the United States in a world where economic and military security increasingly rest[ed] upon technology." The Soviet Union had absorbed the lesson, "matching and outmatching" the United States in space expenditure, notwithstanding the people's dire need for consumer goods. Western Europe, by contrast, was spending only about one-thirtieth as much as the United States on space technology. Their relative lack of interest in space could "lead only to political and economic strains and to weakness" he insisted. It was in America's interest, therefore, that the technological gap in the space sector should be narrowed: "What has stimulated, energized and advanced us, may well stimulate, energize and advance them," Frutkin suggested.[11]

This Cold War agenda, and the relatively paltry investment in space in Western Europe, obliged NASA to step in if it could. As the author of a 1964 CIA report put it, whatever measures the Europeans took to build their capability, "the assistance of the US—both officially and through unofficial commercial channels—has been, is, and will probably remain the critical factor in the success of any European space program in this decade."[12] This was the thinking that lay behind President Johnson's and NASA's support for Germany's $100 million Helios program described in the previous chapter. It also informed the administration's interest in assisting ELDO, though here the thrust to technological cooperation had to contend with a far more complex and contested policy agenda.

Two Barriers to Collaboration: French Ambitions and Intelsat's Mission

NSAM294: Curbing France—Two major hurdles stood in the way of moving beyond scientific to technological collaboration with Europe, notably in the strategically key domain of launchers. The first was enshrined in NSAM (National Security Action Memorandum) 294, signed by McGeorge Bundy and dated April 20, 1964. NSAM294 was a response to de Gaulle's determination to develop an independent nuclear deterrent, including guided missiles, and the French president's dislike for international and supranational institutions (including the North Atlantic Treaty Organization, NATO) that restricted France's sovereignty and autonomy of action. For Washington, to curb proliferation on the continent "it continues to be in this government's interest not to contribute to or assist in the development of a French nuclear warhead capability or a

French national strategic nuclear delivery capability." To that end NSAM294 directed that "effective controls be established immediately" to stop "exchanges of information and technology between the governments, sales of equipment, joint research and development activities, and exchanges between industrial and commercial organizations [...], which would be reasonably likely to facilitate these efforts by significantly affecting timing, quality or costs or would identify the U.S. as a major supplier or collaborator."[13] Closing the technological gap with France in the domain of rocketry would undermine the stability of the Atlantic alliance that the United States hoped to construct to meet the Soviet threat, rather than contributing to the defense of the free world.

NSAM338 Promoting Global Telecommunications—The second major stumbling block to technological sharing was the administration's determination not to launch foreign satellites for "separate systems," which would do "significant economic harm" to an American-led global telecommunications satellite system. US-European tensions over this issue threatened to sabotage the collaborative process in the 1960s and early 1970s, and their evolution over the decade will be treated in detail in what follows.

In his famous speech before Congress on May 25, 1961, President Kennedy not only committed the United States to put a man on the moon before the decade was out, he also called for the establishment of a single, global communications satellite system. Soon thereafter a White House press release dated July 24, 1961, invited all "nations to participate in a communications satellite system, in the interest of world peace and closer brotherhood among peoples throughout the world." The president called for the establishment of a privately owned corporation, Comsat, to handle the American portion of the system. He also made it clear that he wanted Comsat to establish the system quickly to ensure an American space first.[14] The Communications Satellite Act, signed into law by Kennedy on August 31, 1962, authorized Comsat to "plan, initiate, construct, own, manage, and operate itself or in conjunction with foreign governments or business entities a commercial communications satellite system."

Kennedy's proposal challenged the domination by Britain of an international communications network of submarine cables. It also politicized an ongoing business-led communications satellite enterprise.[15] It recognized the cultural importance of an American-led global communications system: a main leitmotif in the May 1961 speech was, after all, the use of space technology to win the hearts and minds of those who were faced with a choice between "freedom and tyranny."[16] It was also a more immediate, pragmatic, and commercially important way of "selling" space to the American public than a remote and immensely challenging lunar landing. These aims—the creation of a commercial entity that would both be profitable and promote an American foreign policy agenda in the context of intense Cold War rivalry—dominated the negotiations surrounding the definition of the role of Comsat. They influenced its relationships with the administration (notably the State Department and the Federal Communications Commission (FCC)), and with European telecommunications entities and governments.[17]

It was assumed from the outset that, to ensure the most efficient utilization "of a very rare resource—the electromagnetic frequency spectrum," Comsat would establish a single global system: competing global, regional, or national systems would be discouraged. This would avoid "unnecessary and wasteful" economic competition, the needless duplication of technologies, and facilitate the standardization of equipment.

The global system foreseen by Comsat distinguished sharply between the financing and management of space and ground segments. The space segment comprised all of the satellites in the system, along with their supporting tracking, control, and command facilities. It would be jointly owned and financed by the participants (nations, groups of nations, or regions), whose capital investment would be proportional to their potential use of the global system. The ground segments would comprise stations transmitting and receiving data from satellites and would be under the control of national private or public telecommunications entities.

The Europeans, organized through a new European Conference on Satellite Communications (CETS), were concerned from the outset about what they saw as American dominance in Comsat. In a key meeting in Rome in 1964 they voiced their hesitations over the dual role of Comsat as having both a major role in defining policy and as managing the system. They were also concerned by a procurement policy that was based on accepting the technically and financially best bid from aerospace industries. This would necessarily favor US high-tech industry, given the huge technological lag between the two sides of the Atlantic at the time.

Italian ambassador Edigio Ortona opened the Rome meeting. He spelt out CETS's position. He suggested that the global comsat system should be "owned and managed by a world organization" to ensure that all the participating countries could have an "adequate voice" in its management.[18] This should be entrusted to a multinational general conference in which each country had one vote "in principle" on most matters. Ortona also urged that in the early stages of the system special provision should be made for European industry so that it could close the technological gap with the United States.

The American delegation rejected these proposals outright. It would proceed at once to raise the capital for the global system with or without the Europeans. Comsat would manage the space segment (while each country would control its own ground segment), and voting would be determined by investment (which in turn was based on a country's use of the system).[19] Financial share would be correlated with projections of traffic data based on the use that countries made of international cable facilities: this meant that the United States' voting weight would be greater than 50 percent, perhaps as much as 65 percent. Contracts would be awarded competitively, thus effectively ensuring that no European firm could hope to participate significantly in the early technological fruits of the system. To rub in the point, in March 1964 Comsat announced that it had contracted with Hughes Aircraft Co. for the design, manufacture, and testing of two synchronous orbit communication satellites "to demonstrate to the Europeans," as one US negotiator put it, "that their refusal to agree would not hold up the system."[20]

The Europeans had little choice but to accept the US conditions. To satisfy their concerns it was agreed that the legal agreements then adopted would only apply in the interim. They would be reevaluated by an Interim Communications Satellite Committee (ICSC) in 1969, when steps would be taken to set up a permanent organization whose operation would be defined by the newly negotiated definitive agreements. The interim voting procedures were the very last to be settled. The United States would (initially) have 61.0 percent of the vote in the Interim Committee, followed by Britain (8.15 percent), and France and Germany

(6.1 percent each). As more countries joined the organization so the percentage shares would shift, but it was agreed that, in any event, the US share of the vote would not drop below 50.6 percent. Decisions would be taken if accepted by the United States along with other countries whose combined vote was "not less than 12.5%." This effectively stopped France and Germany (12.2 percent together) having a veto over all decisions.

The interim agreements establishing the first multinational communications satellite organization were opened for signature on August 20, 1964.[21] The governing body of the consortium was called Intelsat—the International Telecommunications Satellite Corporation. Its executive instrument was the ICSC. Intelsat had overall responsibility for the design, development, establishment, and operation of the single global system that would be managed by Comsat. A related Special Agreement dealt with the ground segment. By the end of 1965 the number of participants in the consortium was 46: no communist country was among them. The initial estimate for the cost of the space segment was $200 million. It would provide voice, telegraphy, broadband data, high-speed data, and television services. Comsat's investment quota, which translated into voting weight on the interim committee, was a little more than 56 percent at the end of 1965. Western European countries had a share of about 28 percent.[22]

The initial structure of Intelsat ensured US preponderance in the system. As Charles Johnson, a senior official in the Johnson administration, put it, Intelsat was "an unusually attractive international vehicle for the U.S."[23] Since the interim agreements stipulated that the United States' voting weight could never drop below 50 percent, "we control."[24] All the same, as more countries sought to exploit the new opportunity, Comsat, as manager on behalf of an international consortium, was obliged to encourage bids from foreign firms for equipment for the space segment. NSAM338 was promulgated on September 15, 1965, to meet this challenge.[25] The "core" of NSAM 338, to quote National Security Adviser McGeorge Bundy was "to use our technological superiority to *discourage* commercial competition with Comsat and/or wasteful investment in several duplicative Free World defense-related systems."[26] The United States would provide technical information, launch vehicles, and launching services to other nations only when they assured the administration that "what we supply is needed to develop or use the global commercial system."[27] For military communications, the administration's "aim was to encourage selected allied nations to use the U.S. national defense communications satellite system rather than to develop independent systems, and to accommodate allied needs within the U.S. system."[28] In short through NSAM338 the United States aimed to use its technological preeminence and its veto powers in the Interim Committee as levers to restrict the proliferation of competitors to the single global telecommunications system that was managed by Comsat.

NSAM338 posed immense problems for NASA and the State Department. On the one hand, through the Intelsat agreements, the United States was encouraged to help other countries build up their communications satellite industries so as to get some return on their investment in the international consortium. NSAM338 instructed them not to provide that support unless they were given guarantees that the benefactor did not build a satellite system that could do significant economic harm to the single global system being established by Intelsat. It also discouraged technological sharing in the domain of launchers, for fear that other countries would eventually seek to develop their own communications satellite system outside the Intelsat framework. NSAM294 added to these restrictions by insisting that technological assistance to Western Europe in the area of rocketry should do nothing to advance the French military program. These requirements bedeviled the efforts made by NASA and the State Department to share technology with Europe in the latter part of the 1960s.

NASA and ELDO: The Early Initiatives

In 1959 the British government of Prime Minister Harold Macmillan decided to cancel an expensive program to build an already obsolete Intermediate Range Ballistic Missile called Blue Streak.[29] Rather than waste the money already spent, and disband the expert teams that had been assembled to work on the missile, the government decided to strip it of its military characteristics (which had been developed in conjunction with the United States) and to offer it to European partners across the Channel as the first stage of a multistage satellite launcher. This gesture not only enabled Macmillan to save face at home: it was intended as an expression of goodwill to the emerging European Common Market, which the British had opposed in the late 1950s. After lengthy negotiations it was eventually agreed to establish an intergovernmental organization called ELDO to develop a three-stage launcher for civilian purposes. Blue Streak would comprise the first stage. The second stage, called Coralie, would be built in France. The third stage, which promised to be the most advanced technologically, would be built in Germany. Italy would provide experimental payloads to measure the environment during launching and in orbital flight. The rocket, called Europa, would be launched from Woomera in South Australia.[30] The convention establishing ELDO was signed in March 1962. It was ratified by the governments of the seven member states (the five already mentioned plus Belgium and The Netherlands) in 1964.

NASA was quick to react to these developments. In December 1962 Arnold Frutkin, along with a few other representatives, visited Britain, France, and Germany for two weeks to get a closer look at the various installations involved in the project. They told the Europeans that "cooperation in the launch vehicle area was possible to a limited extent."[31] Those limits were set by several conditions. The European programs had to be directed to peaceful civilian applications, and be of mutual technological interest to NASA and ELDO. Most important of all, the agreements had to be multilateral and not bilateral. NASA would only collaborate through ELDO and not with individual national authorities in the domain of rocketry.[32] This was supposed to avoid the exploitation of American technology in national military programs. It would also promote European integration. As State Department official Robert F. Packard put it, this change in US policy had to be viewed "as part of the entire spectrum of our national interests in Europe, among which a major U.S. interest is to encourage those developments which promote the interdependence and integration of the European countries such as Euratom and the Common Market."[33]

It was not easy to translate these good intentions into practical action. The enormous lead that the United States had over the ELDO member states severely limited the areas of technological collaboration that could be of mutual interest. In addition, the structural weakness in ELDO that had been evident to many from the start, namely, the lack of a strong centralized system of project management and control, was of increasing concern in Washington.[34] There is a porous barrier between many civil and military technologies used in rockets/ missiles. The ELDO Secretariat had little authority over the people and firms developing the separate stages in Britain, France, and Germany. Thus, contrary to what NASA and the State Department had originally hoped, routing sensitive knowledge and technology through ELDO was no guarantee that it would not eventually emerge in national military projects, notably that in France. A report prepared by the CIA in May 1964 confirmed the danger: "[T]he organization has no enforcement machinery to police compliance, and the possibility is raised that ELDO might contribute to the spread of ballistic missile technology." This was just what NSAM294, promulgated in April 1964, wanted to stop. Indeed, notwithstanding requests from Europe for "propellants, guidance components and other launch-vehicle hardware and technology," the CIA analysis found that export licenses had only been granted for a few select items.[35]

In 1965 the member states of ELDO decided that their launcher should be upgraded to have a geostationary capability (the Europa II program). This required constructing a more powerful third stage than previously planned for. In May of that year a senior engineer in the European organization, Bill Stephens, wrote NASA asking that ELDO and NASA staff discuss together "the more fundamental problems which have been encountered by NASA in designing, testing and launching liquid hydrogen/liquid oxygen upper stages, the development philosophy followed," and the possibility of establishing links between European and American firms in the Europa II project. Frutkin saw this request "as a valuable opportunity to advance our relationship with ELDO as a multilateral institution, to establish a ground for limiting or delaying assistance in the missile field to competing interests in Europe, and to establish a counterweight to National missile programs."[36] In other words, Frutkin was persuaded that the risks of technological leakage into the French military program could be averted and that a way could be found both to assist ELDO and to respect the constraints imposed by NSAM294.

Another impediment emerged even as NASA and the State Department were considering Stephens's request: the restrictions on technology transfer in the telecommunications sector. This issue was given new urgency by the demonstrative success of Early Bird launched in April 1965. NASA was particularly disturbed by a restrictive clause inserted in the draft policy statement being circulated at the time by J. D. O'Connell, the special assistant to the president for telecommunications. That clause suggested that, to impede the development of foreign communications satellite services outside the global Intelsat framework, the United States should deny help with launch vehicles and launching services to foreign governments (unless the necessary guarantees were forthcoming). This was so controversial that NASA administrator James Webb took it upon himself to write O'Connell and ask that the extension of the restrictions to cover not only launch services but also launch vehicles be removed. As Webb put it, "In effect, although perhaps not intended, this [extension] places in a policy paper otherwise exclusively addressed to communications satellites, a blanket prohibition on transfer of technology concerning launch vehicles."[37]

NASA's alternative suggestion was, as Webb put it, to make "detailed and fine distinctions" between the kinds of technology that could be shared and those that could not.[38] An example was that between solid propellants and nonstorable liquid propellants such as liquid hydrogen and liquid oxygen. As Frutkin put it, the latter technology "has not been deemed to accelerate the more advanced solid propellant program which France is developing in connection with strategic delivery objectives."[39] Thus whereas blanket policies made no attempt to distinguish between various types of rocket fuel, and their implications for national security, NASA was at pains to distinguish between the military potential of solid (high security risk) and nonstorable (low security risk) propellants. Their suggestion fell on deaf ears. The broad restriction remained in place in the final policy statement that accompanied NSAM338, promulgated in September 1965.

The Crisis in ELDO in 1966 and the Renewed Pressure for Technological Sharing

In February 1966 the British government circulated an aide-mémoire to its partners in ELDO.[40] It remarked that the organization was unlikely to produce any worthwhile result and that Her Majesty's government saw little interest in continuing as a member of the organization and contributing financially to its program. Development costs of Europa had more than doubled from the initial estimate of about $200 million to over $400 million. The time to completion had slipped from five to seven-and-a-half years. The British first stage, Blue Streak, had been successfully commissioned in June 1965, while the French and German stages were still under development. The British were therefore effectively subsidizing continental industries to produce a launcher that, in fact, would be obsolete technologically and commercially uncompetitive with American heavy launchers.[41]

The timing of this move was deemed most unfortunate in Washington. First, the European integration process was in a very brittle state at the time and even NATO seemed to be on the brink of fragmentation.[42] The French had precipitated a crisis in the European Economic Community (EEC) by boycotting the EEC's decision-making machinery so as to liberate the country from its "subordination" to community institutions and the dilution of sovereignty that that entailed.[43] In this inauspicious climate, everything possible had to be done to sustain the momentum for European unity. As Undersecretary of State George Ball emphasized, European integration "is the most realistic means of achieving European political unity with all that that implies for our relations with Eastern Europe and the Soviet Union...and is the precondition for a Europe able to carry its proper share of responsibility for our common defense."[44] ELDO was not central to European integration. But just when France was challenging the momentum of European unity, the significance of the United Kingdom's threat to leave ELDO risked being amplified by those who were increasingly hostile to supranational ventures on the continent.

The British challenge to ELDO also came at the very moment when her partners were becoming increasingly vocal about the putative "technological gap" that had opened up between the two sides of the Atlantic.[45] President Johnson took this matter so seriously that in November 1966 he personally signed NSAM357, instructing his science adviser, Donald Hornig, to set up an interdepartmental committee to look into "the increasing concern in Western Europe over possible disparities in advanced technology between the United States and Europe."[46] In its preliminary report, the committee concluded that "the Technological Gap [was] mainly a political and psychological problem" but that it did have "some basis in actual disparities." These included "the demonstrated American superiority in sophisticated electronics, military technology and space systems." Particularly important were "the 'very high technology industries' (particularly computers, space communications, and aircraft) which provide a much greater military capability, are nationally prestigious, and are believed to be far-reaching in their economic, political and social implications."[47] For the Johnson administration, then, the technological gap, even if inflated in Europe, was a problem that had to be addressed, and the mutual development of space technology through an organization like ELDO was one way of doing so.

Finally, NASA again emphasized that enhanced international collaboration in space would aid nonproliferation. Quoting James Webb this time, it would be "a means whereby foreign nations might be increasingly involved in space technology and diverted from the technology of nuclear weapons delivery."[48] The United States could use the carrot of technological sharing with ELDO to redirect limited human and material resources away from national programs that were more difficult to control and which might encourage the proliferation of weapons delivery systems.

The continued and spectacular success of the French space program gave this argument for saving ELDO an added urgency. On November 26, 1965, France had become the third space power by launching its own satellite with its own launcher, Diamant-A, from Hammaguir in Algeria. The feat was repeated in February 1966. This three-stage launcher combined "militarily significant solid and storable liquid fueled systems"—just the kind of technology the United States did not want it to develop—in a highly successful vehicle derived from the national missile program.[49] In the light of these achievements and de Gaulle's growing determination to affirm his independence of the EEC and the Atlantic alliance, "[t]he US is concerned that, if ELDO were to be dissolved, France might devote more of its resources to a national, military-related program or that it might establish undesirable bilateral relationships [with the Soviet Union] for the construction of satellite launch vehicles."[50] The United States had to contain this threat and ensure that European institutions emerged "from the present crisis with their prestige, power and potential for building a united Europe as little impaired as possible."[51]

The Johnson administration took two steps to address this situation. First, they let Britain know that they were deeply concerned about the implications of its possible withdrawal from ELDO. In addition, the administration formally undertook to provide technological support to ELDO. On July 29, 1966, Walt W. Rostow, one of LBJ's two national security advisers, signed off on National Security Action Memorandum 354.[52] NSAM354 was a response to a request from the Department of State that the United States "clarify and define" its policy concerning collaboration with the "present and future programs" of ELDO. The document affirmed that it was "in the U.S. interest to encourage the continued development of ELDO through U.S. cooperation." It referred to the results of an ad hoc working group, established by the State Department and chaired by Herman Pollack, that had prepared a statement "defining the nature and extent of U.S. cooperation with ELDO which the U.S. government is now prepared to extend." This statement was to be "continually reviewed by the responsible agencies," above all, the Department of Defense and the State Department, along with NASA, "to ensure that it is current and responsive in terms of developing strategies."

The help that Pollack's working group proposed was extensive. It was divided into three categories: general, and short-range and long-range assistance.[53] The first contained some standard items—training in technical management, facilitating export licenses, use of NASA test facilities—but also suggested that a technical office be established within NASA "specifically to serve in an expediting and assisting role for ELDO." Short-range help included "technical advice and assistance" in items such as vehicle integration, stage separation, and synchronous orbit injection techniques, as well as the provision of unclassified flight hardware, notably the strapped-down "guidance" package used on the Scout that had already been exported to Japan. Long-range assistance was focused on helping with a high-energy cryogenic upper stage of the rocket, as had been requested by Stephens on behalf of ELDO the year before. It was proposed that Europeans be given access to technological documentation and experience available in the Atlas-Centaur systems, that ELDO technical personnel "have intimate touch with the problems of systems design, integration, and program management of a high-energy upper [*sic*] such as the Centaur," and even that the United States consider "joint use of a high-energy upper stage developed in Europe."[54] In short, in mid-1966, the United States was considering making a substantial effort to help ELDO develop a powerful launcher with geosynchronous orbit capability by sharing state-of-the-art knowledge and experience and by facilitating the export of hardware. This support—it should be added—would not normally be available on a bilateral basis to European national launcher programs.

None of this would have been thinkable as long as NSAM294 (denying technology that might help the French military program) and NSAM338 (denying technology that might subvert a single global comsat system) were not revised. Indeed in spring 1966 it was evident that NSAM294 was due for review. European booster technology was advancing rapidly without external help. A blanket denial of export licenses now would unnecessarily harm both US business and foreign policy interests. Even worse, it might encourage a request to a non-US supplier, most obviously the Soviet Union with whom de Gaulle was fostering technological collaboration as an expression of French autonomy. Reiterating NASA's demand that policy for technology transfer should make "detailed and fine distinctions," Richard Barnes, the director of Frutkin's Cooperative Projects Division, insisted (and Webb concurred) that the interpretation of restrictions on technology transfer determined by NASM294 had to be more specific. The guidelines, he wrote the chairman of the NSAM 294 Review Group in the State Department, should deny to a foreign power "*only* those *few critical* items which are clearly intended for use in a national program, would significantly and directly benefit that program in terms of time and quality or cost, and are unavailable in comparable substitute form elsewhere than the US" (emphasis in the original).[55] Correlatively, it should share items that were "of only marginal benefit to the national program" or "were available elsewhere than the US without undue difficulty or delay." This was happening already in sensitive areas. The release of inertial guidance technology to Germany had been officially sanctioned in July 1964 on condition that it was not employed "for ballistic missile use or development."[56] A strapped-down "guidance" package had been offered to Japan. By contrast, and foolishly in Barnes's view,[57] an American company had recently been refused a license to assist France with the development of gyro technology even though gyros of comparable weight and performance were already available in France. In short US policy should take into account the kind of technology at issue, its likely uses in practice, the global state of the market for the technology, and the importance of collaboration from a foreign policy perspective.

While Barnes was putting NASA's case to the State Department, Webb was doing what he could to get the Department of Defense to support NASA's approach. Writing to Defense Secretary McNamara in April 1966, Webb pointed out that although high-energy, cryogenic, or "non-storable" upper stages might conceivably be employed for military purposes, in practice they would probably not be deployed in that way. He argued that anyway the risks of technological leakage into the military program were outweighed by the benefits of promoting a civilian rocket. As he put it, "Even in the case of France it seems likely that encouragement to proceed with upper stage hydrogen/oxygen systems now under development might divert money and people from a nuclear delivery program rather than contribute to that which is already under way using quite different technology." Here, and in general, wrote Webb to McNamara, rather than a blanket restriction, "we might be better off were we to concentrate on a few very essential restrictions, such as *advanced* guidance and reentry systems" (my emphasis). In a supportive reply McNamara reassured the NASA administrator that he strongly favored international cooperation in space and that he had directed the DoD staff "to be as liberal as possible regarding the release of space technology for payloads and other support items."[58]

It was fairly easy to revise the restrictions embodied in NSAM294 to accommodate the changing balance of technological power between the United States and France, particularly once the French had shown that they had mastered launcher technology sufficiently to place their own satellite in orbit. The constraints imposed on sharing booster technology in NSAM 338 were less easily dislodged, and were a serious irritant to US-European relations. Frutkin wrote with some exasperation that the Europeans were persuaded that the United States was "seeking by all means, fair or foul, to maintain political and technical control of Intelsat."[59] Barnes was equally frustrated by "European fixation on comsats and launch vehicles."[60] Of course people in France and Germany may have been exaggerating the situation, but the administration itself recognized that they had some cause to complain. Charles Johnson admitted in an exchange with Walt Rostow that the odds were so heavily stacked in the United States' favor in the (interim) Intelsat agreements that it was "difficult to maintain international cooperation on this basis."[61] Barnes agreed. There had been a "deterioration of 'climate for cooperation' caused by (1) US policies and actions within the Intelsat, and (2) US export policies in support of the 'single global system.'"[62]

NASA's view was that, unless they acted fast, and softened the restrictions in NASM338, the United States would lose all control over the direction of the European communications satellite system, as well as support for American policies in Intelsat. Frutkin was convinced that the United States had to be prepared to provide launch services on a reimbursable basis for (experimental) foreign communication satellites. This would "extend the market for American vehicles, remove some incentive for independent foreign development of boosters, and assure that we could continue to exercise critical leverage in foreign comsat activities rather than lose such leverage." An (anonymous) internal memorandum argued, along similar lines, that technological sharing was the best way to enroll foreign firms and their governments in American comsat policy. By allowing "United States firms to enter cooperative arrangements with the communications and electronics manufacturing industry in other countries," notably in Western Europe, industries in these countries would develop the technical know-how needed for them "to compete effectively for contracts for the space segment of the global communications system." This would "remove a current irritant, primarily expressed by the French but also shared by the British, Italians and Germans, about their inability to supply hardware for the Intelsat space segment." And even if such technological sharing did not irreversibly lock these European countries into the single global system favored by the United States, one could expect them to have a "greater incentive" to collaborate with America in developing that global system. They were also likely to be more cooperative and sympathetic to the US position during the renegotiation of the interim Intelsat agreements scheduled for 1969. Anyway, if the United States did nothing to help these nations, they would eventually develop the technology on their own, without American help, and would be quite capable of establishing separate, regional communications satellite systems in due course.[63] As Frutkin explained,

(a) We do need to improve our situation in Intelsat with specific reference to the 1969 negotiations. (b) We already have a strong technical lead in the comsat field. (c) We already have an adequate voting majority in Intelsat. (d) We can rely upon our technical, moral and financial strength to assure continuing leadership—without seeking to deny technology to our partners in Intelsat.[64]

The proposal from Pollack's working group to help ELDO develop or acquire the kick-stage and propulsion technology needed to place a communications satellite in geosynchronous orbit was entirely coherent with this attitude.

A Missed Opportunity

In September 1966 NASA administrator Webb traveled to Europe to discuss space collaboration with Germany and other potential partners. Frutkin briefed him shortly before his departure. While the "general atmosphere for space cooperation with the United States may have improved slightly," wrote Frutkin, the steps taken to date had done little more than "clear the air somewhat." The Europeans, he told Webb, "know of no progress in easing US restrictions upon communications satellite technology," and "it may be sometime" before the progress that had been made in Washington could be divulged to them. Webb was therefore to repeat the standard answer to the usual request for comsat launch assistance: "that we could certainly give consideration to such a proposition on the assumption that the European countries take their Intelsat commitment to a single global system as seriously as we do."[65]

The damage caused by this reticence was amplified by President's Johnson official offer of support to Germany just before Christmas in 1965. It will be remembered (see previous chapter) that in an exchange of toasts with Chancellor Ludwig Erhard at a state banquet, LBJ suggested that existing scientific cooperation should be extended to embrace "an even more ambitious plan to permit us to do together what we cannot do alone." The president gave two examples of "demanding" and "quite complex" collaborative projects, which would "contribute vastly to our mutual knowledge and to our mutual skills": a solar probe and a Jupiter probe.[66]

This gesture was driven by political concerns: collaboration in space science was being instrumentalized by the State Department not only to recognize Erhard's support for the United States in Vietnam, but to drive a wedge between French and German policies in Europe. Indeed, Erhard was forced to relinquish his post in November 1966, accused of mismanaging the economy and of being too pro-American and anti-French. In addition, LBJ's offer was interpreted by some as a strategy to divert scarce European resources into science and away from applications, notably telecommunications, that is, as a clumsy effort to secure American preponderance in Intelsat. "All in all," wrote Frutkin to Webb in August 1966, "we must say the President's proposal got off to a poor start due to misunderstandings which are inevitable when a proposition of this sort is made in the headlines without preparation of the ground."[67] Barnes put it pithily: because of European "suspicion and distrust," aggravated by President Johnson's spectacular overtures to Chancellor Erhard, there was "no prospect for escalating cooperation with Europe unless (1) US is willing to modify its present export control policies, and (2) we could offer other possibilities for cooperation in areas of interest to them (i.e., comsats and vehicles)."[68]

The opportunity for the United States to shape the European program was, however, slipping away. By September 1966 ELDO had temporarily resolved its crisis: the British had agreed not to withdraw in return for their contribution to the budget being reduced from 38 percent to 27 percent.[69] The organization had also reoriented its program unambiguously in favor of developing Europa II that achieved geostationary capability by adding a fourth, French-built solid-fuel stage to the previous rocket. In parallel, France and Germany decided to fuse their national comsat projects in a joint experimental telecommunications satellite called Symphonie. Symphonie would be launched by Europa II from a new base near the equator in Kourou, French Guiana.[70] ELDO had moved from an artificial political construct to an organization that was working to improve its management structure and that now had a well-defined technical mission. For the moment at least, the Europeans would blaze their own trail into space. They would do so under a new regime led by a Republican president who was sworn into office in January 1969.

Chapter 4

European Participation in the Post-Apollo Program, 1969–1970: The Paine Years

The negotiations over European contributions to the post-Apollo program concerned the biggest single attempt to integrate a foreign nation or region into the technological core of the American space program during the first decades of NASA's existence.[1] These discussions were carried on for about three years, and engaged several NASA administrators: Thomas Paine, from October 1969 until he left NASA in September 1970; George Low, who temporarily led the organization while a successor was found; and then James C. Fletcher. They also engaged multiple arms of the administration: NASA of course, as the lead agency, but also the State Department, the Department of Defense, the Office of Telecommunications Policy, the National Security Council, and, hovering in the wings, the Office of Management and Budget (OMB), which assumed extensive powers in the Nixon administration.[2] They were of deep concern to industry. And they were dominated by issues of technology transfer and launcher policy, here embedded in a framework that touched on matters of international diplomacy, national security, and American technological, commercial, and political leadership of the free world.

In a speech to the United Nations in September 1969 President Nixon called for the "internationalization of man's epic venture in space." Feeling himself mandated to broaden the base of the post-Apollo program, Paine made a concerted effort to seek international partners, and made his case with passion to the Australians, the Canadians, the Japanese, and the West Europeans. It was the last who were best positioned to take advantage of it. European engineers, managers, and policymakers, who had learnt so much from NASA in the early 1960s, were deeply impressed by the Apollo missions: the United States, it seemed, could do anything it wanted in space. The gap in technological, engineering, and managerial capacity that had opened up between the two sides of the Atlantic in the space sector had now become a chasm—and yet here they were being invited to join in NASA's next major program. Their reactions combined awe at American achievements, with pride that they were deemed worthy of inclusion in the next leap forward, and with fear born of uncertainty. Given their limited resources, if they made a major commitment to NASA's post-Apollo program they risked sacrificing an indigenous space program of their own devising, above all an autonomous launch capability. If they rejected the American offer they would be doomed to an inferior position, always collaborating from a position of weakness with the world's space leaders. NSAM 294 and NSAM 398 were suggestive of what that could entail: a vulnerability to the constraints on international collaboration imposed by US commercial, political, and security concerns, which could mean launchers denied, technology and managerial skills withheld, and prime contractors always based on US soil.

The account that follows will flesh out these more general considerations in greater detail. It is divided into three chapters. The first covers the period from the end of 1969 to early 1971, when the budget appropriations for FY1972 were finalized—and much to NASA's distress, post-Apollo did not figure largely in them.[3] The second chapter covers 1971. While some progress was made on defining the parameters of US-European collaboration, the year was dominated by a separate if related concern: the implications of the definitive Intelsat agreements (accepted in principle by 73 governments on May 21, 1971) on the availability of US launchers for European telecommunications satellites. Finally, there is the period inaugurated by President's Nixon's statement on January 5, 1972, that the space shuttle (more precisely the STS, Space Transport System) would be the centerpiece of NASA's post-Apollo program. Poised to move quickly, NASA rapidly took advantage of the new situation. Plans for a major technological collaborative project were refined in a series of meetings with experts from both sides of the Atlantic. A variety of possible platforms for a European contribution were explored, including the construction, under the guidance of an American prime contractor, of parts of the orbiter itself. Alternatives included the European-led construction of a "space tug," an orbit-to-orbit vehicle intended to ferry hardware and people from the shuttle's low-earth orbit to the moon, the geostationary orbit, and so on, and a Sortie Can or a RAM (Research Applications Module), a capsule or a palette for doing space science that would be lodged in the shuttle's cargo bay.

The managerial, industrial, and technological complexities of direct participation in the orbiter soon overwhelmed NASA's wish to have any partners directly engaged in building its new space transport system. The agency also started having grave doubts about the wisdom of developing the tug, which had emerged as Europe's preferred contribution to the program. Taking the bull by the horns, in June 1972 it was announced, to the distress not to say anger of many of its partners, that the United States could only support a European effort to build a "sortie can" for space science experiments, while encouraging international participation in the *use* of the shuttle system. Germany decided to take advantage of this offer, and took the lead in developing what later became known as Spacelab. The French, by contrast, were now even more emphatic that meaningful technological collaboration with the United States was impossible. The withdrawal of the tug, and the conditions under which the United States would launch foreign communication satellites, played into the hands of those who were seeking political justification for an independent European launcher program. The French authorities, yielding to pressures from engineers in their national space agency and the Gaullist wings of the political elite, took prime responsibility for developing a European heavy launcher called Ariane, which made its first successful maiden flight on Christmas eve, 1979.

The Post-Apollo Program

Soon after taking the oath of office in January 1969 President Nixon established a Space Task Group (STG) chaired by Vice President Spiro Agnew. Its three other members were NASA administrator Tom Paine, Lee A DuBridge, the president's science adviser, and Robert C. Seamans, the secretary of the Air Force and former deputy administrator of NASA. The STG's aim was to find ways of making cuts in the space program, and to come up with a "coordinated program and budget proposal" that factored in "international implications and cooperation."[4] It submitted its report to the president on September 15, 1969, and met the press two days later.[5]

The STG proposed three alternatives programs having different budgetary levels. All shared the same goal, "and I emphasize the word 'goal,'" said Agnew, "and not a commitment—a manned landing on Mars before the end of the century." Each offered a different path to that goal depending on how quickly it was achieved.[6] At the core of the STG's program lay an orbiting space station and a space transportation system. The station, envisaged for the mid-1970s, would be initially designed to house 6–12 astronauts. It could be expanded by the subsequent addition of modules to accommodate 50–100 people. Paine emphasized that since "a substantial reduction in the cost of space transportation [was] essential...a new and truly low-cost space transportation system [was] an integral part of the space station concept."[7] Three components were foreseen for this system: a reusable space shuttle that could access low-earth orbit from a terrestrial launch pad, reusable space tugs to move people and equipment from the shuttle's cargo bay to various other orbits as well as onto the moon, and, third, a reusable nuclear engine, derived from the Nerva project then well under way.[8]

The original shuttle concept made maximum use of existing aeronautical technology.[9] As described by Paine, the shuttle, which would hopefully make its maiden flight in about 1976 or 1977, would "look like one of these giant new 747 intercontinental jets, but instead of being on the airstrip horizontally for a takeoff it will take off vertically, so it will be racked up sitting on its tail. Instead of having jet engines slung under its wings," the NASA administrator went on, "it will have rocket engines, of the type that power our Saturn 5 rockets, clustered in the tail." The second "orbiter stage," mounted on the nose of this "booster stage," would also be a spacecraft with wings "about the size, weight and appearance of a big transcontinental Boeing 707." Both were fully reusable, had a crew of two (plus passengers in the orbiter), and would be piloted back to earth at the end of their missions, where they would land horizontally, like airliners. It was hoped that the reusability of the space transport system could reduce the cost of injecting one pound of payload into orbit by at least an order of magnitude, from some $500 with a Saturn launch vehicle in the 1960s, to something below $50 per pound of payload in orbit in the 1970s. Seamans was quick to emphasize that the Department of Defense (DoD) was particularly attracted by this feature.[10]

While DoD support was obviously an asset in Congress it had important technological implications. Apart from requiring a large payload bay and extremely powerful motors, the DoD insisted on a high cross-range capability (on the order of 1,250 nautical miles) for the orbiter.[11] The Air Force wanted the shuttle to be able to recover an orbiting payload and return to the Vandenberg Air Force Base in Southern California after a single 110-minute shuttle orbit. The landing strip would have moved about 1,250 miles east as the earth rotated during this time. The operational flexibility required by these kinds of missions required sacrificing payload weight for the added weight of the Thermal Protection System (TPS) that would be needed to protect the orbiter in the hypersonic maneuvers called for. It also required NASA to replace a straight-wing configuration with a delta-wing.[12]

What of international collaboration? The STG identified it as one of the five principal program objectives of the post-Apollo program. During the month prior to the release of its report the Nixon administration issued two National Security Study Memoranda, NSSM 71 and NSSM 72, signed by the national security adviser, Henry Kissinger. NSSM 71, dated August 14, 1969, established an interagency committee to review policies "governing the access by foreign countries to certain advanced technologies vital to our national security." It had to "give full consideration" to the administration's commitment "to international cooperation in the peaceful application of nuclear and space technologies and to the necessity for the free exchange of scientific knowledge when national security is not impaired."[13] NSSM 72, dated September 4, 1969, called for the creation of a small ad hoc group on International Space Cooperation to report on possibilities for cooperation "with friendly countries as well as the Soviet Union."[14]

In a letter to the president in August 1969 Paine welcomed the policy review authorized by NSSM 71, which he hoped would "clear away unnecessary restrictions which could seriously obstruct the increased international activity" Nixon had called for.[15] He saw possibilities for collaborating in planetary exploration with the Soviet Union, and for closer collaboration with Japan, Australia, and Canada. But it was Western Europe that particularly interested him. The possible scope of cooperation reflected NASA's judgment of where European scientific and technological strengths lay, and what they could afford. The emphasis was placed on applications satellites, planetary missions (along with the Soviets), and the inclusion of foreign astronauts in post-Apollo manned flight programs.[16] The State Department echoed these sentiments.[17] Indeed, at this stage of planning, no one saw much scope for Europe doing more than being involved in science and applications, and in *using* the space station and the space transportation system. Participation in hardware development as such was not seriously considered.

Europe Is Invited to Join

On October 13–15, 1969, Paine met with the ministers of science and senior space program officials of France, Germany, and the United Kingdom. He also described to a distinguished committee of senior officials of the European Space Conference (ESC) the details of what he called the "President's new space program," which would be presented to Congress for funding in FY1971.[18]

The ESC was a gathering of ministers or their delegates in the several countries in Europe interested in defining a European space policy for the 1970s. It first met in 1966. The national representatives got together when needed, and very frequently in times of crisis. The ESC was superseded when the European Space Agency (ESA) came into being in 1975, when its key functions passed to the ESA Council.

Paine spelt out the STG proposals in considerable detail to his European audience (as well as to authorities in Australia, Canada, and Japan).[19] He suggested that NASA could achieve its goals within its then-current levels of funding ($4–6 billion annually). And he welcomed European participation. He had come to Europe, he said, to "personally make it as clear as I can that it is the desire of America not only to continue but indeed to expand the cooperation which from our standpoint has proved so fruitful and which we hope, from your standpoint, has also been significant."

It is important to realize how radical Paine's proposals were. Space was no longer being defined primarily as a strategic resource to be deployed in a competitive struggle for global technological supremacy with the Soviet Union. It was rather being seen as a new frontier to be explored and colonized, a place to live and work. "For us in America," Paine said, "which has been called the new world, we feel that space may represent another new world, a seventh continent, which is now opening to mankind in the region 100 miles above the surface of the globe."[20] Europeans realized the revolutionary implications. A report to a committee of the ESC written by Jean-Pierre Causse (of ELDO) and Jean Dinkespiler (of ESRO) noted that "[t]his really does mean *a total metamorphosis of space activity*" (emphasis in the original).[21] The delegates to the ESC meeting "expressed the hope that European countries would soon have the necessary data to enable them to give as positive an answer as possible to the offers of cooperation made by the American authorities."[22]

Paine was careful not to oversell cooperation. He avoided giving definitive schedules or firm commitments, talking instead in general terms of program directions, plans, hopes. This was not simply because the program was still somewhat schematic, and would surely be implemented piecemeal, as Congressional and presidential approval was forthcoming. The most significant reason was that NASA did not want to steer Europeans down particular paths at the outset. Participation was not to be imposed from above but something that bubbled up from below because the Europeans wanted it. As Frutkin put it,

> We would not wish to constrain imaginative European thinking and initiative regarding the structuring of participation, i.e., we want to give the fullest and freest reign to European proposals. [...] Europeans must determine for themselves *whether* they are interested in participation and *what* is the nature of their interest. It would then be a short and logical step for them to give thought to *how* that interest should be pursued and structured. (Emphasis in the original)[23]

To improve communications, it was agreed in February 1970 that ESRO and ELDO would together station a representative in Washington on a permanent basis. An ELDO team, headed by Causse, would make periodic visits to NASA and its contractors to keep abreast of developments in both the space shuttle and station. They would be invited to regular NASA "internal" three-month briefings, and NASA would provide for "full observation and participation opportunities in the planned summer study activities on the space station in 1970–71."[24]

Classification was another important obstacle that was quickly removed. Deputy Administrator George Low and Robert Seamans agreed at once "that the space shuttle program should be conducted on a generally unclassified basis" in the same sense that the Apollo program was unclassified, bearing in mind "the international flavor of the program."[25] In mid-February 1970 Paine and Seamans signed an official agreement between NASA and the Air Force establishing a joint NASA/USAF committee whose task was to ensure that the shuttle "be designed and developed to fulfill the objectives of both the NASA and the DOD" and confirmed that it "will be generally unclassified."[26]

NASA was emphatic that collaboration would be pointless if Europe did not reciprocate, above all by increasing its space budgets. In 1969 ESRO's annual budget was slightly over $50 million, ELDO's was about $90 million, and the entire European effort, including that of individual nations, was about $300 million.[27] Frutkin was quite blunt about it in his briefing for Paine before the administrator's trip to Europe in October. It was imperative, he wrote, for Europe to increase its level of financing several-fold if it had "substantial space ambitions and wishes to take hold of the opportunities of the future." In any event, "significant participation in planning for future space exploration and use cannot really be considered, and would even be a waste of time," he added, "if there is not an intention to seek much larger funding."[28]

Europe's Response

By May 1970 the ELDO Council had voted $500,000 for conceptual studies of the tug, while the ESRO Council had voted a similar sum for a modular element of the space station.[29] The full-time ELDO and ESRO representatives to NASA in Washington had been nominated. Industries on both sides of the Atlantic were exploring ways of working together. It was being suggested that Europe would contribute up to about $1 billion over the next decade to a $10-billion post-Apollo program by providing both discrete elements such as the space tug and highly integrated elements such as parts of the shuttle.

In response to European requests, NASA arranged for briefings on the station and shuttle in Europe in the summer 1970. Speaking in Paris and in Bonn early in June, Frutkin once again emphasized the agency's enthusiasm for European participation, and identified five basic principles that would underpin it: "(1) self-funding of participation, (2) management integrity, (3) adequate exchange of technical information, (4) equivalent access to space facilities, and (5) the broadest possible participation."[30] Participation could take four forms—studies and R and D; developing a separate element like the tug; developing an integral part, element, or subsystem of the shuttle itself; and utilization by foreign experiments or foreign astronauts. Frutkin stressed that the sooner Europeans became engaged in the program, the greater would be the scope for participation.

Europeans could not act fast, however. Their own internal uncertainties and divisions over the future directions of the European space program were amplified by the need for certain assurances from the US authorities regarding the space transportation systems and the space shuttle. On the industrial side they hoped for "technical access to the space shuttle and space station projects," along with a "European role in the production as well as the development phase of any items Europe undertake." On the political side, they wanted guaranteed, reimbursable access to American launchers and launch facilities both before and after the shuttle was operational. Both of these requests—for meaningful technological collaboration, and for guaranteed access to the shuttle—raised serious policy issues. It will be remembered that NSAM 294 specifically excluded foreign access to ballistic missile technology. What guarantee was there that STS technology, and above all the development of the technologically advanced tug, would not leak into national missile programs? As for the question of shuttle availability, this was potentially subject to the restrictions imposed by NSAM 338. NSAM 338 specifically disallowed NASA to launch telecommunications satellites that could undermine the single global telecommunications system being put in place by Intelsat (to be described in detail shortly). As a major NASA policy statement explained in May 1970, "in its 'worst case' form," the demand for launch guarantees "raises the question of whether Europe should in principle be permitted to buy US STS launch services to establish commercial communications satellite systems which the United States might regard as competitive with Intelsat. The European view," it went on, "is that Europe cannot be expected to contribute to the development of a key Space Transportation System whose use would be subject to U.S. 'whims.'"

To sum up. In the months after Paine had enthusiastically promoted NASA's new vision and program for space in Europe, the negotiations over European participation had become intertwined with a number of other related issues that complicated the decision-making processes enormously. Europe's resources were limited. They were willing to invest more in space. But they faced a stark choice. Paine summed up the alternatives in a letter to Nixon. Europe "must choose either an independent European space effort of a limited and retrograde character or commit to a much bolder joint program that will be dominated by the United States."[31] The NASA administrator had gone to the heart of the dilemma as seen by many abroad: independence along with technological obsolescence, or cooperation at the risk of domination.

The First Major Barrier to Participation: Data and Technology Transfer

Export licenses were required for sharing sensitive technology in the shuttle program. This included items that facilitated national comsat abilities (specifically picked out for tight control in NSAM 338), including the acquisition of launchers, or that contributed to independent national strategic weapons systems (NSAM 294, under review in terms of NSSM 71). The limits to what was permissible were quickly tested by McDonnell Douglas in April 1970. The firm wanted an export license allowing it to share "A Proposal to Accomplish Phase B Shuttle Program" with potential industrial partners in England, France, Germany, Italy, the Netherlands, Belgium, Sweden, and Japan.[32] NASA wanted the Office of Munitions Control in the State Department to accord a license to the firm. The Department of Defense wanted the export license withheld.

NASA recognized that McDonnell Douglas's Proposal contained technical material in a number of potentially sensitive areas—"thermal protection, aerodynamics, avionics, structures, cryogenics, materials, propulsion, flight control, and so forth." However it claimed that most of this material was available in the open literature. It would not significantly contribute in any way to either strategic delivery or comsat capability "beyond that already existing in the receiving country, nor would its release be prejudicial to the interests of the United States."[33] It insisted too that the mere release of this documentation had to be distinguished from the possible transfer of hardware in any subsequent phase of active foreign participation. A license now was essential if a foreign entity was to be able to evaluate whether or not to participate in the post-Apollo program at all.

The Department of Defense felt that McDonnell Douglas's proposal provided a "broad display of advanced technological capability" that seemed to be "well in excess of what might be required to secure international participation in the space shuttle." It recognized that the shuttle itself had "no significant strategic delivery implications at this point." Yet it feared that "an unrestricted flow of the best US technology in a number of areas would in time lead to concern regarding the development of independent strategic delivery capabilities." It wanted NASA to secure government-to-government agreement on areas of international participation *before* industry was involved. That agreement would provide a legal umbrella under which the flow of technology from US contractors to foreign industrial partners could be "properly managed on a case-by-case basis."[34] As NASA's legal counsel explained, the DoD did not dispute the fact that the information contained in the McDonnell-Douglas proposal was available in the open literature, and it agreed that the shuttle itself had no strategic delivery implications: "it objects on the grounds that the mere flow of technology would be harmful."[35]

The restrictions on technology transfer frustrated Frutkin. He pointed out to Paine that these objections overlooked the fact that "military missile technology has been widely exported under various rationales and with certain assurances." He compiled a list of "controversial" technology transfer cases with Europe that were still pending in the Office of Munitions Control, one going back to September 1969 for technical assistance for Helios (see chapter 2).[36] In July 1970 he wrote, despairingly, that "the present attitudes and practices of the DOD and Department of State fraternity concerned with the export of unclassified technology in the space area would present virtually insuperable problems for us and make it extremely difficult to get satisfactory solutions in such a time frame as to establish credibility with the Europeans."[37] He called for a procedure that "should be as automatic as possible, should not be established on a case-by-case basis since this would entail unacceptable delays and uncertainties, should be premised on the establishment of adequate safeguards (guarantees and assurances) rather than on excluding particular items of technology, and should apply to NASA as well as to its contractors."[38] This framework for collaboration required, in Frutkin's view, a radical shift in perception by those who drew back from technological sharing: rather than see the risks to the United States they should focus on the benefits. These were several.

First, Frutkin stressed that "our principal objective will be to *obtain* a foreign technical contribution, rather than to *provide* technology. Thus we do not mean to export tug technology; we expect foreign cooperation to develop it for us." Of course he recognized that "*anything* we do by way of serious cooperation with other countries will inevitably enhance their capabilities should they wish to divert them to military purposes."[39] The choice was therefore between no technological exchange at all, or some degree of exchange with safeguards built into it.

Frutkin was emphatic that even though European participation would involve a degree of technological exchange, it would be biased in favor of the United States in two ways. The United States would get hardware from Europe, say, contributions to the avionics system of the tug. Europe, on the other hand would mostly have access to US technology through documentation and visits, a "second-order kind of exposure" that would not "produce the level of know-how that working in the technology imparts, nor that receiving actual operational hardware does." In any case, the "total technology base" of foreigners was "so much smaller than ours that they have much less opportunity to assimilate and apply technology they get from us" than is the converse.[40] In sum, wrote Frutkin, "the risk we take in broadening the flow of technology to our Western alliance partners is a very, very small risk and totally appropriate to the gain that we make in international participation and in direct program contribution."[41]

To strengthen the case NASA studied the extent of technology transfer in different collaborative programs, canvassing the views of experienced officials in the agency, the DoD, and in major US aerospace corporations.[42] The study considered two extreme cases: certain parts of the shuttle, and the separable tug. Regarding the first, it concluded that Europeans could usefully contribute to the development of the vertical tail of the shuttle and to some elements of the attitude control system to the advantage of both partners. The ensuing "transfer of critical technology to Europe would be a relatively small percentage of the program value." At the other extreme, the tug, even though a more independent system, would call for considerable technology transfer. It comprised two main components, a propulsion module and an avionics module. Europe lacked experience in some aspects of the former (e.g., cryogenic storage for long periods). It had only "limited experience and know-how in navigation, guidance, power distribution, instrumentation and data management systems" as required by the avionics module. To control technology flows here the United States could furnish subsystem components rather than state-of-the-art technological know-how. Reviewing the situation, NASA suggested that technology transfer to Europe, be it through "integrated" or "coordinated" participation, was relatively unimportant and controllable. What Europeans sought most of all was program management and systems engineering experience rather than specific tasks. In any event before it was agreed to collaborate on such tasks the United States would need to make a more refined study of European capabilities so as to identify strengths, from which domestic corporations could benefit, and weaknesses, where technology transfer had to be carefully controlled.

Aerospace leaders agreed that technology transfer posed few dangers.[43] They favored partnership on the grounds that it would stabilize the program absent an "assurance of adequate and steady funding" from Congress and that it would curb the "stimulation of independent and competing programs in Europe." They were also persuaded that the program would be so challenging that the US aerospace industry would benefit far more from its 90 percent share than the Europeans would from their 10 percent effort, emerging "from the post-Apollo enterprise even further ahead of the Europeans than when we started."[44]

On July 17, 1970, National Security Decision Memorandum (NSDM) 72 was released.[45] It addressed the "Exchange of Technical Data between the United States and the International Space Community." It established an interagency group, to be chaired by a NASA representative, to review policy and procedure for technical data exchange between the United States and foreign governments and agencies, beginning with Europe. And it specifically asked that those guidelines and procedures "be designed to provide for timely and effective interchange of technical information between the parties, while at the same time insuring the protection of U.S. national interest."

In response to this directive, an "Ad-hoc Interagency Group on NSDM-72" was established to formalize policy. It had the important role of both strengthening NASA's leadership, and of providing a forum for interagency consultation that could help avoid internal "polarization" and protect NASA from being "pictured as advocating giving away national values while the other agencies strive to protect them as required by Statute and Executive Order."[46] The group met weekly in August, it was chaired by Frutkin, and it included representatives from the DoD, the State Department, and the National Security Council.[47]

The group agreed that technological exchange should be handled in two phases. Phase A was a period of consultation before an intergovernmental agreement was signed. Phase B was implemented after such an agreement was adopted. "In Phase A we are releasing information so that foreign governments can assess whether or not to participate with us; the information is not too sensitive and the risks are not very great."[48] Requests for material would be authorized by NASA, after consultation with the Department of Defense when appropriate. If no objection was raised by the DoD within 72 hours the requested information would be released to foreign governments interested in participating in the post-Apollo program.[49] (This "Phase A" mapped onto Phases A and B in the NASA project development schedule.) In Phase B, by contrast (corresponding to Phases C and D of NASA's project schedule), the United States "would be dealing frequently with hard security and sensitive data," involving "definitive designs, know-how or hardware," and a more restrictive regime would be implemented to control it.[50]

This section has provided a quick look at how the flow of scientific, technological, and managerial knowledge between NASA and its contractors, on the one hand, and foreign entities on the other, was handled in the early days of the post-Apollo program. The object of the exercise was to streamline the procedure for obtaining licenses or assistance agreements *before* any partner had even committed to participate. NASA wanted the definition of the post-Apollo program, and the areas in which collaboration was possible to be as transparent as possible, so enabling foreign entities to decide for themselves where they might best contribute to the American program. As Frutkin stressed, foreign participation was being sought, above all, where there was an existing indigenous technological and industrial strength abroad. The aim was not to give technology away, but to creatively combine what others had to offer into NASA's effort. The implementation of this philosophy in 1970 and 1971, and the growing conviction in 1972 that it was impracticable, will be described in subsequent chapters.

The Second Major Barrier to Participation: Launcher Availability and the Constraints of the Intelsat Agreements

Launcher availability, and its relationship with the ongoing negotiations on the definitive Agreements establishing Intelsat, overshadowed the post-Apollo negotiations.[51] It was raised at a joint meeting of senior representatives from the State Department and NASA early in 1970.[51] U. Alexis Johnson, the deputy secretary for political/military affairs, headed the delegation from State. Johnson was an enthusiastic proponent of international collaboration who had just pushed through a very controversial agreement to share rocket technology with the Japanese (see chapter 10). Paine and Frutkin represented NASA. Those present noted that

> [w]e had anticipated, and the Europeans have now informally but unmistakably confirmed, that they cannot be expected to participate in the development of the shuttle unless they can be assured that they will be able to purchase shuttle launchings for any peaceful purpose. Put bluntly, this means they will not participate if they could be denied access to US launching capabilities whenever their purposes are judged undesirable by the US on narrow grounds of US national interest.[52]

As NASA saw it the entire collaborative project with Europe hung in the balance, its success crucially dependent on this one issue. Robert Packard in the State Department put it thus: "However ungrateful or disingenuous or misguided the reactions of many of our partners in Intelsat, we have managed to evoke their hostility and distrust in comsat matters to a greater degree than in most other areas of our relationships with them. This includes many of our closest allies and foreign associates."[53] A positive response to European demands "could well cause Europe to abandon large launcher development programs." It would also make the Europeans more willing to trust the United States in the renegotiation of the Intelsat interim agreements. A negative reaction, by contrast, would not only mean losing a contribution of up to a billion dollars to the post-Apollo program. It would "confirm the position of those Europeans who preach the need for non-dependence on the US," which would in turn "provoke decisions in Europe to channel funds into competitive and independent, as well as wasteful, European space programs." It would also complicate the United States' position in Intelsat, "strengthen[ing] those forces which argue that the US continues to seek by every means to dominate space telecommunications into the 1980's."

The sentiments expressed here do of course have a familiar ring about them. As we saw in the previous chapter, the combination of US dominance in Comsat and the administration's support for their position expressed in NSAM 338 had been a source of constant friction between NASA and Western Europe, and indeed between NASA and other arms of the US administration. Yet, while both Frutkin and Barnes were immensely irritated by European distrust of US motives in Intelsat (Europeans seem to think, Frutkin wrote, that the United States was "seeking by all means, fair or foul, to maintain political and technical control of Intelsat"[54]), it was also recognized that it was difficult to maintain international cooperation on the basis of the interim Intelsat agreements. In 1966–1967 the debate was dominated by worries over the "technological gap" and NASA strongly favored a proactive attempt to encourage technological sharing with European industries so that they could compete more effectively with US firms for comsat contracts in the space segment. The debate in 1969 had moved beyond this. As the negotiations over the definite agreements got under way, the entire structure of Comsat itself was being challenged, and questions were being asked regarding the United States' willingness to provide launch assistance to foreign countries that wanted to establish their own comsat systems separate from the global system.

The negotiations over the Intelsat agreements brought home to the Europeans how vulnerable they were to American technological leadership. As we saw earlier,

NSAM 338, promulgated on September, 25 1965, denied American assistance in the development of foreign communication satellite capabilities, and further emphasized the determination of the US authorities to leverage that leadership, and the dynamism of their industry, to shape the contours of an international communication satellite system to their commercial and political advantage. Within a few years the evidence of a fruitful collaboration between the state and private industry was there for all to see. Between 1964 and 1970, NASA and the DoD (which developed its own system) had invested $207 million and $377 million, respectively, in research and development for communication satellites: Comsat had invested $143 million.[55] The sophistication and capacity of four generations of Intelsat satellites increased apace. Intelsat I (*Early Bird*), the first satellite built by Hughes, was an experimental-operational satellite, with 240 two-voice channels, that inaugurated commercial transatlantic communication in June 1965. The Intelsat II series added a television channel. The first successful Intelsat III launch, on December 18, 1968, had 1,200 two-way voice circuits and four-color television channels. This series ensured global coverage by 1969, ushering in the first and only global commercial communications satellite system, and so achieving Intelsat's prime mission objective. The first Intelsat IV satellite was successfully placed in orbit in January 1971, as the final negotiations on the definitive Intelsat agreements were being settled. Positioned over the Atlantic, Intelsat IV had 3,000–9,000 two-way voice circuits and up to 12 color television channels.[56]

Economically speaking, the United States was the main investor in, and benefactor of this system. As the State Department reported to Congress in 1970, "Since 1964 the 76 members of the Intelsat consortium have invested $350,000,000 in the system and America's share (and voting power) is currently about 52% or $226 million. Ninety-two percent of the total spent ($323,500,000) went to American contractors."[57] There was no foreign procurement for Intelsat I. It was 2.3 percent in Intelsat II and 4.6 percent in Intelsat III. It rose to about 26 percent in the first four of eight in the Intelsat IV series, for which Hughes Aircraft subcontracted $19.6 million abroad. It then fell back to about 10 percent for the next four in the series.[58]

The benefits reaped in the space segment were supplemented by the dominance of US firms in the ground segment. Congress was told that, by 1970, 28 countries had invested some $250 million in 50 earth stations, in which at least 50 percent of the hardware had been provided by American manufacturers.[59] Indonesia was one such country. A ground station built near Jakarta that was carrying 95 percent of Indonesia's international communications was equipped with hardware provided by an American company. It was operated by an American firm that employed only relatively uneducated Indonesian technicians, and that was slated to reap all the profits from the operation for the first 20 years.[60]

Situations like this were obviously not tenable in an "international" organization. France, which firmly believed in the technological, commercial, political, and cultural value of communications satellites—and who realized that increased investment and improved technology were keys to improving Europe's situation in Intelsat procurement—took the lead. As mentioned in chapter 3, in 1967 it agreed with Germany to build an experimental communication satellite called Symphonie as a first step toward the acquisition of the technological and industrial know-how needed to compete meaningfully with American firms for Intelsat contracts. Symphonie was a small (185 kilograms) satellite that was to be launched with ELDO's Europa II rocket in 1973. Its capabilities were marginally better than Intelsat I and II. It would be placed in geostationary orbit at longitude 15°W, where it could provide the overseas territories and provinces of France "with cultural television programmes and commercial or governmentally run telephone services," as one French engineer put it in September 1970.[61] The first flight model would be a "technical test phase." It would be followed by a "non-commercial experimental operational phase," in which interested countries could build the ground stations needed to receive its signals. In 1970 there were still some doubts about the more contentious "operational phase" that would follow.

The negotiation of the definitive Intelsat agreements got under way in January 1969. The Europeans moved forcefully to ensure that they had a greater say in the running of the new body that would emerge. After long, difficult, and bruising discussions the definitive agreements were adopted in May 1971 by a vote of 73–0, with 2 member countries absent and 4 abstentions including France.[62] The legal and political anomaly, whereby Comsat was both the representative of an international organization, the defender of US corporate interests, and the manager of the space segment was abolished. The definitive arrangements unambiguously split the administrative and financial side of Intelsat from the technical operations. It vested authority for the former in an Assembly of Parties, the prime political organ of Intelsat, in which all Member States had one vote. Technical matters fell into the domain of a more restrictive Board of Governors, the extension of the ICSC, and in fact the heart of power in the Intelsat system.

The European demand that Intelsat should help develop technology in those countries that lagged behind the United States—another issue that had been raised in 1964—was again rebuffed by the American delegation, now with the support of the developing countries. The latter saw no reason to subsidize the economy of an industrialized region by placing contracts for hardware that was more expensive than the best bid. All the same a compromise wording was found that left the door open for some concessions in procurement policy. Article XIII of the definitive agreements not only specified that procurement would be open to "international invitations to tender." It also allowed that, if there was more than one bid that offered the "best combination of quality, price and the most favorable delivery time," the contract would be awarded primarily so as to stimulate worldwide competition, thus nominally breaking the grip of US firms on the supply of goods and services.[63]

Article XIV(d) was one of the most controversial items in the Intelsat definitive agreements, and one of particular pertinence here. It dealt with the rights of Intelsat member countries "to establish, acquire or utilize space segment facilities *separate from* the Intelsat space segment facilities to meet its international public communications services requirements" (my emphasis). Comsat went into the negotiations demanding that "the definitive arrangements should contain an absolute prohibition against any Intelsat member's participation in the establishment or use of any communications satellite system other than the Intelsat system for international telecommunications purposes." The draft agreement put forward by the US delegation suggested "sanctions for the breach of this obligation by way of suspension and eventual expulsion from Intelsat."[64]

The US delegation led by Comsat won little if any support, either from government authorities at home or from other nations in Intelsat. The compromise hammered out, against strong American opposition, allowed for separate regional comsat systems under two conditions. First, such a system had to be "technically compatible [...] with the use of the radio frequency spectrum and orbital space by the existing or planned Intelsat space segment." Second, and more ambiguously, the new system would be allowed if it did not cause "significant economic harm to the global system of Intelsat."[65] These two conditions (technical compatibility and absence of significant economic harm) would first be voted on by the Board of Governors, where votes were weighted, and a complex formula was established to curb the powers of any single major country, or bloc of industrialized countries.[66] The board's recommendation would be passed on to the Assembly of Parties, where each country had one vote. To be implemented its recommendations required the support of two-thirds of those present and voting.

The final agreements were not specific as to how "significant economic harm" was to be established, other than saying that two-thirds of the delegates in the Assembly of Parties had to agree on it. On one interpretation, to proceed with a separate system required a *positive* vote from member states, that is, two-thirds of them had to agree that the new system would *not* do the global system significant economic harm. On another interpretation, comsat systems could proliferate unless Intelsat made a *negative* finding, that is, unless two-thirds of those voting agreed that the system *would* cause significant economic harm to Intelsat.

This distinction was crucial for the proponents of a separate comsat system. On the first interpretation, a so-called positive finding, the onus was on the candidate for a new system to persuade two-thirds of the member states that it would not do significant economic harm to the global network. On the second interpretation, a negative finding, the onus was on two-thirds of the Intelsat Assembly of Parties to show that it would. The assumption was that the requesting member was confident that the separate system would not be at the expense of the global system. Countries like France that wanted to develop separate regional systems obviously preferred the "negative" interpretation, since this placed the onus on their opponents to muster widespread support to stop them.[67]

The definitive agreements did not empower Intelsat to make *binding* determinations on any of its parties. It was reduced to a consultative body, which could only make advisory recommendations. Thus a member or group of members could proceed with developing a separate system even if the Intelsat Assembly voted that it would do significant economic harm to the international organization.[68] The only factor stopping a member state defying Intelsat would be the opprobrium of those in the body who sought to defend and respect its procedures, and who would be incensed to find revenue-producing traffic diverted to a separate system to the exclusive benefit of a limited number of (necessarily industrialized) countries.

For NASA, Comsat's attempts to protect Intelsat from competition were unnecessary and counterproductive. Europe was a decade behind the United States in the development of communications satellites—a decade in which US business would still dominate the market. Anyway US industrial support would be essential for Europeans to build advanced communications satellites for the 1980s, thus ensuring a net dollar inflow into the country for both the payload and the launcher. In short NASA was emphatic that the advantages of foreign participation in future space programs "far outweighed any benefit we could hope to get from continuing to restrict launch services so as to protect Intelsat."[69] Hence the conclusion:

> What is required is a positive internal policy directive, short and clear enough to be unequivocal, permitting the Department of State and NASA to make clear as appropriate that the US will agree, in any broader agreement on major foreign contributions to the post-Apollo program, to make STS launch services available on a reimbursable basis for any peaceful purpose. The internal statement should come from the White House so all agencies can and will reflect it.[70]

The same assurances would have to be provided for the supply of reimbursable launch services in the 1970s, before the shuttle was operational.

By mid-July 1970 the State Department and NASA had devised a satisfactory formula, in their view, as regards the availability of launchers (the DoD having withdrawn from the issue).[71] They were persuaded that some proliferation of communications satellites systems was unavoidable if the United States was to maintain credibility in the negotiations over the final version of the Intelsat agreements. To meet this new situation NASA declared that NASM 338 should be revised to allow for reimbursable launchings or to sell shuttles "to those who will have participated substantially in the development of the shuttle," like the Europeans. New international agreements being finalized in Intelsat would no longer express President Kennedy's proposal for a single global comsat system; they would have to allow for "domestic, regional or specialized communications systems." In this more relaxed regime it was suggested that the United States would launch foreign comsats unless "the appropriate organ of Intelsat reached a negative finding with respect to such a system." In this situation the United States would be "obliged to withhold our own collaboration," that is, it would not collude with a petitioner that wanted to override a majority vote in Intelsat.[72]

NASA was aware that while this was as far as they could go at the moment, it would not satisfy Europeans. As Frutkin explained to Packard in the State Department, "[T]oo many interests in Europe would wish to regard such a qualified assurance as negative at this point, exactly as was done with the Symphonie launch correspondence some years ago."[73] On that occasion the United States had stipulated that, to respect the Intelsat agreements, it would only launch the Franco-German satellite if it was to be used for experimental, not operational purposes. Imposing conditions on launching comsats to satisfy Intelsat was anathema to the Europeans, who felt that, if they judged that they were respecting the agreements, it was not up to the United States to use their monopoly on access to space to thwart European ambitions. A Congressional rapporteur on an ESC meeting held in Bonn early in July reported back to Washington that he was told "time and again" that European participation in post-Apollo required an "iron-clad agreement by the United States to make available launch vehicles and services, unconditionally, for any peaceful purpose."[74] In sum, the launcher issue was never going to be resolved as long as some influential members of the European space community demanded unconditional access to American launch services. The stark choice between a major contribution to the post-Apollo program and the maintenance of a costly and putatively obsolete European launcher was also seen as the choice between deriving advanced technological knowledge from the United States and being at the mercy of the United States' interpretation of the definitive Intelsat agreements.

The (Brief) Honeymoon on Launcher Policy: The Lefèvre Mission in September 1970

In July 1970 the European Space Conference took a major step forward in what came to be known as the first package deal.[75] At the core of this deal lay the decision to reorient ESRO's mission away from scientific satellites toward applications, with top priority being given to a communications and an aeronautical satellite. ELDO would continue to work on the completion of the Europa II launcher. Belgium, France, and West Germany, along with the Netherlands (to the end of 1971 only) were also willing to go it alone and to embark on the development of a more powerful launcher, Europa III.[76] The delegates also agreed to establish a working group that would draft a convention establishing a single European space agency, similar to NASA. Collaboration with the United States was not neglected. Not only did the ESC commit itself to an immediate start on the project definition phase of an aeronautical satellite (Aerosat) in cooperation with NASA, it also wanted "all possibilities" for European participation in the post-Apollo program to be studied by a working group of ESRO and ELDO officials in consultation with NASA. It voted $2.5 million through June 1971 for studies of the space tug.

The American authorities were heartened by these developments. The Europeans would gain technical, managerial, and industrial benefits, would be able to "avoid investment in the development of redundant European launch capabilities," and would be given additional assurances regarding American launchers and launch services.[77] For the United States, a major European contribution could be of substantial domestic value. It would bring in financial resources and technology, and enhance the use of the systems, so strengthening the justification for developing them. It would improve NASA's political hand as it battled for its budgets. As George Low wrote in a memo intended to "emphasize [his] own enthusiastic and strong support" for "wide and meaningful cooperation in the post-Apollo program"—if successful, such a project would "have a strong influence on support for our post-Apollo program objectives both in Congress and within the Executive Branch."[78] It would also provide a template for further partnerships of this type, and contribute to the North Atlantic alliance. On the down side, a joint project would be more complex to manage, for "although the Europeans will be heavily dependent on us, we will become dependent in some measure on them." But only in some measure: this was not a level playing field, as emphasized in a position paper prepared for the State Department:

> In view of the preponderance of U.S. resources and effort which will be put into the development of these systems and the far greater use which the U.S. will have for them, when operational, this collaboration with the Europeans would be very asymmetrical [...]. There will be no credible basis on which this collaboration could be viewed as an equal partnership. The responsibility and control will necessarily be American.[79]

The challenge then was to give the Europeans a meaningful stake in a program in which the balance of power was tilted heavily toward their partner.

On September 16–17, 1970, a delegation led by the Belgian minister of state in charge of scientific policy and planning, Théo Lefèvre, was hosted by the Department of State in Washington, DC. Lefèvre headed the mission in his capacity as chairman of the ESC. He was assisted by J. F. Denisse, the president of the French national space agency, and Lord Bessborough, the United Kingdom's minister of state in the Ministry of Technology. Causse and Dinkespiler were among the very few scientists and engineers present. The American delegation was headed U. Alexis Johnson for the State Department. The other principal members were George Low in his capacity as acting NASA administrator, Edward David, the science advisor to the president, William Anders, the executive secretary of the National Aeronautics and Space Council, and John Morse, deputy assistant secretary of defense for European and NATO Affairs. Senior staff members from these various arms of the administration attended as advisors, including Arnold Frutkin and Dale Myers, NASA's associate administrator for Manned Space Flight. This was then a discussion at a very high level of a "preliminary and exploratory character," to sound out the "political, financial and other implications of an eventual European participation" in the post Apollo program.

The meeting moved beyond the usual formalities and exchanges of views and tried to make concrete progress on fundamental matters of policy.[80] During the opening session on September 16 the members of the American delegation made brief statements expressing their enthusiasm for European participation in post-Apollo, while taking care to add that the program had not yet been officially adopted. They indicated possible areas where the Europeans may like to cooperate, from building a discrete element of the orbiter to participating in an integrated system. They remarked that the collaboration would be guided by no exchange of funds and management integrity. A number of other items that were of concern to Europeans—access to information and facilities and participation in decision-making—were addressed. However, the burning issue, and the one almost immediately raised by Lefèvre, concerned the availability of launchers. Europe, Lefèvre said, did not have the financial means to maintain an independent capability in satellites and launchers and to participate in the post-Apollo program. Faced with this dilemma, it had to have US launchers available "without political conditions, and on a commercial basis." There was a preliminary exchange of views on these issues the next morning.[81] Two weeks later, on October 2, 1970, the US undersecretary of state officially replied to Lefèvre. His 14-page letter carefully described the administration's thoughts on three key European concerns: acquisition of launch services and launch vehicles, the extent of European involvement in decision-making, and European access to US information and facilities.

Johnson reassured the Europeans that they would have a role in decision-making and management commensurate with the extent of their participation. They would be consulted in the development of the shuttle and the space station whenever matters arose of "significant, mutual concern to both parties." There would be an "extensive role" for Europe in the management of those areas in which its contractors were involved, even if they worked under an American prime. Europe would also have to be "a partner in reaching any decisions which have a measurable impact upon European costs or European tasks." As regards the use of either the shuttle or the space station, "we would expect Europe to take part in mission planning and experimental programs in generous proportion to their use." That said, given the preponderance in the United States' contribution to both development and use, "overall responsibility for management of the post-Apollo program would necessarily rest with the U.S."[82]

As for access to information and facilities, Johnson noted that the aim was to make optimum use of resources and skills on both sides of the Atlantic. In doing so one had to distinguish between general and detailed access to technical data and facilities. All countries would have general access, meaning access through visits and published information, to all technology and facilities in the overall development of the program. Detailed access—meaning "access to design, development, and production data to the level of commercial know-how"— would be allowed to participating countries "commensurate with the measure and character of their participation."[83] It would be released by the United States or by Europe "on a need-to-know basis necessary for the accomplishment of their specific tasks under the agreed collaboration," and in phase with their progress with those tasks.[84] Access to technological know-how was thus tied directly to the extent of investment and participation, and was not a generalized right that could be acquired with a minimum of effort by the foreign partner. As Frutkin put it in a briefing document, Europe can "determine the extent of its access to commercial know-how in the program by increasing its contribution, and through it the number of interfaces it will be involved in, and through its requirements for such information."[85]

Sensitive information, classified or unclassified, was not directly dealt with in the meeting on September 16–17, but the State Department had prepared itself for the question if it arose. Only individuals or teams clearly identified as requiring it would be granted access to this knowledge, it would be restricted to the location where the work was done, and it could not be transferred or applied in strategic military weapons systems. If by chance guidance or reentry technology was involved Europeans could only be allowed access to such knowledge "if it could be clearly demonstrated that (1) better technology and know-how exists in the prospective contributing country or, (2) in the case of only equivalent technology and know-how, there are over-riding reasons to seek foreign participation in these areas, and (3) neither the U.S. technology nor end products resulting from it would be transferred to any third party."[86] In sum, if there was little to lose the need for international collaboration could trump national security, but only under the strictest need-to-know regime and with appropriate safeguards.

The launcher policy described by Johnson was effectively that agreed between NASA and the State Department in July (see earlier).[87] It was conditional on Europe making a substantial contribution to the program, meaning "at least 10%" of its estimated cost of some $10 billion over ten years. This share could be met by contributing significant new technology to the system, or by developing a major system or subsystem, or by a combination of these. If the Europeans were willing to make this 10 percent-plus financial engagement, the United States "would no longer determine the availability of launch services for European payloads on a unilateral case-by-case basis": American and European interests would be on an "equal footing" with regards to the supply of launchers "for possible commercially competitive purposes." This "blanket assurance" to launch had to be "consistent with relevant international agreements," however: the United States would respect the decisions of the Intelsat Assembly of Parties. In particular, unless two-thirds of the Intelsat member states voted that a proposed separate system did do significant economic harm to the global system (a "negative finding"), the United States would launch for Europe.

Johnson did not want to provide substantive criteria for "significant economic harm," as requested by the Europeans. The Intelsat negotiations were drawing to a close, major concessions had been made, and this was no time to reopen the debate on the highly contested Article XIV of the definitive agreements. However, he did stress that the United States "would provide the requested launcher facilities [...] even if it had voted against the project."[88] On the other hand, if there was a negative finding, the United States would still "consider their position, without saying that under no circumstances could they provide launchers"[89] In short Johnson assured the Europeans, however significant economic harm was defined, the United States would not apply "the principle of being consistent with Intelsat arrangements" "in a narrow way."[90]

The general philosophy underlying Washington's position is clear. It was no longer trying to "help" a weaker ally, as in the 1960s. Europeans had a financial, technological, and industrial contribution to make to post-Apollo. Once they had decided what they wanted to do, the United States would determine how best to meet their requirements, consistent with Washington's desire to foster international collaboration and to protect its national interests broadly defined. Of course the relationship would, of necessity, be dominated by the United States. The asymmetries in contributions of all kinds were evident, "Nor will it be in our interest to attempt to enhance the benefits for the Europeans artificially."[91] Thus time and again when the Europeans sought to be treated as "equal partners"—in decision-making, in access to technology, in negotiations with third countries—they were reminded of their subordinate position. Europe's ability to influence events would be proportional to their share in the program and restricted to the areas in which they were directly engaged.

The position on launchers followed the general pattern: greater US flexibility was tied to substantial European participation. The State Department made it clear that, if that participation was forthcoming, the United States had no intention of using its power in Intelsat to indiscriminately protect American interests. The willingness to interpret voting majorities in terms of a negative finding, which favored the petitioner, was indicative of this flexibility.

Of course, there were still areas of uncertainty. How binding on the United States was a "negative finding"? How did one measure "significant economic harm"? Johnson recognized Europe's fears of being held hostage to American launch policy if they did not retain independent access to space. He was willing to give near-blanket assurances of launcher availability: after all, there was a difference between launching foreign payloads "subject to case-by-case determination on the one hand and, on the other, offering an assured, on-going commitment to do so for all European space projects (so long as they are for peaceful purposes and consistent with international agreements.)"[92] However, he was extremely reluctant to commit the United States to launch European telecommunications satellites "unconditionally," and in defiance of a "negative finding" by two-thirds of the appropriate Intelsat organ—a situation that, he thought, was most unlikely to arise anyway.

While the US authorities played down the difference between the two parties, the Europeans tended to emphasize them. Lefèvre insisted that an adverse recommendation in Intelsat was not legally binding, and that Europe could legitimately defy it if it had its own launcher. Europeans also wanted to interpret "economic harm" so widely that they could reconcile their commitment to Intelsat with "projects which could be competitive with Intelsat rules without jeopardizing its existence." Johnson was emphatic that the credibility of the United States as an international actor demanded that it respect the decisions taken by Intelsat (even if it had voted against them). Europe could not be treated differently to any other petitioner. The United States, Johnson wrote to Lefèvre on October 2, 1970, "would adhere to the language and intent of article XIV, and would expect other countries to do the same."[93]

The First Setback: The ESC Meeting on November 4, 1970

In his report back to the ESC early in November, Lefèvre began with a very positive account of the gathering in Washington, and with an enthusiastic endorsement of the technological novelty of the post-Apollo system and of the United States' "desire to internationalize the conquest of space, for the benefit of humanity as a whole."[94] He spelt out clearly the shift in US policy on launchers—from a case-by-case decision to a (near) blanket assurance—and stressed that it was conditional on Europe making a substantial contribution to the American-led program. And he emphasized that, in the event of a negative finding in Intelsat, even if the United States were reluctant to go against an internationally-sanctioned decision, it might still "exercise its freedom of decision" on whether to accede to Europe's request for a launch. This did not bother Lefèvre unduly. He put a positive light on the launcher issue, emphasizing that *"Europe will have a large availability of American launching devices"* even if a few "uncertainties" needed ironing out. He was satisfied that enough progress had been made in Washington for Europe to enter "the negotiation phase proper" in a program that would "give a new dimension to European efforts and a greater responsibility vis-à-vis international cooperation." In July, said Lefèvre, there had been general agreement in the ESC that telecommunications satellites and the means to launch them should provide the backbone to a European space effort run through a single agency. Residual doubts on this program should now be put aside. After today, he concluded with determination, "[w]e must know exactly which countries are willing to continue and organize a joint effort, meaningful and reasonable, so that Europe will efficiently participate in the development of space techniques with the twofold purpose of promoting technological progress and keeping its cultural and political independence. [...] The time has come to act."

Lefèvre's hope of pulling the collected ministers together behind a unified policy was soon shattered. The British led the opposition. They had already voiced two concerns in the meeting held on September 16–17. First, they hoped that participant countries would be granted *"full access to, and unrestricted use of, all know-how, design rights, etc* generated by *any* part of the post-Apollo program"* (my emphasis).[95] This idea had been killed at once. The United Kingdom had simply not come to terms with the asymmetry in the partnership, nor with the implications "of the obvious preponderance of U.S. investment and use," which undermined any "credible basis" for the level of sharing that the British hoped for.[96] Second, Britain was extremely reluctant to make a "substantial" contribution to a program whose content and cost was still not defined. In September Johnson's reply reflected the difficulty NASA was having in getting Congress to support Paine's original program.[97] This ambiguity was picked up by the new British minister of aviation supply, Freddy Corfield. Corfield had taken up his post in September 1970 after the Conservative Party ousted Harold Wilson's Labour government at the general election in June. As he put it, "There have been considerable changes in the form of the proposal since it was first suggested and at the present moment there is no specific programme approved by the American government. The timescale is uncertain and the cost estimates and incidence of expenditure remain to be clarified."[98] The new British government was engaged in a comprehensive review of public expenditure. It could not accept "a commitment to share the costs of 10% participation [in post-Apollo], running to as yet unquantifiable but probably very large sums of money, and this in a context of a project too loosely defined to enable any assessment to be made of the benefits in relation to resources involved." Nor was this necessary to secure US launchers for applications satellites, in Corfield's view. He said that 10 percent participation may be needed for *blanket* assurances. But the United Kingdom did not seek them. The British government was persuaded that, as in the past, "for all purposes for which Europe is likely to require launchers, we can expect to be able to rely on a reasonable American response."[99]

Corfield's opposition was given added traction by the Gaullist minister for industrial development and scientific research, François-Xavier Ortoli, who put a different twist on the uncertain situation across the Atlantic.[100] Whatever the costs of post-Apollo—and current estimates were likely to escalate—a 10 percent European share would probably far exceed the costs of developing a European launcher. In return the benefits were dubious: the US guarantees for launchers were not watertight, and access to technology was too restricted. On balance, therefore, it was cheaper and more advantageous technologically and industrially for Europe to go it alone.

The discussions were finally suspended at 2 AM in the morning of November 5, a day earlier than anticipated. Belgium, France, and Germany agreed to pursue the possibilities of post-Apollo collaboration with the United States and invited others who were interested to join in the next round of discussions.[101]

These negative reactions to Washington's proposals infuriated Frutkin. Already at a meeting in Florence organized by Eurospace (an industry lobby group) in September 1970, he had emphasized the obvious—that "it would be extremely unrealistic to assume that there would be total access to the technology of the programme, at the know-how level, if the U.S. is contributing 90% and Europe 10%."[102] "Equal" partnership (Ortoli) or "full access" (Bessborough) were inconceivable granted that asymmetry in commitment. As for the fluid state of the post-Apollo program, in Florence both he and Dale Myers confidently asserted that the post-Apollo program would be adopted, and stressed that the advantage of its content not being settled was that the Europeans could participate in the definition phase, so helping structure its shape in line with their interests.[103] Instead of approaching post-Apollo collaboration in this spirit, Frutkin wrote Low, the meeting in November was conducted "with high emotion and political pre-judgment, with little reference to the available facts which should determine European interests, and with persistent unrealism on trade-off possibilities, conditions, risks and benefits."[104]

Domestically Frutkin did try to turn one complaint made in November to NASA's advantage. A few days after the abortive ESC meeting he wrote to Robert Behr of the National Security Council to tell him of Britain's reluctance to commit to post-Apollo participation "because of the uncertainty of the US commitment to the space shuttle and to continuity in our major programs."[105] He added, somewhat menacingly, that "[w]e would have to be prepared, in the event we do not move the shuttle forward, to find Europeans concluding that we provide a very poor foundation for international enterprises and that we have seriously delayed and diverted their own regional programs, perhaps deliberately."[106] A few days later he met with Johnson and Pollack in the State Department to discuss the NASA budget for FY1972 and the "need for a clear and credible signal to the Europeans that the United States is moving ahead with the space shuttle program."[107] This in turn led to both State Department officials drafting memoranda for Kissinger affirming that while the post-Apollo program did not stand or fall by virtue of international participation, it was imperative to offer Europeans an "assured alternative" if the United States expected them "to forego independence."[108] Johnson's memorandum was particularly explicit about what was at stake: the benefit of European know-how, a contribution of about $1 billion, national security concerns ("there are obvious advantages to having the Europeans as partners in the United States program, as compared to their developing a separate and independent space launching capability over which we might have little or no influence"), and political considerations (success in post-Apollo would promote intra-European cooperation and further major scientific and technological projects; it would also strengthen the capability of Washington's NATO allies and of the alliance).[109]

These arguments amplified an appeal made by Low to Kissinger at the end of October, and reinforced by him after the ESC meeting early in November. NASA's acting administrator explained that the agency was now willing to defer a start on the space station in favor of the shuttle. This was not only because the shuttle was "the correct next major step in the United States space program." It was also because "a go-ahead on the space shuttle, in FY1972, is of crucial importance in relation to the possibilities for very substantial international contributions to and participation in our major space undertakings of the future."[110] Another round of interdepartmental discussions was held, and the need for clear directives from the top was emphasized if European support was not to drain away.[111] On January 4, 1971, the national security adviser replied to Low indicating that no definite policy directive could be expected at this time.[112] NASA's final budget resubmitted to Congress for FY1972 was slightly below that of FY1971. It would take another year before the president eventually endorsed the space shuttle.[113]

Paine's Departure

Just before these delicate and complex negotiations got under way the Europeans lost one of their most trusted allies: NASA administrator Tom Paine.

Paine was convinced that if NASA was to going to sell its post-Apollo program, it had to adopt what he called a "swashbuckling, buccaneering, privateering kind of approach."[114] He tried to enroll the White House in his ambitions plans by writing several letters to the president, encouraged by Nixon repeating publicly in March 1970 that he hoped for "greater international cooperation."[115] His energetic advocacy was not, however, matched by the administration's support for NASA.

NASA's ambitions were reined in by transformations to the decision-making process on the budget. Nixon elevated the Bureau of the Budget (BOB) into the Office of Management and Budget (OMB) in July 1970, and gave it wide-ranging powers to evaluate program performance and budgetary requests *before* they were submitted to Congress. This arm of the administration was thus a cardinal player in the assessment of budget requests coming up from the various government agencies (only the CIA and the DoD were apparently able to override their strictures), and its officers had a crucial role in transforming general policy statements into concrete programs with a realistic (in their eyes) dollar amount attached to them.

These changes had palpable effects on NASA's budget. Indeed between the time that Paine made his ebullient speech in Europe in October 1969 and the Congressional debate on the budget in the first six months of 1970, he saw NASA's future funds cut by over 25 percent. His proposed budget for FY1971 was $4.25 billion, a sum that he had already reluctantly reduced by about $0.25 billion. The BOB lopped $0.8 billion off that. Senior White House staffers Peter Flanigan and Thomas Clay Whitehead pruned it further. Flanigan was an investment banker who had been Nixon's campaign manager in 1968 and who had been given oversight responsibilities for space. Whitehead was a systems analyst from RAND who was asked by Flanigan to assess NASA's budget and planning procedures. Flanigan and Whitehead reduced the BOB figure to $3.53 billion, and then, even as Paine was announcing this to the press, cut it by a further 2.5 percent to $3.3 billion as part of an across-the-board reduction to present a balanced budget to Congress for FY1971. Paine's budget proposal thus suffered a massive reduction of some $1.2 billion in a few months.[116]

Then there was the situation in Congress. A survey of Congressional opinion covering the first 11 months of 1970 remarked that "[i]nflation, increasingly pressing domestic social problems, urban decay, environmental pollution and growing popular disenchantment with Federal programs that could possibly be called technological luxuries" had pushed space well down the list of national priorities. This was exacerbated by the success of Apollo 11 and 12, which suggested that the United States was well ahead of the USSR in the "space race."[117]

The Cold War rationale for a major space program had lost its bite, and the satisfaction of domestic social needs was uppermost in the minds of both Congress and the Senate.

President Nixon was also less committed personally to space than was President Johnson, who had, of course, made the conquest of space his signature item in the run up to the 1960 presidential election. What is more when Nixon spoke of international collaboration he had the communist bloc foremost in mind. European matters took second place to his concern to establish east-west détente. This was translated into the signature of major international agreements intended to stabilize the international order, including a collaborative space venture that involved the docking in space of an Apollo and a Soyuz spacecraft (see chapter 7).[118]

In a climate where interest in space was rapidly declining, where financial restraint was imperative, where the old Cold War arguments for a major space program had lost their punch, and where the budget process was dominated by people who were determined to clamp down on expenditure and were very reluctant to authorize new open-ended projects, support for a major post-Apollo program was anything but assured. Paine was not particularly good at adapting his proposals to this political reality: he rather naively believed that his enthusiasm and the self-evident (to him) merits of NASA's proposals would persuade the White House to endorse them and Congress to fund them. He was even less able to manage the internal dynamics in the White House and the power that the BOB had over preliminary budget estimates, nor the hostility felt by people like Tom Whitehead to his ambitions.

In August 1970 Thomas Paine decided to return to private life. One of the last things he did was to thank Henry Kissinger for his "strong and effective support" in their "joint efforts to increase international participation in the space programs of the United States." He also expressed his "deep appreciation" to U. Alexis Johnson (State Department) for his "help and encouragement" in the past, and urged his continued "strong support [...] to increase substantially participation by other nations in our space program."[119]

As Joan Hoff puts it, Paine's departure from NASA on September 15, 1970, "came as a welcome relief to both the legislative and executive branches of government."[120] The reaction in Europe was just the contrary. Paine's "conviction and enthusiasm," his "friendliness and open-mindedness," would be missed. So would his recognition, not generally shared in Washington, that "we cannot have significant international cooperation without some real dependency, each side upon the other."[121] The secretary general of ELDO spoke for all on that side of the Atlantic when he wrote to Paine that "[w]e will [...] be deeply affected by your leaving NASA which will mean the break in an important personal link which has been of the greatest value at this still rather provisional stage of our common enterprise."[122]

Chapter 5

European Participation in the Post-Apollo Program, 1971: The United States Begins to Have Second Thoughts—And So Do the Europeans

Bumps on the Road to Engaged Collaboration

On February 5, 1971, the Apollo XIV Lunar Module touched down on the surface of the moon. This was followed a few weeks later by the release of the President's Report to Congress on Foreign Policy in the 1970s.[1] The report used the successful completion of the Apollo XIV mission to reiterate that the achievement was not simply a reflection of American scientific and technological capability. "It is equally a measure of an older American tradition, the compulsion to cross the next mountain chain. The pressurized space suit is, in a very real sense, today's equivalent of the buckskin jacket and the buffalo robe. Apollo XIV is the latest packhorse, and its crew the most recent in a long line of American pioneers." It ingeniously introduced the international dimension by stressing that "mutual help and cooperation" was "essential to life on the American frontier." In a reference to the new climate of détente it noted that NASA and the State Department had been instructed to pursue broader collaborative projects with Moscow "with the utmost seriousness." Congress was also advised that while "substantial participation" was being sought in the post-Apollo program, "the result is uncertain, for there are very real difficulties to be solved."[2] Two of those concerned the scope of NASA's international commitments.

White House staffer Tom Whitehead was particularly outspoken in this regard.[3] In a memo to Peter Flanigan, who had oversight responsibilities for NASA's budget, he wrote that the agency was failing to "make a transition from rapid razzle-dazzle growth and glamour to organizational maturity and more stable operations in the long-term." Its overheads were too high. The agency lacked direction. Above all its pursuit of European funding for post-Apollo had not been thought through. The White House had not yet decided what the shape of the program would be, yet if the Europeans were to commit $1 billion to it, "the President and the Congress will have been locked into NASA's grand plans because the political cost of reneging would be too high." What is more, "the kind of cooperation now being talked up will have the effect of giving away our space launch, space operations, and related know-how at 10 cents on the dollar," to the disadvantage of US business. Whitehead, in fact, thought that it would be better to take space operations out of the political realm and anchor them more firmly in the commercial area, where they would be free from "international bickering" and better serve the needs of American high-technology industry. What NASA needed now, he wrote, was "a new Administrator who will turn down NASA's empire-building fervor," and present the OMB and the White House with "broad but concrete alternatives." "In short," Whitehead wrote, "we need someone who will work with us rather than against us, [...] and will shape the program to reflect credit on the President rather than embarrassment." The man eventually chosen for that job was James C. Fletcher, who took over as NASA's administrator in April 1971.

Comsat leveled additional criticism of NASA and the State Department's handling of European collaboration. In a sharply worded letter to U. Alexis Johnson dated December 29, 1970, Comsat president Joseph Charyk spelt out his concerns.[4] Charyk noted that, in the negotiations over the definitive Intelsat agreements that were drawing to a close, "we had assumed that the United States would refuse to provide launch services for a separate regional system unless the Assembly of Parties, with the concurrence of the United States, found that the proposed system would be technically compatible with the Intelsat system and would not do significant economic harm to that system" (a positive finding, as explained in chapter 4). From what he had heard, however, it seemed that the United States would be prepared to launch regional satellites for Europe under a quite different set of conditions (a negative finding). Comsat's entire strategy and, in particular, its willingness to retreat from its initial negotiating position—that no separate system should be tolerated at all—was being undermined by the kind of concessions Johnson was making to the Europeans (see table 5.1).

For Charyk the *only* condition under which the United States should launch a separate system would be if the Assembly of Parties, *by the required two-thirds vote and with the concurrence of the United States, made a positive finding.*[5] Anything else would "appear to us to be indefensible" (table 5.1). He ended by asking the government "to clarify its intentions," and to provide Comsat and the US delegation with the "clearest possible assurances" on the conditions for launcher availability. This would have a "direct bearing" on the US delegation's willingness to accept the very diluted version of Article XIV(d) in the definitive agreements due to be signed soon. Put bluntly, what Comsat could not achieve at the negotiating table it wanted the State Department to achieve by exploiting the United States' monopoly of access to space to deny launcher availability to regional comsat systems unless they could be shown to do no economic harm to the single global system.

Comsat's "attack," as Pollack called it, placed NASA and the State Department in an acutely difficult position.[6] It took several weeks for Johnson to work out his position in discussion with Low, Charyk, Frutkin, and Whitehead. To draw closer to Charyk, Johnson decided to reverse the position he had discussed with Lefèvre in October, and to align himself (partially) with Charyk (table 5.1). As Johnson explained, this meant that if earlier Intelsat had to prove that the separate system did do it economic harm (i.e., the presumption was that it did not),

Table 5.1   Changing State Department position on implications for Intelsat if United States launches a comsat for a foreign entity

Position taken by	United States will launch a separate comsat system	Pertinence of US vote
Johnson to Lefèvre, Sep–Oct 1970 (negative finding)	Unless two-thirds majority finds that the separate system would do significant economic harm to Intelsat (and may even launch if it does)	Need not have voted with the majority
Charyk to Johnson, Dec 1970 (positive finding)	If two-thirds majority finds that the separate system would not do significant economic harm to Intelsat	Must have voted with the majority
Johnson to Charyk, Jan 1971 (positive finding)	If two-thirds majority finds that the separate system would not do significant economic harm to Intelsat	Need not have voted with the majority
Johnson to Lefèvre, Feb 1971 (positive finding)	If two-thirds majority finds that the separate system would not do significant economic harm to Intelsat	Need not have voted with the majority

now "the proponent(s) of a regional system [would] bear the burden of persuading two-thirds of the Assembly that the proposal will not cause significant economic harm to Intelsat and will not prejudice the establishment of direct links to the global system."[7]

Johnson would not go all the way with Charyk, however. He insisted that the international structure of Intelsat obliged the United States to accept an affirmative vote that achieved the required majority, even if the United States was in the minority position (table 5.1). Nor would the State Department yield on this point: a two-thirds positive finding, with or without US support, was "absolutely necessary in order to reach any agreement with the Europeans."[8] To reassure Charyk and the Comsat Board, Johnson pointed out that it was very unlikely that a regional system could achieve a two-thirds favorable finding if the United States was opposed to it.

NASA was not happy with this concession to Charyk. The Europeans would obviously be furious. Low feared that the reversal of the more flexible position previously suggested to Lefèvre "will effectively kill the chances for post-Apollo participation by Europe."[9] The only way to "soften the blow," he said, would be to make an advance commitment to launch Europe's planned operational satellite system, Eurosat, foreseen for the early 1980s. This decision had to be taken before the next Lefèvre mission to Washington, scheduled for early February. Low felt so strongly about this that, according to Frutkin, "if we could not arrive at a policy decision and so inform the Europeans, he would feel obliged to tell the President that he could not expect to carry out the President's charge to NASA to develop post-Apollo participation."

Eurosat, to be situated in a geostationary orbit at longitude 5°E, would have 3,000–5,000 circuits by 1980, and 8,000–20,000 circuits by 1990.[10] It would carry part of the intra-European traffic in telephony, telegraphy, and telex of the CEPT (European Conference of Postal and Telecommunications Administrations) and Eurovision TV programs on behalf of the European Broadcasting Union

(EBU). Coverage would include Western Europe and the Mediterranean basin, and extend to the five Nordic countries as well as Turkey. NASA concluded that Eurosat would do significant economic harm to Intelsat only if it provided television as well as voice, record and data services between all of these countries. If, however, it provided television to the Mediterranean basin exclusively, and a full range of services to the remaining countries, it would cause "measurable but not significant economic harm."[11] This was the configuration of the satellite that, in Low's view, Johnson would have to launch for Europe if he did not want the reversal of his position to sabotage all hope for post-Apollo negotiation. Unfortunately, Johnson made no mention of Eurosat in his conciliatory letter to Charyk.[12]

The Lefèvre Mission in February 1971 and Its Aftermath

A European delegation led by Lefèvre met again on February 10 and 11, 1971, at the State Department. They had prepared the ground with a lengthy letter sent the month before.[13] They wanted, as the Belgian minister put it in his opening statement, to participate in post-Apollo in ways that "would facilitate mutual dependence," "co-management," in a "joint venture" in which the partners would have "equal rights" to information, even though Europe only contributed 10 percent of the budget.[14] They sought associated benefits in terms of launcher availability and access to technology—Europe wanted to buy or build under license American launchers that could be launched from their new equatorial base in Kourou, French Guiana. They also insisted that once the shuttle was built, or rather "jointly developed," as they put it, it would be "available without restrictions to each of the partners for peaceful uses." Lefèvre reiterated that the Europeans sought access "to all the technology developed within the framework of the post-Apollo program, and not just that part of it which is necessary from [sic] executing the tasks accorded to Europe."

These requests were strategic rather than realistic. A State Department briefing document emphasized again that "[t]he very marked asymmetry in the partnership and the very advanced stage of US planning leave no alternative but to regard the post-Apollo program as a US program, not as a joint program."[15] As for the related request for technology sharing, the State Department emphasized that "[i]t is not possible in the world of commercial competition, congressional overview, and US industrial self-interest, to provide Europe full access to the commercial know-how developed in the post-Apollo program in return for a 10% contribution to that program."[16] As for launchers, the United States had no objection to Europe launching American rockets from foreign soil, or building Americans launchers abroad under license—but only if they respected "Intelsat-linked conditions" wherever they were launched.[17] As NASA feared, U. Alexis Johnson's new interpretation of those conditions was the biggest single blow to European hopes. And NASA was not alone. At a meeting of the senior staff of the National Security Council on the eve of the European visit, National Security Adviser Henry Kissinger doubted whether the United States was being "reasonable" in refusing to give an "unequivocal commitment" to provide launch services for European communications satellites.[18]

Lefèvre was incensed.[19] Europe needed launchers "without political conditions," he fumed; it could not participate in the post-Apollo program otherwise.

He reminded the State Department of its original interpretation of the Intelsat vote. He could not see why the United States was now demanding a positive finding in the Intelsat Assembly before it would agree to launch a regional European comsat. This new interpretation was against the spirit of cooperation that had prevailed until then.[20] Resenting the insinuation that Europeans were behaving irresponsibly, Lefèvre also pointed out that the Europeans were just as concerned as were the Americans to respect the definitive Intelsat agreements— but since the Assembly of Party's recommendations were not legally binding, they could not stop a country or region launching a rival system even if the assembly made an adverse finding. In European eyes Johnson was reinterpreting a consultative recommendation as a binding determination. They were treating a relatively weak legal finding as a non-negotiable political constraint.

Johnson could not budge: his hands were tied by his commitment to Charyk. The meeting inevitably ended on a sour note on February 12. Lefèvre "stated that the results of the discussion had been very disappointing," and affirmed that "if the US position remains unchanged, Europe would have to have a negative view toward post-Apollo participation."[21] A stream of telegrams from embassies abroad confirmed Europeans' puzzlement and anger. The member states of CETS, meeting on March 22, were unanimous in agreeing that the "proper interpretation" of Article XIV(d) was that enshrined in the "negative finding" (see table 5.1).[22] One European speaker after the other, including those who were sensitive to the dilemmas faced by the United States, expressed their disappointment at the new turn of events.[23] It was too much for Frutkin. Why was there so much criticism of the United States in this forum when it had done so much to promote international collaboration in space? Should one expect the leader in space technology to remedy the technological gap? How could one expect parity in technological exchange when the levels of contribution to a collaborative venture differed so greatly? Given the enormous benefits derived from collaboration over going it alone, could the Europeans not be "a little more relaxed about pressing for national advantage"?[24]

The Benefits of Collaboration: What the United States Could Offer

The uproar over launcher availability crowded out ongoing discussions at the technical level over the modalities of European collaboration in the post-Apollo program. Three possibilities were on the table: (1) the space tug (figure 5.1) that would ferry satellites from the shuttle's low-earth orbit to other, notably geostationary orbits; (2) experimental modules for the station or the shuttle (Sortie Cans or RAMs); and (3) the construction of components of the orbiter itself.

Frutkin and other senior NASA personnel discussed these matters on February 1, 1971.[25] They concluded that a reusable tug was "the most valuable and desirable element the Europeans could contribute to the post-Apollo program." It was "an essential element which cannot be undertaken directly by NASA for a number of years." For financial reasons, in the short term the agency would probably have to use expendable adaptations of the Centaur or Atlas rockets for tug missions. If Europe built a tug and had it ready by 1979, the United States could take advantage of that alternative. Even though the tug

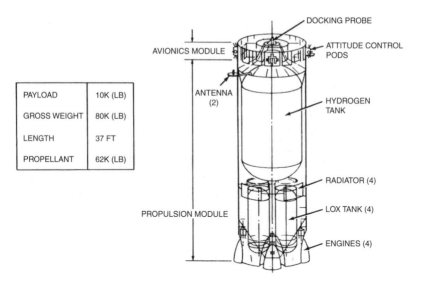

PAYLOAD	10K (LB)
GROSS WEIGHT	80K (LB)
LENGTH	37 FT
PROPELLANT	62K (LB)

Figure 5.1   The space tug concept.

Source: Technology Transfer in the Post Apollo Program. NASA HQ MF71–6399, 7-27-71, Record Group NASA 255, Box 14 Folder II.H, WNRC. Permission: NASA.

was a big step forward, the advanced technology that it required—in structure, propulsion, and controls—was probably within European capabilities, and could productively feed back into NASA's work. Interfaces would be clean, management simplified, and, in the event of failure, delays, or overruns, the impact on shuttle development would be minimal. The USAF's attitude was the only "major uncertainty," but it was felt that this was not an "unmovable obstacle," if only because the Air Force might not get funding for its own tug and could probably manufacture a tug developed abroad if it needed one.

Frutkin and his colleagues viewed the manufacture Sortie Cans or RAMs as the next best task, for the same reasons as the tug. The least desirable contribution was selected elements and structures for the orbiter itself. Technology transfer was a major concern here, even though US industry had identified excellent possibilities for subcontracting elements of the orbiter to European sources. If a single European firm made a critical item, like the vertical tail, it would obtain "proportionately more in general knowledge about the STS system than could be justified by the depth and amount of contributions to the program."

These ideas were presented to a joint meeting of experts from February 16 to 18. The leaders of the European delegation, Causse and Dinkespiler, were extremely impressed with the clarity of the presentations made by NASA.[26] The cross-range requirements for the shuttle, and their implications, were spelt out in detail. The plans for the station were explained, and the importance of RAMs emphasized. A mission model for the use of the shuttle covering all payloads was also presented. Some 60 flights per year from 1980 onward were foreseen; the tug was needed for about two-thirds of them. NASA's preference for Europe to build the tug or a RAM that could be used with the station or with the shuttle alone was stressed. Its concerns about subcontracting out parts of the orbiter were emphasized.[27] One issue on which all agreed—Low, Johnson, and Lefèvre—was that collaboration should be in a multilateral framework. This was to simplify management, to pressure European nations to work together, and to stop individual countries from signing bilateral agreements with NASA to the benefit of their home industry.

Post-Apollo Collaboration and High Politics

Europe's insistence that post-Apollo collaboration was tied to "guaranteed" launcher availability meant that little progress could be made at the technical level until the political problems were resolved. On February 24 the State Department briefed all pertinent embassies regarding the debates that had been held with the Lefèvre delegation two weeks before and the uproar that it had provoked. It insisted that the "distortions" and "misunderstandings" that had occurred in the European press should be countered by stressing the important new concessions that the United States had made. They were willing to consider supporting regional European telecom systems. They would define the specifications of the system that the United States could support in Intelsat, with particular reference to the proposed Eurosat system that had been suggested for the 1990s by the Europeans themselves. They guaranteed that Europeans would not only have preferential access to the shuttle, but could acquire one for use at their own launch suites for launching their own payloads.[28] None of this satisfied Lefèvre. He stressed once again that the whole launcher question was tied to Europe's wish to engage in a long-term telecommunications satellite program, and needed assurances that launchers would be available before deciding to abandon autonomous access to space. The "final decision power" vested in the United States by the new terms for launching comsats made it impossible for Europe—in Lefèvre's view—to embark on any "medium or long-term programming of our space activities."[29]

Frutkin was determined that NASA's views should be heard through the diplomatic cacophony. He spelt them out to Herman Pollack in the State Department and to Robert Behr on the staff of the National Security Council. He insisted that a bold step was needed to undo the damage done by Johnson's reversal of the United States' position in Intelsat. Nothing less could insure a 10 percent/$1 billion European commitment to post-Apollo, which would be at the expense of Europe developing an indigenous launcher and make them dependent on the United States for access to space.[30] He noted that a recent agreement brokered with Japan by the State Department offered launch support for comsats subject to Japan itself deciding whether they satisfied Intelsat's conditions: there was no additional override clause linking US launcher availability with Intelsat recommendations. Europe could not be treated differently. If it was, he told Behr, NASA and the United States would lose credibility as partners in international collaborative projects in science and technology. The agency would not be able to carry out the "personal and repeated directive of the President" to engage foreign partners in the post-Apollo venture. And Europe would be backed into "an independent launcher development program which would proliferate missile technology and win Europe total independence with respect to Intelsat initiatives." Behr added a further argument in a memo to Kissinger of March 4, 1971: "Many large US firms have discovered that their international business depends upon the existence of foreign capabilities and skills to which they can relate." For them even a 10 percent participation in the STS program would build the kind of "framework" that would narrow the technological gap and facilitate transnational and transatlantic collaboration at the level of the firm.[31]

Pollack agreed with Frutkin that the United States would have to make American launchers available to the Europeans for their own use, but insisted that Washington should decide whether or not the payload would do significant economic harm to Intelsat.[32] Behr included in his memo for Kissinger the arguments against a strong European presence in post-Apollo and against a relaxation of the US line in Intelsat. At the top of the list was Tom Whitehead's claim that "we are giving the Europeans too much technology for too little return," and that they were demanding a "disproportionately large share of program management responsibilities." He noted that the savings to the United States would be considerably less than $1 billion because of the increased managerial complexity. And he suggested that any concessions to Europe would alienate Third World countries in Intelsat, who might feel that once again the industrial powers were arranging concession among themselves that were denied to the less-developed world. "There was no need for you to get involved in this hassle at the moment," Behr assured Kissinger. But he did suggest that the national security adviser meet with the president's science advisor, Ed David, to discuss "whether it is in the interest of the U.S. to continue the development of the space shuttle/station and, if so, to what extent do we wish to engage foreign participation?"[33] His comment serves as a reminder that no firm decision had yet been taken on whether to proceed with the shuttle at all, let alone with a European contribution at it.

In fact momentum was growing to exclude Europe from the post-Apollo program. On February 22, 1971, Ed David met the president along with White House staffers Peter Flanigan and John Ehrlichman—but no one from the State Department. David informed Kissinger about a month later that the meeting had concluded that "a joint effort with the Europeans is not in our best interests."[34] He mentioned that this seemed to be Nixon's view too. The president's science adviser listed five reasons against European participation. Topping the list was "substantial high technology transfer" in return for a 10 percent contribution that was of dubious value anyway, since it would be offset by the increased costs of a cooperative program, and by the United States probably having to "undertake back-up programs for those elements or systems being developed abroad." Management would be complicated by the need to satisfy partners. Collaboration would strengthen Europe's capacity to compete "for our own commercial exploitation of our satellite technology." And, finally, the "proposed arrangements [would] lock us into the shuttle program, depriving us of the flexibility to tailor the program to our evolving needs." Kissinger was not prepared to take these remarks at face value: as he cautioned Ed David, "[O]ne should not attempt to deduce Presidential decisions from casual conversation." The upshot was that all agreed it was necessary to continue discussions with the Europeans but to "to slow the pace," not to soften the US position on launchers, and to produce a "technical cost/benefit analysis of the various alternative ways of cooperating in space with Europe"[35] in case it was needed to exclude them from post-Apollo.

Toward the end of April another high-level attempt was made to explain the European view in Washington, this time by a German delegation led by the minister of education and science, Hans Leussink. The Germans had been strong supporters of post-Apollo participation from the outset. As Behr told Kissinger, the reaction in Bonn to the "failed" Lefèvre mission in February had been one of surprise and shock, "by what they universally described as the 'hard line' taken by the US" at the meeting. "They expressed fear that it will be impossible for Europe to participate in the post-Apollo program and dismay that Europe will find it necessary to develop an independent launcher capability which would be wasteful and also a divisive element in US-European cooperation."[36] In short, Bonn felt that it was being forced by Washington to support a program in Europe that it did not like at the expense of a collaborative venture with the United States that it had always sought.

Leussink met with Low, Frutkin, and others at NASA on April 21. Low repeated the offer that Europe do the tug that would be required for "the majority of the shuttle missions," and contribute, on a subcontracting basis, "some portion of the shuttle air frame." As regards launcher availability, Low emphasized that he understood the concerns expressed abroad and hoped "that we can work out our problems on this subject in the near future."[37]

The next day Leussink met with Ed David.[38] He insisted that if post-Apollo cooperation failed it would undermine the drive to European integration. He emphasized that Germany was willing to collaborate on any reasonable venture—shuttle, tug, or space station module—but that it needed launch assurances (failing which it would be "forced, with other European countries, into development of Europa III, leaving no money for post-Apollo cooperation"). Leussink also pointed out that he would be quite satisfied if the Europeans—like Japan—could buy American boosters or build them under license abroad.

The internal divisions within the administration emerged full-blown at a special meeting on post-Apollo cooperation held immediately after Leussink's visit. It took place in the White House Situation Room on April 23, 1971.[39] The meeting was chaired by Kissinger, and attended by White House staffers Peter Flanigan and Tom Whitehead, the new NASA administrator, James Fletcher, and his assistant, George Low, Edward David and Norman Neureiter from the Office of Science and Technology, and Robert Behr from the National Security Council.

Kissinger explained what was at issue: "[W]hether a program of co-operation with the Europeans is desirable. If the President wishes to have a program of some substance," he went on, "the US will have to provide launch services to the Europeans. If he decides against a program of co-operation, we will hold to a hard line on the provision of such services." Flanigan was particularly outspoken, objecting to the dangers of technology transfer and charging that NASA was advocating a program that was "nothing like what the President wants." Flanigan claimed that for Nixon "a symbolic gesture like flying a European astronaut in space" would suffice. Whitehead supported him, arguing that what really mattered for the president was cooperative ventures that resulted in a net foreign policy benefit for the United States: the "grandiose program" being pursued by NASA was likely to have the opposite effect.

Kissinger, on balance, showed more sympathy for NASA than for David, Flanigan, and Whitehead. He was more sensitive than they to the foreign policy aspects of post-Apollo cooperation. He disliked the presumption by the White House staffers that they had privileged access to the president's wishes. And he implied that the debate was being driven by muddled preconceptions rather than by reasoned argument. NASA was asked to provide within two weeks, in cooperation with David's office, a paper "(1) defining technology transfer and analyzing its implications and (2) describing the various possibilities for space cooperation with the Europeans in addition to the shuttle." Kissinger told Behr that a reply from the State Department to Lefèvre would have to wait until this paper had been prepared and evaluated, and a program direction had been selected by the president.[40]

NASA's Approach to Technology Transfer with Europe

It took much longer than anticipated to prepare the reports called for by Kissinger at the end of April. Both were eventually available in June, and evolved further as the meeting with Kissinger was postponed time and again: it was eventually held on or around August 9.

NASA's report on "Alternatives to Post-Apollo Participation" identified various areas where collaboration might be possible, and discussed the costs and benefits of each: flying a foreign astronaut to Skylab, a joint Air Traffic Control preoperational system, a joint experimental applications technology satellite (ATS). However it effectively dismissed these as significant alternatives to post-Apollo.[41] Indeed the original report was so one-sided that Herman Pollack summarily rejected it as "essentially a contentious paper reciting the dire consequences that would follow from backing out of the post-Apollo proposals. It denied that there are any suitable alternatives."[42] A revised version took a slightly more balanced approach but drew very similar conclusions. In short, as far as NASA was concerned, there was no way the United States could now deny the Europeans participation in post-Apollo, no alternative that could compensate for the drastic foreign policy setbacks of such action

The paper on "Technology Transfer in the Post-Apollo Program" expanded on the text produced the year before. It also went through two versions, a detailed one in anticipation of a meeting with Kissinger on June 8, and a punchier alternative prepared when the meeting was postponed to the end of the month, before being postponed again.[43] The core of the argument was developed at a meeting on May 5 with senior personnel in the Office of Science and Technology, including Ed David, and with input again from representatives of the American aerospace industry, notably MacDonnell Douglas, North American Rockwell, Grumman Aerospace, and General Electric.[44]

As we saw in chapter 4, NASA's case against its critics was intended to turn on its head the charge that the agency was ready to give away American aerospace technology at ten cents on the dollar. Thus

The thrust of foreign participation in such a post-Apollo program will be to contribute to the US effort rather than set up a flow from us. Such a direction of flow will be further supported by our practice of a policy already fully communicated to potential European participants, namely, our commitment to select projects for European development only where the capacity is already substantial in Europe.

That capacity was described by US firms already collaborating abroad. Grumman Aerospace wrote that the French firm Marcel Dassault was "one of the most capable manufacturers of high performance aircraft in Europe," and "should be able to contribute any portion of the Shuttle prime structure that France might undertake," except perhaps the main cryogenic tankage. Their work with Grumman on metallic and polyimide thermal protection materials and design should put them in a position to be able to manufacture some portion of the Thermal Protection Subsystem (TPS) that was "so crucial to the shuttle's cross-range capacity." Grumman also claimed that the German firm Dornier was "well capable of handling structural sub-assemblies" for the shuttle, and had excellent research and test facilities that could be used during the development of the shuttle.[45] North American Rockwell was contracting with the British Aircraft Corporation for shuttle phase B participation in "structural elements, aerodynamics, flight test instrumentation, and data handling."[46] The McDonnell Douglas Aircraft Corporation, for its part, was actively pursuing international collaboration with ERNO in Germany, Hawker Siddeley in Britain, and SNIAS (Société nationale industrielle aérospatiale) in France.[47]

It was noted that Europe could also make major contributions to the tug and the RAM. In summer 1971 Messerschmitt- Bölkow-Blohm gave a final presentation to ELDO and NASA of its pre-Phase A study of the tug, as did Hawker Siddeley Dynamics leading a group of ten European companies.[48] The Convair Division of the General Dynamics Corporation, which had been selected to perform a Phase B RAM study, was subcontracting parts of the work out to MATRA in France (systems design and analysis, guidance, and control), ERNO in Germany (material science and manufacturing in space), SAAB in Sweden (phased array, data processing, image compensation), and Selenia in Italy (bulk data handling, millimeter wave communication system, etc.).[49] In sum, even if Europe's ability still had to be tested in practice, its capacity to make major contributions to significant parts of the post-Apollo program was not in doubt, at least not to US aerospace corporations.

On the basis of their experience US aerospace firms were convinced of the quality of the work done in European industry, and were not particularly concerned that significant technology would be leaked to them. On the contrary since Europe's contribution was limited to 10 percent, they repeated that "the US should come out further ahead of Europe than when we started." In fact, the industrialists' experience suggested that what Europeans lacked above all was not technology but "general management and systems engineering knowhow." Though they would acquire some insight into this through post-Apollo collaboration, "the American companies consider that such know-how would be directly applicable in Europe in only very limited ways." The risks were more than outweighed by the benefits of a European presence, which, the companies argued, would "stabilize" the American program, and "avoid a stimulation of independent and competing programs in Europe."

NASA backed these claims with a more detailed analysis of the nature and scope of technology transfer in the four areas it had identified. It again took the vertical tail of the shuttle to be typical of an area where Europe might be included in the orbiter program, to the advantage of both partners (e.g., by the provision of test data for the supersonic plane, Concorde). Special materials were the most advanced technologies here: titanium for the main structure, carbon/carbon for the Thermal Protection System on the leading edge of the orbiter, which could reach temperatures as high as 2000°F. NASA's study indicated that only the latter posed any risks of advanced technology transfer. Accordingly the agency suggested that this part could be separated out from the rest of the tail and manufactured entirely in the United States before being integrated into the tail on either side of the Atlantic

As before, the tug, which would be constructed by a European prime contractor, was broken down into two major elements, the propulsion module and the avionics module.[50] This chapter stressed again that much of the propulsion system—the main engine, cryogenic propellant insulation to permit long-term storage, and so on—posed novel challenges to the Europeans and would undoubtedly advance their technological base, though not with NASA's direct help. By contrast, the avionics module called for a different approach. The Europeans only had limited experience and know-how in areas such as navigation and guidance (see chapter 4). This was not seen as a major impediment, however. It could be argued that it was in NASA's interest to provide certain elements to make possible a larger foreign participation, and the flow of technology could be controlled if US firms could provide the subsystem components the Europeans needed as integrated technological units. In particularly sensitive cases they could simply supply a "black-box" for integration by the European prime.[51]

At the end of July NASA prepared an extensive presentation summing up its findings on the nature of technology transfer in the post-Apollo program.[52] Some of the 50-odd viewgraphs, for the shuttle orbiter tail and wing, and the space tug, are presented in figures 5.2–5.4. NASA stressed the depth of technological capacity in the European aerospace industry, much of it due to the development of military aircraft and Concorde. It also drew attention to the positive attitude of US industry to collaboration (figures 5.2 and 5.3). The overall conclusions bring together in one image NASA's and US industry's arguments intended to allay fears that post-Apollo collaboration with Europe would lead to significant technology transfer (figure 5.4).

Frutkin's (and US industry's) efforts to reassure the higher echelons of the administration that there was no serious danger of significant technological leakage to Europe were to little avail. The State Department was not persuaded. Hermann Pollack was delighted that NASA's report did "not substantiate those who in the February 22 meeting with the President argued against the post-Apollo cooperation program on the grounds of unwarranted or uncompensated technological transfer." All the same he felt that the technology transfer issue was still not definitively settled.[53]

NASA's arguments did not satisfy Ed David either. Writing to Kissinger in late July, the president's science adviser remarked that even though "the NASA study (concurred in by Jim Fletcher) suggests that the technology transfer question as well as management complications are not of significant proportions, my personal concerns on these points have not yet been answered to my full satisfaction, nor can they be answered until there is a better understanding of the potential European contribution."[54] He agreed to continue technical discussions with the Europeans but only to define more clearly "without any

ITEM	COMPANY	PROJECT
LARGE DELTA WING DESIGN, FABRICATION, TESTING	BAC	CONCORDE
	SNIAS	CONCORDE
	HSA	VULCAN AIRBUS
SUPER ALLOY STRUCTURES CAPABILITY	SNIAS	VERAS
	DASSAULT	MYSTERE
TITANIUM AIRCRAFT STRUCTURE MANUFACTURING	HSA	TRIDENT HARRIER
	BAC	A300 CONCORDE
	ERNO	EUROPA II
	MBB	F-4
	SNIAS	CONCORDE
	DASSAULT	MYSTERE

Figure 5.2   European experience in shuttle-related wing & tail structural hardware.
Source: Technology Transfer in the Post Apollo Program. NASA HQ MF71–6399, 7–27–71, Record Group NASA 255, Box 14 Folder II.H, WNRC. Permission: NASA.

- EUROPEAN AEROSPACE TECHNOLOGY IS GOOD
    - U.S. FIRMS WOULD NOT BE DISADVANTAGED BY POST-APOLLO TECHNOLOGY FLOW
- EUROPEANS NEED AND DESIRE SYSTEMS ENGINEERING AND MANAGEMENT EXPERIENCE
    - U.S. FIRMS BELIEVE SUCH TRANSFER HAS LIMITED APPLICABILITY
- EUROPEAN PARTICIPATION WOULD REQUIRE SIGNIFICANT U.S. MANAGEMENT SUPERVISION
    - THIS COULD REDUCE THE MONETARY VALUE OF EUROPEAN CONTRIBUTION
- MAJOR EUROPEAN GAIN WOULD COME FROM PERFORMANCE OF TASKS
    - U.S. WILL GAIN PROPORTIONALLY MORE
- EUROPEAN PARTICIPATION WILL HELP STABILIZE THE POST-APOLLO PROGRAM

Figure 5.3   US industry views on post-Apollo cooperation.
Source: Technology Transfer in the Post Apollo Program. NASA HQ MF71–6399, 7–27–71, Record Group NASA 255, Box 14 Folder II.H, WNRC. Permission: NASA.

precommitment, the potential interests and contributions of both sides." They should be undertaken recognizing that "we are not committed to *agree* to foreign participation [but only] to give positive *consideration* to foreign interest in participation," be it from Europe or from Japan.[55]

- • EUROPEAN DEVELOPMENT OF SHUTTLE COMPONENTS FEASIBLE

  - • TECHNOLOGICAL BENEFITS TO BOTH U.S. AND EUROPE
  - • SMALL TRANSFER OF TECHNOLOGY

- • EUROPEAN DEVELOPMENT OF SPACE TUG FEASIBLE

  - • GREATER TECHNOLOGY TRANSFER, CONTROLLED THROUGH U.S. PERFORMANCE OF CERTAIN CRITICAL TASKS
  - • SIGNIFICANT SYSTEMS ENGINEERING SUPPORT

- • EUROPE PRIMARILY INTERESTED IN PROGRAM MANAGEMENT AND SYSTEMS ENGINEERING GAINS RATHER THAN DISCRETE TECHNOLOGIES

- • CRITICAL VALIDATION OF EUROPEAN TECHNOLOGICAL CAPABILITY NECESSARY BEFORE SPECIFIC TASK COMMITMENTS

- • U.S. INDUSTRY BELIEVES EUROPEAN PARTICIPATION DESIRABLE

Figure 5.4   Summary evaluation of European participation in post-Apollo.

Source: Technology Transfer in the Post Apollo Program. NASA HQ MF71–6399, 7-27-71, Record Group NASA 255, Box 14 Folder II.H, WNRC. Permission: NASA.

One strong argument remained to tilt the balance: foreign policy. NASA administrator Jim Fletcher told U. Alexis Johnson in May that, in his view, "the decision as to whether to engage the Europeans in the Post Apollo program rests essentially on foreign policy considerations."[56] Pollock rehearsed the argument again for Johnson two days later.[57] The program would strengthen ties with Europe, which were still "the cornerstone of our efforts to build a peaceful world." The region's industrial, economic, and technological strength were critical to the existing balance of power and closer ties in advanced industry and technology would consolidate the Atlantic Alliance. Space programs in particular were ripe for "meaningful cooperation," and would "offer highly visible and dramatic symbols of the fruits of partnership." On the other hand, without post-Apollo cooperation, "Europe's program will be organized around France, and we would have little input and little influence on such a program." Failure to work together in space might also impact Western European decisions in other high-tech sectors—such as the development of "breeder" reactors and uranium enrichment facilities. In short for Pollack, as he put it to Undersecretary Johnson, the decisions now being taken on both sides of the Atlantic as regards the future framework for advances in high technology would "have a profound, long-term effect on Europe and on its relations with the USA," and "successful multi-national cooperation in post-Apollo [would] improve our ability to influence Europe's decisions in these other fields."

Early in August 1971, then, there was general agreement between Fletcher, Johnson, David, and Flanigan that the United States should continue the technical discussions with Europe, making it absolutely clear that this did not commit either party to participation of any kind in the post-Apollo program. Kissinger accepted this position. On August 18, 1971, he wrote to Secretary of State Rogers to tell him that Nixon had confirmed "his support for continued pursuit of opportunities for international space cooperation in general, and specifically with the Europeans."[58] He asked the State Department to prepare a reply to Lefèvre that suggested that technical discussions be continued with a view to defining "possible cooperative relationships between Europe and the U.S. in the program of STS development." He was emphatic that no commitment of any kind should be made that might later impede "an independent decision by the U.S. on the desirability or schedule of STS development." With the future of the STS and of Europe's participation in it thus left open, Kissinger asked that the scope of the discussions be extended to include "an exchange of views with the Europeans regarding the content of space activities" in the post-Apollo era, as well as to consider "other potential areas for cooperation in space exploration, operations and launches." This also meant, of course, that from now on "U.S. launch assurances for European payload will not be contingent upon European participation in a joint STS program, but will be treated separately to the degree possible." The results of these technical discussions were to be made available to the president by January 15, 1972.

As it happens, the week before Kissinger wrote this memo to Secretary of State Rogers the president had seemingly taken a major step forward as regards the content of the post-Apollo program. In a famous memorandum to Nixon dated August 12, 1971, the deputy director of the OMB, Caspar Weinberger, had proposed that NASA's annual budget be stabilized at $3.3–3.4 billion, and that it should make provision for a shuttle. The American people and the world needed to be reassured, Weinberger wrote, that the United States was not "giving up our super-power status, and our desire to maintain our world superiority." Jobs also had to be protected in an aerospace industry made vulnerable by the wind-down of the Vietnam War. Nixon scrawled "I agree with Cap" on this memo.[59] But this position was not yet formal, and in any event was not communicated to the Europeans.

The New "Solution" to the Launcher Problem: September 1971

On September 1, 1971, Undersecretary of State U. Alexis Johnson replied to the letter that he had received six months earlier from Theo Lefèvre.[60] He plunged directly into the launcher issue. The United States, said Johnson, had reviewed its position in an attempt to meet European concerns. He noted at once that that new position was "not conditioned on European participation in post-Apollo programs," and he hoped that it would provide "a basis for confidence in Europe in the availability of U.S. launch assistance." Johnson reaffirmed that, of course, American launch assistance would still only be for satellites that were for peaceful purposes and consistent with its obligations under relevant international agreements and arrangements. However, it would be available both from American territory, and "from foreign launch sites (by purchase of an appropriate U.S. launch vehicle)." As regards the interpretation of the thorny Article XIV(d) of the Intelsat agreements that had been signed on May 21, 1971, Johnson proposed three possible scenarios, presented in table 5.2.

Johnson went on to say that to avoid Europe investing heavily in a satellite system only to find that the United States would not launch it, the American

Table 5.2   Revised US policy on launching comsats that could do "significant economic harm" to Intelsat's global system

Case	Intelsat finding	US position on launcher availability
I	A favorable recommendation	Will launch
II	Absence of favorable recommendation, but United States supported the system	Will "expect" to launch, provided petitioner acted in "good faith"*
III	Absence of favorable recommendation and United States had not supported the system	May launch, if system modified to meet Intelsat objections

Note: * More precisely: "[S]o long as the country or international entity requesting the assistance considers in good faith that it has met its relevant obligations under Article XIV of the definitive agreements."

authorities would consult with the European Space Conference in advance of it embarking on any major program to evaluate its consistency with the Intelsat agreements. A concrete example of such a system had been suggested by the Europeans earlier in the year (the Eurosat system, see earlier). The United States judged that this system would do measurable, but not significant economic harm to Intelsat. If it were officially presented to the organization, "we would expect to support it in Intelsat."

It is clear that the State Department had been persuaded that it should be as flexible as possible over the launcher question now that the definitive Intelsat agreements had been signed. Certainly, the cornerstone of US policy remained the same: that it would launch a separate communications satellite system if two-thirds of those voting in Intelsat agreed that that system did not do significant economic harm to the global system (and this whether or not the United States had been one of those voting in favor). What was new, however, was that now Johnson was prepared to take the Intelsat vote as a "finding" or "recommendation," and not as a legally binding directive. In other words, if the requisite two-thirds majority was not obtained he was willing to consider launching a separate system. Such willingness was further nuanced depending on whether the United States had been in favor of the system or not. By accepting to launch absent a two-thirds positive finding on a system that the United States favored, Johnson was effectively willing to risk criticism of the American position in Intelsat to placate European fears. He was suggesting that the US authorities would take upon themselves the responsibility of demanding changes which, in the view of their their experts, would make the separate system acceptable, without having recourse again to Intelsat. This was a major reorientation indeed. It was also no longer conditional on European space agencies making a major commitment to post-Apollo participation: launcher policy was now completely distinct from whatever framework for US-European space collaboration was jointly adopted for the 1970s.

Experts Meet and Work Packages Are Defined

In his letter to Lefèvre Johnson suggested that a technical working group led by Charles W. ("Chuck") Matthews, the deputy associate administrator in NASA's Office of Manned Space Flight, should soon meet with Europeans to discuss both participation in the Space Transportation System and other possible modes of collaboration (as required by Kissinger). Fletcher was determined to focus the discussions on the orbiter, however. At a high-level meeting on October 6 he insisted that the "ensuing technical discussions [with the Europeans] should concentrate initially on defining tasks and working relationships for the space transportation system project, since time is catching up with us." NASA would define the concepts and the configuration of the shuttle in the next two or three months, and select a prime contractor by spring 1972: it would be "imperative to know the extent of European participation by that time."[61] Thus Matthews's objectives were clear when he came to Paris later in the month. The American delegation was to engage in "technical discussions aimed at a definition of candidate areas for possible European participation in the space transportation system, viewed in the broader context of program requirements for the 1980s," including payloads in which Europe may like to participate. Political, managerial, and contractor-to-contractor arrangements would be temporarily bracketed.[62] After a successful meeting on October 22, 1971 with about 40 space officials it was agreed that a joint group of experts meet in Washington at the end of November to study detailed proposals from NASA.[63]

Charles Donlan, assistant director on the Space Shuttle Program, visited a number of European aerospace companies before the expert meeting to validate again the technological capability of major European aerospace contractors in Britain, France, and Germany (see figure 5.2). He ended his trip persuaded that European firms could carry out any part of the post-Apollo program that was contracted to them, and realized that they had more experience than their American counterparts in working in different languages across technical interfaces.[64] There was not going to be one-way flow of technology from the United States to Europe in the post-Apollo program.

The first "Technical Conference on US/European Cooperation in Future Space Programs" opened in Washington on November 30, 1971. Four main areas were identified and disaggregated into separate work packages where needed. These were the space shuttle itself, the space tug, orbital systems, which included Sortie Cans, Pallets, and RAMs, and general technology development (which will not be dealt with here). It was assumed that the shuttle would have an American prime contractor, so here the work packages had to reflect elements that could be subcontracted. The tug was different, in that if Europe undertook the project the US role would probably be limited to minor contributions such as furnishing some components and providing data on shuttle interfaces. Various scenarios of US-European collaboration were foreseen for the orbital systems and for technology development.[65]

The discussions were held in the context of an ongoing battle between NASA and the Office of Management and Budget (OMB) over the agency's budget for FY1973, and the configuration of the shuttle that could be afforded. Indeed as late as September 1971 Low and Fletcher seriously considered whether they should forgo a shuttle altogether for something less expensive like a reusable glider.[66] By November, however, the Europeans were assured that the most attractive design for the shuttle that was emerging featured a delta-wing orbiter with an external hydrogen/oxygen tank fueling its main engines along with two reusable boosters to provide sufficient thrust for lift-off. The fuel system for the boosters was still under discussion but would be settled when the call for proposals for shuttle development was issued in spring 1972. The selection of candidate work packages for collaboration was thus restricted to the orbiter and its main engines. The orbiter was disaggregated into separate parts (see figure 5.5), and a corresponding list of work packages was drawn up; this excluded components already developed in the Apollo program, which would be transferred to the shuttle. This list fell into four main areas—airframe, propulsion, instrumentation, and aerodynamic testing—for which a total of 14 individual work packages were identified "along with their general description, schedule and approximate costs." They were essentially illustrative, and ranged from small elements that could be accomplished by a single firm to larger elements that would engage a European consortium.

The airframe attracted most attention. It was believed to "provide an opportunity for European participation because of the ability to identify areas that have relatively simple interfaces and therefore would prove to be more straight-forward to manage and integrate into the total vehicle."[67] The nine work packages discussed here were tail assembly ($20–30 million), main wing ($65–75 million), elevon ($20–25 million), center fuselage, forward and aft ($100–125 million), cargo bay door ($30–40 million), radiator ($10–15 million), landing

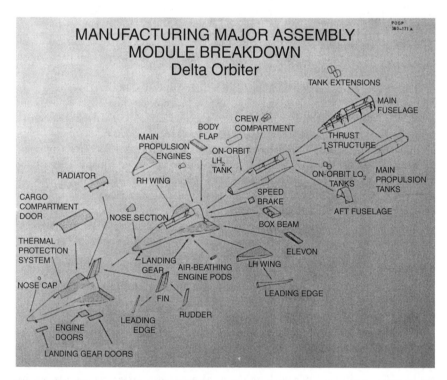

Figure 5.5 Disaggregated orbiter, showing subsystems that were potential candidates for collaborative projects, as proposed by McDonnell Douglas.

Source: Presentation made by the firm on May 7, 1971, attached to memo from Sh (Sam Hubbard) dated May 19, 1971. Record Group NASA 255, Box 17, Folder VI.D.3, WNRC. Permission: NASA.

gear and door ($20–30 million), nose section ($3–5 million), and ejection seat ($10–12 million).

The tail assembly package consisted of the orbiter fin and rudder/speedbrake (see figure 5.6), and could easily be integrated into the vehicle by the prime contractor. NASA was careful to exclude from possible consideration the leading edge and thermal system. Sharing technology here risked upgrading European capabilities in domains with a commercial impact, for example, high performance aircraft. The pattern was repeated with the offer for technological collaboration with the main wing, which was somewhat more difficult to integrate than the tail. Like the tail, the wing had a beam and rib arrangement with an aluminum skin. Again the work package on offer included the design, fabrication, testing, and certification of the primary structure. Yet as with the tail, the leading edge and thermal protection system installation was excluded.

Four propulsion packages were identified: the orbital maneuvering system (OMS) pods ($40–50 million), the reaction control system pods, the air-breathing engine pods ($20–30 million), and the auxiliary power unit. This list deliberately excluded engines because of the criticality of integration and the considerable development experience already available in the United States.

Not much detail was provided on the instrumentation work packages, where it was thought that Europe might be able to contribute $20–30 million. NASA officials were wary of encouraging close technological collaboration here because

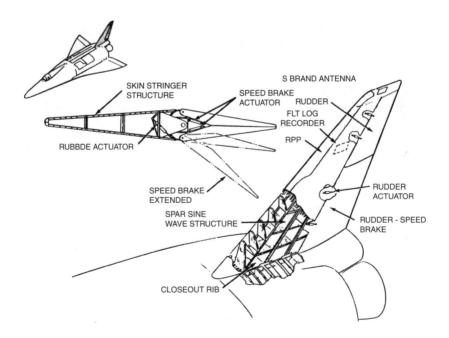

Figure 5.6   Disaggregated orbiter vertical tail.

Source: Technology Transfer in the Post Apollo Program. NASA HQ MF71–6399, 7–27–71, Record Group NASA 255, Box 14 Folder II.H, WNRC. Permission: NASA.

of the "multiple interfaces involved and the critical interdependency with other systems." Subcontracting out parts of the instrumentation would "prove too costly and difficult to integrate and manage."[68] Table 5.3 summarizes the work packages on the orbiter that NASA thought might be suitable for international cooperation.

The second major element in the post-Apollo package was the space tug, which would transfer payloads from the shuttle's low-earth orbit to higher orbits, most notably the geostationary orbit (see figure 5.1). As we saw earlier, at this time the tug was a central feature of the shuttle system and "a key element of the system economics."[69] In April 1970 the European Space Conference voted 500,000 MU (1 MU or monetary unit, was about $1) for pre-Phase A studies (i.e., conceptual studies) of the tug, and early in 1971 two lead European contractors, Hawker Siddeley Dynamics (HSD) and Messerschmitt-Bolkow-Blohm (MBB), presented their findings to NASA. Further impetus toward European engagement in the tug came toward the end of 1970, when NASA decided that it could not afford to develop the tug concurrently with the shuttle, and began to look at the possibility of using existing Agena and Centaur upper stages in its stead as a stopgap measure. In February 1971 another 400,000 MU were allocated to HSD and MBB to further study the tug in the light of changes to the shuttle configuration. They presented the results of their new studies in Huntsville, Houston, and Washington, DC, in October.[70] Thus, when the experts met on November 30, 1971, considerable progress had already been made in Europe on the definition of a tug.

The experts meeting in November strongly endorsed this work. They were persuaded that the tug was "a logical area for European participation in the post-Apollo activities since it is an easily 'separable' item with a relatively clean set of interfaces. In addition," the *Report* on the meeting went on, it provided "significant technology challenges in a number of areas and represents a key element of the Space Transportation System."[71] Encouraged, on January 5, 1972, ELDO issued requests for proposals for two Phase A space tug studies costing $1.4 million. Additional funds would be sought later that month for what Causse told NASA was Europe's "main area of interest" in post-Apollo.[72]

Table 5.3   Work packages on the orbiter proposed to Western Europe for collaboration in November 1971

---

1. Tail assembly (fin, perhaps rudder/speedbrake) NOT leading edge and thermal protection
2. Main wing NOT leading edge and thermal protection
3. Elevon
4. Central fuselage, fore and aft
5. Cargo bay door
6. Radiator
7. Landing gear and door
8. Nose section
9. Ejection seat
10–13. Propulsion (without engines)
14. Instrumentation (difficult to integrate)
TOTAL COST ~ $400 million

The Europeans liked the tug. NASA now estimated that it be would be needed on 40–50 percent of the foreseen 60-odd annual shuttle flights. If each reusable tug could be used ten times, this implied a production rate of three–five new tugs annually over a ten-year period. This was a rate of industrial production higher than that foreseen for the anticipated European rocket program. A reusable tug, in particular, was also technologically challenging. As one European report put it, it represented "a new type of space vehicle partaking both of the launcher, by its propulsion and structure, and also to a great extent of the satellite, by reason of its launching in the orbiter, its life in orbit and its intelligent systems (attitude control, rendezvous, docking)."[73] For Europe the tug prefigured the kind of space vehicle that would be common after the 1980s. It also gave Europeans direct access to the heart of NASA's space flight planning and operations.

The variety of orbiting systems under discussion exploited various possible uses of the shuttle as a platform for short observations and experiments in space. This idea evolved as the funding for a possible space station was pushed into the 1980s. It was inspired by the use of a converted Convair 990 airplane as an airborne laboratory that scientists were using to make astronomical experiments and earth observations. If an aircraft cargo bay could be transformed into a useful scientific platform for space research, why not use the shuttle in a similar way?[74] On one variation, called a Sortie Can or Sortie Module, a shirt-sleeve experimental environment was lodged in the Shuttle's cargo bay and connected to the orbiter crew compartment by an access tunnel. It could be combined with various external pallets attached to the back end of the Sortie Module (and could also be replaced entirely by such pallets). The pallets would only be humanly accessible by EVA (extravehicular activity) and could serve as platforms for large instruments such as telescopes. In a second variation, called a RAM (Research and Application Module), the entire module would be tailored to a specific discipline or human activity, such as materials science and space manufacturing.[75] When the experts met on November 30, 1971, NASA had not committed itself to going further than Phase B studies (Definition) of a RAM, and was thinking of beginning an in-house Phase B study of a Sortie Can six months hence. Both were candidates for European participation, being reasonably autonomous systems that could be used by both parties. NASA also stressed that in the field of payloads to be carried by these systems, all areas were open for European cooperation. It was concluded that Europe should start at once on a Phase A study on the sortie can and pallet, in close cooperation with NASA, so as to be up to speed when NASA began its in-house Phase B effort in about May 1972. Steps would also be taken to define jointly a candidate experiment program for European researchers on the two systems.

The position of the European delegates to the Washington meeting was an unenviable one. The European space program had been dealt a serious blow between the exploratory meeting with NASA in October and the more extensive gathering to define work packages at the end of November. On November 5, 1971, the much-awaited F11 test flight of the Europa II rocket failed catastrophically when the rocket exploded three minutes after lift-off.[76] ELDO immediately established a committee of enquiry, and requested American participation in the seven working groups set up to investigate the accident: Frutkin responded positively, if somewhat gingerly.[77] With their technological and managerial confidence shattered, the attraction of working closely with the United States in a post-Apollo program, no matter what the cost, had an increasing appeal in European capitals.

An Alternative Collaborative Project: Aerosat

Kissinger's insistence that NASA should not restrict its options for collaborating with Europe to the post-Apollo program brought the ongoing negotiations over a jointly developed aeronautical satellite system into focus at the end of 1971. The idea of a NASA/ESRO suite of satellites to handle air traffic over the Atlantic was one of the collaborative ventures promoted by NASA administrator Tom Paine in his early enthusiasm for international projects in the post-Apollo period.[78] By August 1971, and notwithstanding the multiple stakeholders and conflicting interests involved, it was agreed in Madrid (to cite the European report of the meeting) that a preoperational aeronautical satellite system would be "jointly developed, funded, managed, implemented and evaluated" by Europe through ESRO and by the United States through the FAA (Federal Aviation Administration), along with other interested governments.[79] Europe was prepared to assume 50 percent of the full program cost, and although the prime contractor would be chosen by open competitive bidding (and might well not be European), it was stipulated that European partners be included in the scheme and would obtain a "fair and reasonable" share of the contracts. Liberal provision was made for technology sharing. "For the first time," wrote the science correspondent of the prestigious French daily *Le Monde*, "co-operation with the United States in the field of application satellites seems to be getting under way under conditions of equality."[80]

The assault on this project was again spearheaded by the OMB, along with Whitehead and Flanigan.[81] They wanted American industry to drive space activities, and they were hostile to the idea that the FAA and ESRO would be co-owners of the system. They wanted it to be owned privately and leased to governments. In line with the associated concern to restrict technology transfer, they would have no truck with the idea that if the Europeans paid half the program costs they should be entitled to their fair share of industrial work. There was to be no constraint on US industry's competitive advantage and no transfer of technology from the United States to Europe, as would be inevitable in a joint program. They also objected to the idea that the satellite should be restricted to aircraft, citing economic efficiency: the Office of Telecommunications Policy wanted a single system for both maritime and aviation services. Thus armed, Whitehead and his allies demanded an in-depth policy review before ratification of the FAA-ESRO memorandum of understanding (MoU) that had been drafted in Washington on August 20, 1971. As Frutkin noted in his diary in October, "European confidence in cooperative projects has been dented by the long delay in our responding to the Lefèvre letter, by the obvious uncertainty of the shuttle's future and by US behavior on an aeronautical satellite."[82] In November Johnson warned Kissinger that if the United States withdrew from Aerosat at this stage it would have serious repercussions "not only our future co-operation in post-Apollo and other space related activities, but on overall US-European relations."[83]

In 1971 the enthusiasm for post-Apollo cooperation that Tom Paine had injected into US-European relations began to wane. The gap widened between the considerable progress made at the technical level between joint groups of experts spearheaded by NASA and the increasing doubts raised at the level of high policy. Whitehead and Flanigan, with the support of David and Fletcher, became increasingly and effectively vocal in their opposition to close collaboration. Their fear that the United States would sacrifice its technological lead, and that US industry would be harmed, was mingled with the White House staffers' distrust of NASA. In their eyes NASA wanted Europe in the shuttle program to protect it from domestic political cuts, even cancellation, and was accordingly willing to give away American technology at ten cents on the dollar, as Whitehead put it. Their pressure on NASA was amplified by Charyk's demand on behalf of Comsat that Johnson tighten up the conditions under which the United States would launch foreign telecommunication satellites before the definitive Intelsat agreements were signed in May.

The European position was summed up by Secretary of State Rogers in a memo to the president in January 1972. As he put it, "[T]he prospects for substantial European contributions to the post-Apollo program are clouded [...] by residual European doubts about whether our offer of launch assistance is sufficiently adequate to permit Europe to forgo the development of its own large and expensive rockets." Delays in reaching agreement on the Aerosat project were also being read "as an ominous sign concerning our future intentions on space cooperation."[84] In fact the Europeans had now realized that they would not be treated as privileged partners under the Intelsat framework. The State Department would be flexible, but it would not give them a formal cast-iron guarantee to launch a separate European comsat system. European space policy for the next decade was further complicated by the disastrous failure of the Europa rocket in November. Divisions were emerging between those who felt it was important to work with the United States in an advanced technological project come what may (led by Germany) and those who saw little or no advantage in it (like Britain and France, for different reasons). An important policy initiative was needed to energize the decision-making process. The official authorization of the shuttle program in 1972 by President Nixon did just that.

Chapter 6

European Participation in the Post-Apollo Program, 1972: Disentangling the Alliance—The Victory of Clean Technological Interfaces

The Shuttle Is Authorized...and the Options Shrink

On January 5, 1972, President Nixon announced that the United States should proceed at once to develop "an entirely new type of space transportation system designed to transform the space frontier of the 1970s's into familiar territory," readily accessible to humans in the decades to come. The space shuttle would "revolutionize transportation into outer space." It would "take the astronomical costs out of astronautics." It promised to become "the workhorse of our whole space effort, taking the place of all present launch vehicles except the very smallest and the very largest" (the Scout and the Titan-III rockets) soon after it became operational at the end of the 1970s. The economic benefits of reusability, which promised to "bring operating costs down as low as one-tenth of those for present launch vehicles," would allow the shuttle to transport humans safely, routinely, and relatively cheaply. The shuttle would take America "out from our present beach-head in the sky to achieve a real working presence in space." It would also secure the "pre-eminence of America and American industry in the aerospace field" by engaging the talents of thousands of highly skilled workers and hundreds of industrial contractors who would ensure that the United States maintained its leadership in "man's epic voyage into space."[1]

Nixon did not refer to the military use of the shuttle in his public statement. He authorized NASA administrator James Fletcher, and his deputy George Low, to mention military applications, however. And indeed this aspect was one of several emphasized by the two NASA officials at the San Clemente White House immediately afterward.[2] In the press conference Fletcher claimed that the low cost and the ability to launch "on a moment's notice, when something strange happens," to be in space within "24 to 48 hours," would certainly interest the Department of Defense. The NASA administrator said that he was "sure the military will be using the shuttle routinely for most of their payloads," though he did not specify that this would involve the development of the space tug.

The president only made passing reference to international collaboration in his official January announcement. In conversation just beforehand, however, he told Fletcher and Low how important international collaboration was to him, particularly the flying of astronauts from all nations, East and West.[3] He affirmed that it would also be valuable to encourage "meaningful participation" in experiments "and even in space hardware development."[4] Foreign participation, he said, could reduce the development cost of the shuttle, now estimated to be $5.5 billion, by some 10–15 percent.

The group of NASA/ELDO/ESRO experts were scheduled to meet again early in February to narrow down the options for collaboration discussed at the end of 1971. Now that the shuttle was authorized, there was a flurry of activity inside the administration intended to adjust the US position, and the scope it allowed for transatlantic collaboration to the new, more stable political and budgetary situation. A subcommittee of the International Space Cooperation Committee met four times in the latter half of January. The meetings in January 1972 were chaired by John Walsh, a senior staff member of the National Security Council. They dealt with various technological, managerial, and foreign policy aspects of post-Apollo collaboration with Europe that had the shuttle at its core.[5] One meeting was devoted to presentations from senior businessmen from the Aerospace Corporation, Hughes Aircraft Company, Lockheed of Georgia, and McDonnell Douglas, all of which had experience in working with European firms. A summary report of the findings of the Walsh subcommittee was submitted to Herman Pollack in the State Department on February 18, 1972, and received in NASA on February 23.[6]

The European delegation to the joint experts meeting, held in Washington from February 8 to 10, 1972, was again led by Causse and Dinkespiler.[7] The US delegation was led this time by Philip E. Culbertson, the director of advanced missions in the Office of Manned Space Flight. The first striking development since the previous expert meeting a little over two months before was the reduction in candidate work packages on the shuttle (see table 5.3). There had been fourteen such packages in all in December. Now there were just five which had a "high probability of being suitable for development in Europe": the tail assembly, elevon, cargo bay door, nose cap, and the landing gear and door. Europe's potential financial contribution to the shuttle program had also dropped sharply, from about $400 million for the original fourteen packages to $100–115 million for the five items on offer (see table 6.1).

Several technology-related concerns drove this reduced offer. Culbertson and Frutkin assured John Walsh's ad hoc committee that the five work packages offered to the Europeans were limited to "subsystems which require least transfer of technology,"[8] and to tasks that their firms could carry out "substantially on their own, thus minimizing European need for US technical assistance."[9] NASA also stipulated that anything Europe built should not have a significant impact on critical US schedules. If the Europeans failed to deliver as expected, there had to be "reasonable recovery options" on the US side. It was also important to choose elements whose design was more or less frozen, and not likely to change. National security concerns provided an added twist. Indeed the entire propulsion package was withdrawn at the second experts meeting, probably to reduce the risk of proliferating missile-related technology. The net result was a

Table 6.1   Change between late 1971 and early 1972 in work packages offered by NASA for European collaboration

Orbiter Work Packages Suggested for Collaboration in November 1971	Orbiter Work Packages Suggested for Collaboration in February 1972
1. Tail assembly	1. Tail assembly
2. Main wing	2. —
3. Elevon	3. Elevon
4. Central fuselage	4. —
5. Cargo bay door	5. Cargo bay door
6. Radiator	6. —
7. Landing gear and door	7. Landing gear and door
8. Nose section	8. Nose section
9. Ejection seat	9. —
10–13. Propulsion (without engines)	10–13. —
14. Instrumentation (difficult to integrate)	14. —
TOTAL COST ~ $400 million	*TOTAL COST: $100–115 million*

package that was relatively simple and "exclude[d] the most interesting tasks" (Frutkin). Indeed, he added, "they have already been termed uninteresting by Europeans involved."[10]

Two other major difficulties beset a joint effort on the shuttle even with the work packages simplified to reduce technological transfer to the minimum. One was the problem of project management. As we saw earlier, everybody agreed that the prime contractor on the shuttle would be an American firm. The subcontractors would be European, and they would be paid for their work by the appropriate European funding authorities. While this mechanism respected the principle of "no exchange of funds," it raised problems of its own. Among the more serious management problems identified by the joint expert group were "source selection, the negotiations of out-of-scope changes, limitations on the control by the prime contractor over the subcontractor and the relations between subcontractor and its own government authority."[11] In an international project such strains could rapidly escalate to the government level since the American prime would have to negotiate their resolution with the European funding agencies that were supporting the subcontractors. As one document put it, "There is a high probability that the contention and acrimony of the subcontracting relationships will degrade, rather than improve, our relations with Europe."[12]

The different rhythms of the decision-making process on both sides of the Atlantic complicated technological collaboration on the shuttle even more. NASA was calling for bids for the orbiter in mid-March and expected phase C/D development to begin by July 1, 1972. The expert meeting agreed that if European subcontractors were to be included in the bids by the American prime contractor, a draft government-to-government agreement had to be settled, at least "in principle," by this date. Frutkin rejected this timeline outright two weeks later. "A working level draft is not adequate for the confidence level we need to defer *US* subcontractor negotiations and to authorize instead extensive interplay between US primes and foreign subs," he wrote." The American authorities negotiating with Lefèvre and the ESC should insist upon a "*commitment in principle or letter(s) of intent signed at the ministerial* level in interested countries" by July 1 (emphases throughout in the original). It was necessary, wrote Frutkin, "to shock senior European officials into a recognition of the magnitude of the commitments they would have to undertake and the very inadequate time for negotiating them."[13]

The Collaborative Effort Is Reduced to a Sortie Can

By the end of February 1972 Frutkin was persuaded that NASA should strongly discourage European participation in the shuttle. He was deeply concerned by the management difficulties, cost overruns on the European side, and the risk of delays involved in having Europeans subcontractors build integral parts of the main orbiter.[14] His sentiments were confirmed at the meeting of an interagency group on March 17, 1972, reported on by Pollack to Secretary of State Rogers.[15] Pollack noted that "Kissinger, Flanigan and David each had representatives on this group, and they were unanimous in reflecting the prevailing spirit in their home offices as one of deep skepticism as to the desirability of European participation in the development of the hardware for the space shuttle or other elements of the post-Apollo space transportation system." Their underlying reasons for this attitude, Pollack added, centered on "protecting the technological position of the U.S., maximizing balance of payments and employment benefits for the U.S., and avoiding managerial difficulties that may be encountered in international cooperation in technological activities." In their view, the only reason to continue with the Europeans now was that "we have gone so long and so far in our discussions with the Europeans as to be 'stuck' with their participation." Ten days later, on March 27, Deputy Administrator George Low confirmed that, in his view, only pressing foreign policy concerns could now keep Europe in the post-Apollo program. In a memo to Fletcher that was transmitted to Flanigan in the White House Low wrote that

> our position is that from a programmatic point of view we would like to develop the Shuttle and all of its ancillary equipments domestically. It would be NASA's view to seek foreign participation in the use of the Shuttle, but not in its development. (When I say Shuttle, I also mean tug, sortie module, etc.). However, it is also NASA's position that if there are overriding international reasons to invite foreign participation into the development of the Shuttle, we would be willing to do so provided certain conditions are met [to be specified in separate paper].[16]

By the end of March 1972, then, it was clear that NASA was no longer willing to fight for direct European participation in the STS system, notably the shuttle. Throughout 1971 it had struggled valiantly against those who argued that there would be a serious leakage of technology to Europe. It had devised managerial schemes that, it thought, would both contain technology transfer and be practicable and efficient. It had never persuaded David, Flanigan, or Whitehead of the merits of its case and, now that the shuttle had been authorized, it did not have the will to go on. Only the State Department, by appealing strongly to foreign policy concerns, could save significant post-Apollo cooperation. Johnson was persuaded that such participation "would be damn useful and valuable from a foreign policy and public-relations point of view."[17] Low implied that NASA would be cooperative. Fletcher concurred. "NASA is a service," he told U. Alexis Johnson in January 1972. "We'll do whatever the people want us to do."[18] But which "people"? Whose voice would prevail? Was a consensus possible?

On April 29, 1972, Secretary of State Rogers turned once again to his president detailing the foreign policy situation.[19] Summarizing the history of US-European exchanges he pointed out that until recently the American authorities had "provided the Europeans every reason to believe that the U.S. was seriously interested in having them participate in the development of certain parts of the Shuttle, in one or more of the RAMs and especially in the Tug." In response Europeans had spent or committed $11.5 million on preliminary technical studies. Now all this was in jeopardy. He repeated the arguments that Pollack had reported to him as regards participation in the shuttle. He noted that there were objections that the tug was too difficult technically for the Europeans. That left the RAM. To reduce European involvement to one or two RAMs, however, would be "judged by them as a clear reversal of our previous policy," and would "buy more trouble with the Europeans than can be justified by the ephemeral domestic advantages that we may gain by denying them participation." America must not be seen to change tack now. Rogers suggested that the United States "accept, but not encourage" European participation in the five shuttle tasks identified by NASA—on condition that they made a "prior commitment" to "undertake the subsequent development of one or more RAMs." He also insisted that there was no need for the United States to reverse its position on the tug, since it would require several more years of design study. Instead, what the United States should do was to create an exit strategy for itself, in the event that one was needed, by bringing "the Europeans to agree that consideration of their undertaking of the development of the Tug will be deferred pending further study."

NASA administrator Fletcher wrote to Kissinger a few days later commenting on Rogers's memo. He took a harder line than did the secretary of state. As regards the shuttle work packages, "we continue to feel such European participation is highly undesirable and that it would complicate our shuttle management problems." These concerns could be overridden if the president insisted, but only on condition that the Europeans were responsible for both estimated costs and overruns and also built a sortie module. Fletcher also confirmed to Kissinger that, even if further studies established that a tug was feasible in Europe, NASA wanted "to reserve the right to escape from an agreement," and did not anticipate "technical support of the European study" unless directed by the president to do otherwise. "For all of these reasons," Fletcher wrote, NASA did "not recommend European involvement in the tug."[20]

And then the president's science adviser, Ed David, stepped in.[21] He insisted that the Europeans understood American reservations about technology flow and management difficulties, and were pragmatic enough not to let these concerns in Washington drive them to abandon cooperation. He claimed that the French were going to propose anyway that Europeans give priority to Sortie or RAM modules to be carried in the shuttle payload bay, and that they may abandon plans to develop the tug or contribute to subsystems of the shuttle. In short the United States could drastically reduce the scale of its offer of post-Apollo cooperation without creating the foreign policy blowback that the State

Department feared. With that fear removed, David insisted that negotiations on participation in the orbiter and discussions of the tug should be terminated at once on the grounds that the United States now believed that "they would lead to excessive additional costs and managements complications that the U.S. is unwilling to accept." The United States could accept European participation in the shuttle program, he added, but only "if limited to RAM and Sortie payload modules."

David's view prevailed. Pollack conveyed what was now official policy to a high-level ESC delegation that had come to Washington on June 16 to discuss post-Apollo cooperation. He informed his visitors that European participation in the development of the shuttle "can no longer be encouraged by us even on the limited scale we are still discussing." He also killed "consideration of mutual development of the Tug," which, he said, had "of necessity been set aside." European participation in the development of Sortie modules and in the use of the shuttle system were, by contrast, warmly welcomed.[22]

Why the Tug Was Withdrawn

The withdrawal of the orbiter was not too difficult to swallow; participation was of limited importance anyway. By contrast the unexpected removal of the tug came as a bitter blow. Thirty-five years later Causse still remembered the announcement as coming as a "shock."[23] Nothing could have been more indicative of the asymmetry in power between the two sides of the Atlantic, and of the still-massive disparity in the financial, technological, and industrial capabilities in space between the two "partners." Pollack emphasized at the meeting on June 16 that it was "important that both sides keep in mind the basic, enduring nature of the ties that bind the United States and Europe."[24] He surely wanted to calm ruffled feathers: in reality, he probably only made matters worse.

To put this in perspective we must remember the history. In February 1971, and again as late as February 1972, the joint meeting of experts had made a number of decisions to promote phase A tug activity.[25] NASA let it be known that, for financial reasons, it would only have limited funds available for tug studies and technology development. The preliminary mission model, on the other hand, indicated that the tug should be available soon after the shuttle became operational, as it was required for "over 50%" of the missions. The management of the tug would be left to Europe, with NASA in a "supporting role." An informal version of the proceedings in Paris in February 1972 by NASA's European representative recorded that the agency was "very interested in having Europe consider undertaking the Tug as a Post-Apollo cooperation effort both for the over-all program needs and from the increased international cooperation that such a program would bring."[26] The joint experts group had decided that funds would be allocated to two phase A studies in European industry, that a technology development program would be started as soon as possible, and that the economics of the tug and the mission model would be refined.

Just a few weeks later Frutkin moved sharply away from this position, notwithstanding the advantages noted by the expert group. "The tug is given second place after the sortie module because it is far more difficult to develop and could conceivably give rise to performance difficulties which might impair relationships," he suggested. The risks here were amplified by NASA's decision not to devote substantial resources to the tug, even in the most challenging technological areas. As a result, Frutkin feared that the tug "could also stimulate European advances in technology beyond those of the sortie module."[27]

The nature of those advances was specified in a report prepared by Causse and Dinkespiler for the European Space Conference in March in which they emphasized how important the tug would be for Europe. "The tug by its mission partakes of the nature of a launcher, but by its ultra-light structure, big flight autonomy and automatic rendezvous capability is akin to a space vehicle and actually makes use of highly sophisticated satellite techniques," they wrote. "It pushes propulsion techniques well beyond what is currently envisaged in Europe," and by virtue of "its far-reaching integration with the shuttle and with the payload during operations will afford Europeans effective participation in most American missions."[28] They explicitly told NASA in mid-April that they saw the tug as "a very critical development which, maybe in the future, could be a stage in Europa III."[29] In other words by encouraging the tug NASA not only risked being charged with irresponsible technology transfer but, even worse, of proliferating booster technology.

Then there was the problem of use. Causse and Dinkespiler also sought reassurances that NASA and the Air Force would not build tugs under license in the United States for their own use, and would at least undertake to buy European-built tugs for a certain period of time. NASA had certainly been open to this early in February. Going into the meeting of the joint expert group, Culbertson had written that "[i]f there is a European decision to develop the Tug, Sortie Can or RAM, NASA would expect to commit to use providing it meets our specifications."[30] By mid-April, however, NASA was posing the question differently. If before it was willing to buy tugs as needed—unless Europe failed to deliver—now "we were basically concerned about uncertainties in the definition of a tug, the difficulty of producing one, and the multiplicity of approaches to orbit-to-orbit capability."[31] There were also concerns in NASA about the safety of having a tug powered by cryogenic fuel lodged in the Shuttle's cargo bay. In short, upstream of the question of use, NASA was now having doubts about the safety and the technological feasibility of the tug concept itself.

The Air Force's evaluation of the costs and benefits of developing the tug abroad also struck a blow at European aspirations.[32] It was recognized that contracting out the tug to Europe would save dollars. On the other hand, the Department of Defense was concerned about the dangers posed to national security by having foreign powers develop one of their key technologies. They would have to reveal the nature of their missions. Their requirements might be jeopardized by unilateral decisions, technological and industrial deficiencies, and a lack of operational support by the Europeans. Building the tug abroad would also undermine the domestic industrial base in an already-weakened sector that was crucial to national defense. Summarizing the situation, it seemed to NASA that the Air Force would be willing to use a tug developed in Europe if one were available, but would "undoubtedly" manufacture it under license in the United States. In addition, to secure its supply lines the Air Force would "also likely support development of an alternate, expendable stage [that could perform the tug's missions], based on Centaur or some other existing vehicle."[33] In short, there was no hope that the Air Force would only procure tugs built in Europe, so boosting the production lines of European firms with orders for US "military" technology. By June the tug was dead; indeed it was never built. Studies were terminated in mid-August. That left the Europeans to do the Sortie Module that was later called Spacelab.[34]

There was more to come. On the last day of the June meeting (June 16) Undersecretary of State U. Alexis Johnson finally expressed the official American position on launching the Franco-German telecommunications satellite Symphonie. The Europeans had long sought clarification on whether the United States would be willing to support the Symphonie proposal in Intelsat, and by extension launch it for them. Johnson replied that he could only do so if the proposed satellite was shifted to a different orbital position to that foreseen, and if its geographic coverage was more restricted than planned.[35] This was the last straw for many people in France who were keen to develop an independent launch capability in Europe "to maintain the base of their ballistic missile technology capability and [...] to maintain European independence of the US in space operations."[36] Washington's pared down offer of restricted technological collaboration in the post-Apollo program, the cancellation of Aerosat, and the determination to place launch conditions on Symphonie played into the hands of those who were determined that the region needed to develop its own independent access to space. The new European launcher called Ariane rose from the ashes of the explosion of Europa II in November 1971, and was nourished by the hard line taken by the American negotiators in June 1972.

Spacelab

The offer by NASA to build a Sortie Module as Europe's contribution to the post-Apollo program was taken up by West Germany. On August 14, 1973, NASA administrator James Fletcher and ESRO director general Alexander Hocker signed an MoU for a "Cooperative Programme Concerning Development, Procurement and Use of a Space Laboratory in Conjunction With the Space Shuttle System."[37] The project, which would be spearheaded by the Federal Republic, foresaw the construction of a human-rated laboratory and a number of pallets that would be housed in the shuttle's payload bay. NASA agreed to provide technical support, to manage the operational uses of the laboratory, and to develop essential technological items such as the access tunnel between Spacelab, as it was called, and the orbiter's cabin.

The history of Spacelab has been written from various angles: by Douglas Lord (NASA) and the many European and project managers and engineers who built it[38]; by historians Lorenza Sebesta and Arturo Russo from the point of the general policy framework and the user community, respectively[39]; by industrialists who were engaged in it[40]; and by some of the scientists who actually exploited it.[41]

Though the user communities were not enthusiastic about the venture, industry was more interested. Niklas Reinke has noted the "enormous value in terms of contracts: German industry anticipated no less than DM625 million in turnover from the transatlantic cooperation project."[42] The prime contractor, ERNO (MESH) in Bremen, also gained considerable insight into American methods of systems management. All the same, these were just "add-ons, designed to make [Spacelab] more convincing," according to Wolfgang Finke, a senior official in the Federal Ministry for Research and Development. For him, political considerations dominated Germany's willingness to continue with an "appendix" to post-Apollo after NASA had drastically reduced the scope of collaboration. Central to Bonn's thinking was the need to reassure her Western allies that she was a reliable ally even as she opened up toward the Soviet Union and Eastern bloc (the so-called *Ostpolitik*). "The German government looked for opportunities to demonstrate her attachment to the Western camp and especially her reliance on the United States," said Finke, "without jeopardising her new policy toward the USSR and her neighbours in Eastern Europe. Among other things cooperation in space technology seemed to offer such an opportunity."[43] German officials also hoped that participation in Spacelab, "an endeavour that was at the time considered to be the most advanced," would reduce the technology gap and contain the brain drain.[44]

Spacelab was a collaborative success at the working level. On the other hand it cost far more than expected—$750 million, more than twice the 1973 estimate. The laboratory only flew 16 times between 1983 and 1998 and the overall scientific return was disappointing.[45] ESA and Germany also came to regret the terms of use agreed with NASA in the MoU. Europe agreed to build the first module, to fly one of their astronauts on it along with sharing the payload with the United States, and then to hand it over to NASA who could use it free of charge. NASA only had an obligation to buy one more Spacelab—and it did that and no more. This was far less than Europe had originally hoped for (the sale of four–eight units).

Whatever the disappointments, Spacelab was a major technological project that involved considerable industrial learning and that enabled Europe to engage directly for the first time in human spaceflight. It provided an essential platform for subsequent participation in Space Station Freedom and the International Space Station (see chapter 13). The last word is best left to Reimar Lüst:

> International cooperation does indeed depend a lot on the actual balance of power, but the benefits of cooperation cannot always be explained solely in figures. Just as many European firms today [1989] spend a lot of money to buy themselves into joint ventures with American and Japanese high-tech companies, in order to get knowledge on new technologies transferred into their firms, so ESA had to pay the price of Spacelab to acquire the basics of manned spaceflight.[46]

Conclusion

The idea of foreign participation in the post-Apollo program moved through two quite distinct phases. The first was dominated by NASA administrator Tom Paine, and lasted for about a year from October 1969 through September 1970. Paine's enthusiasm for including other countries and regions—Western Europe, but also Australia, Canada, and Japan—in NASA's ambitious schemes for the 1970s and 1980s, produced a flurry of activity on both sides of the Atlantic. Frutkin took the lead in exploring, along with interlocutors representing the European Space Conference, the financial, industrial, technological, and managerial possibilities of a major contribution to the Space Transportation System. In the second phase, which lasted from the end of 1970 to the middle of 1972, new and extremely determined actors played an increasingly important role in shaping the contours of collaboration. White House staffers Peter Flanigan and Tom Whitehead, with the support of the president's science adviser Edward David, led the charge. They were hostile to Paine's "swash-buckling" approach, believed that NASA had to completely rethink its role and redefine its demands on the public purse, and were deeply concerned about technology transfer to Western Europe. New NASA administrator Jim Fletcher, while willing to fight for the STS, shared their grave doubts about international collaboration. His sentiments permeated through NASA once the president had authorized the shuttle program, and were adopted by Deputy Administrator Low and by Arnold Frutkin. By March 1972 only the State Department was still prepared to make a strong case for European participation in the core of the program, but it was too late.

Senior negotiators on both sides adopted entrenched positions that were immune to argument. For the Europeans, it was the Belgian chairman of the European Space Conference, Theo Lefèvre, with strong backing from France, who demanded watertight launch guarantees before he would fully commit Europe to post-Apollo cooperation. The horizons of his thinking were dominated by the fear that the United States might use its monopoly on access to space to impede the development of a strong European telecommunications satellite industry. The short history of Intelsat in the 1960s had convinced him, and many people in Europe, that America would only relinquish its control of this lucrative market with great difficulty, and that the State Department would work along with Comsat to undermine meaningful competition from separate, regional telecom systems.

On the US side it was Whitehead, along with David, who sought cast-iron guarantees that there would be no significant technological leakage to Europe through its participation in the shuttle program. When Paine launched post-Apollo cooperation in 1969 the argument that the United States should help close the technological gap with Europe still had considerable traction in Washington, DC. By 1970 it was technological competition, not collaboration that dominated the thinking of many in the White House. Whitehead, who could barely conceal his contempt for Paine, was convinced that NASA was recklessly giving away US technology to Europe. Endless reports and analyses failed to change his position, which found resonance with Ed David. When one adds this unrelenting hostility to technological leakage with the problems of managerial organization, the dangers of cost-overruns, and the fears that the Europeans were not quite up to the technological tasks they wanted to undertake, one has a bundle of arguments that was immensely corrosive to any collaborative project.

The president had the authority to bring his White House staffers to heel. The image of Nixon that emerges from these debates, however, is one of a president whose policy pronouncements were often vague, imprecise, and off the cuff—and open to manipulation and self-serving interpretation by his closest advisers. There is no doubt that Nixon was genuine in his desire for international space collaboration, above all with the Soviets and the eastern bloc (see chapter 7). This was central to his geopolitical strategy of détente, a strategy that sidelined Europe in the interest of improving relations with both the Soviet Union and Mao's China. Within that broad policy framework the president was usually vague about the scope and intimacy of technological collaboration. This left considerable room for officials in NASA and the White House—doubtless in good faith—to justify policy agendas, even conflicting policy agendas in the president's name. Though Kissinger was extremely frustrated by these gambits, Nixon apparently ignored them: certainly he did little to clarify his position.

American industry did not share the concerns over technological transfer that so preoccupied senior members of the administration. All of the major American aerospace contractors were positive about incorporating European firms as sub-contractors in various parts of the shuttle system. They had identified European technological strengths, which complemented their own. They were convinced that a foreign contribution would provide greater long-term stability to the program, especially before Congress. And they had no doubt that, even if the Europeans did acquire new and significant knowledge from them, they would still emerge superior in the long run, both because of what they picked up themselves from Europe, and because they were confident that US industry was far more dynamic, entrepreneurial, and innovative than its sluggish, bureaucratized European partners.

The Department of Defense was another actor that had little influence on the trajectory of post-Apollo collaboration. It was deeply implicated in the shuttle design by virtue of its demand for an orbiter cross-range capability of some 1,250 nm. That demand, in turn, made major technological demands, particularly regarding the Thermal Protection System on the leading edge of the delta wings. Europe's experience with Concorde was a potential asset here, however. If the Air Force eventually took little interest in the course of deliberations it was because it rapidly concluded that it would need to build its own tug anyway—and of course as the technological feasibility of that element waned so too did the Department of Defense's engagement in the negotiations.

From the European perspective, the departure of Paine and the arrival of Fletcher turned out to be a serious setback to post-Apollo collaboration. Paine's enthusiasm was infectious, yet his optimism was misguided, even irresponsible. He simply did not have the support in NASA, and certainly not in the Nixon White House for an ambitious collaborative project. Of course Frutkin and Low did what they could to carry out the administrator's wishes. Their efforts were truly Herculean. They had to contend with European negotiators who sought to be treated as equals in a massively asymmetric financial, industrial, and technological project. They found attempts to move forward on discussing concrete sites for collaboration constantly thwarted by European demands for launch guarantees. On top of this they found the ground cut away under their feet by senior officials in other sections of government. Johnson's willingness to yield to Charyk's last-minute demand to reinterpret the meaning of votes on what counted as significant economic harm in the Intelsat Assembly of Parties infuriated Europeans and isolated NASA. So too did the collapse of the negotiations over Aerosat, in which once again Whitehead and Flanigan's concerns about technological leakage played an important role. Fletcher, for his part, seems to have had no stomach for a fight with the White House staffers. More precisely, perhaps, he agreed with their concerns. He too was concerned about the multiple complications that would ensue on giving the Europeans a large technological role in post-Apollo, and quickly came to the conclusion that the only possible merit of post-Apollo collaboration was its foreign policy aspects. It was a definitive move in a climate in which, as we have said, Europe was no longer a major concern in the president's foreign policy agenda, and its growing technological maturity—and not concerns about the "technological gap"—was shaping the contours of policy thinking by senior White House staffers. Rogers and Johnson—the latter already discredited for having made a too-generous deal with Japan over launcher technology (see chapter 10)—could not hope to bring off a major collaborative project with Western Europe under these circumstances.

If the Sorie Can/Spacelab survived this lengthy process at all it was because Germany remained determined to keep the collaborative ball in the air, because the State Department saw considerable interest in working closely with a traditional ally that was itself reevaluating its relationships with the eastern bloc and the Soviet Union, and because this technological element embodied Frutkin's two cardinal principles of "no exchange of funds" and "clean interfaces" in their pure form. In March 1972 George Low wrote that NASA sought foreign participation in the use of the shuttle, not in its development. Spacelab satisfied that requirement.

Part III

NASA and the Soviet Union/Russia

Angelina Long Callahan

Chapter 7

Sustaining Soviet-American Collaboration, 1957–1989

Beginnings

The relationship between the United States and the Soviet Union in space is quite accurately portrayed as one of fierce competition. The launch of the Sputniks in late 1957 and Gagarin's flight in 1961 were deep blows to American pride. They challenged preconceptions about the superiority of American science and technology, even about the superiority of the capitalist system itself. Thus, the global struggle for "the soul of mankind" inscribed itself upon a multitude of scientific instruments, launch systems, institutions, and individuals.[1] For many years, historians have labored to reconcile the paradoxes of Soviet-American cooperation in space with the space and missile races of the mid-twentieth century.

Such histories commonly open with speculation centered on the likelihood of a joint lunar mission proposed by President Kennedy to Premier Khrushchev.[2] Indeed, Kennedy's famed May 1961 "Moon Speech," announcing the United States' "race to the moon" was bookended by both covert and public invitations to collaborate.[3] In so doing, Kennedy unwittingly set up audacious expectations for astronauts and cosmonauts to explore the moon and beyond. With human spaceflight as the agency's signature activity, scholars have struggled to assign some sort of reason to the two nations' rocky progression from (what was apparently) an utter lack of intercourse to the stilted Apollo-Soyuz Test Project and finally the interdependence of the International Space Station.[4] Geopolitics became reified in human spaceflight: cold shoulders through the dire years of missile and space races; détente's climactic 1975 handshake in space; and finally, the Cold War denouement in the International Space Station agreements.

Beginning with the Kennedy-Khrushchev moon flirtations, historians have characterized US offers for cooperation as meeting a "rhetorical goal" and functioning as a "benign hypocrisy." Operating as such, the US space program appeared open to Soviet contributions, but at the same time participated in implicit competition to outdo their rival in hardware and soft power performances. Such narratives explain the complex motives and political economy of major commitments such as a joint lunar expedition, the ASTP, or the ISS.

Well-publicized, expensive, and demanding years of lead-time, these projects were carefully orchestrated under the watchful eyes of presidential administrations and Congress (whose interests at times conflicted with one another and/or NASA administration).

On the flip side of the coin, the many years spanning Kennedy's joint lunar base offer and the Apollo-Soyuz Test Project as well as those years separating ASTP and the International Space Station Agreements are commonly explained by intractable negotiations on diplomatic fronts: wrangling over nonproliferation treaties, controversy over interventions in the developing world, or the uncompromising political will of heads of state. Collaboration seems impossible at these times.

These two chapters aim to add breadth to that presumption, exploring Soviet-American collaboration through the following questions. To what degrees did representatives of NASA attempt to sustain collaborative activities since the 1957–1958 IGY? To what degree might collaborative activities have been shaped by the interests of researchers and policymakers representing state, national, and transnational scientific organizations?

It remains something of a paradox that the United States and the Union of Soviet Socialist Republics/Russia have cooperated in space exploration for more than half a century. While their relations have been strained by fears of technology transfer, threatened by executive posturing, and reshaped by fiscal considerations, to fluctuating degrees individuals making up these research communities have labored steadily to share resources and exchange information.

Khrushchev and Kennedy: Talking About the Weather

Perhaps we could render no greater service to mankind through our space programs than by the joint establishment of an early operational weather satellite system.
— President Kennedy to Premier Khrushchev, March 7, 1962

It is difficult to overestimate the advantage that people would derive from the organisation of a world-wide weather observation service using artificial earth satellites. Precise and timely weather prediction would be still another important step on the path to man's subjugation of the forces of nature.
— Premier Khrushchev to President Kennedy, March 20, 1962[5]

The history of formalized Soviet-American cooperation in space might well be traced to letters and public pronouncements between President John F. Kennedy and Soviet premier Nikita Khrushchev in 1961. Over time (and following occasional lapses in correspondence), the two superpowers narrowed fields of potential cooperation to those outlined in a June 8, 1962, Agreement on peaceful bilateral cooperation in space. Made one year after the orbiting of the first human in space (Soviet cosmonaut Yuri Gagarin) and Kennedy's subsequent announcement that the United States would place a man on the moon by the close of the decade, this agreement to cooperate "for the benefit of mankind" introduced new philosophies to what some have characterized as the "space race."

Therein, the two nations agreed to four fields of cooperation: geomagnetic mapping, experimentation with communications satellites, sharing of biomedical data (for the emerging field of human spaceflight), and exchanging weather satellite images through what came to be referred to as the "Cold Line" facsimile network. Early on, representatives of the two nations agreed to limit work to that which may be characterized primarily as data exchange or even coordinated observation—rather than designing or building instruments together, they agreed simply to share limited amounts of information.

Arnold Frutkin, noting that the content of the 1962 agreement (and the 1963 Memo of Understanding) had on occasion been grossly misrepresented, explained: "They provide for *coordination* rather than *integration* of effort, in other words for a kind of arm's length cooperation in which each side carries out independently its portion of an arrangement without entering into the other's planning, design, production, operations, or analysis. [In unequivocal terms, he assured possible critics,] No classified or sensitive data is to be exchanged."[6] In spite of the relatively low expectations entailed by data exchange and coordinated observations, Soviet participation in 1960s projects tended to be disappointing: their contributions to meteorology came late and were incomplete; their cooperation in the Echo-II satellite less than generous; their exchange of biomedical and geophysical data curt, if not truncated. Following a nine-month delay, waiting for the Soviets to simply name their Joint Working Group (JWG) candidates, one official remarked publicly that it was time for the Russians to "get off the dime." Relations did not become particularly warmer once the JWG began meeting. Of the four aforementioned projects, the Soviets refused to take part in the telecommunications satellite system (opting instead to construct their own system with political allies), cooperated half-heartedly on Echo-II, and in the end, engaged in sustained cooperation in only one field: meteorology.

NASA's administrator Hugh Dryden was particularly critical of Soviet contributions to the Echo-II experiments, detailing what appear to have been half-hearted gestures toward cooperation. His remarks before the Senate Committee on Aeronautical and Space Sciences have been quoted frequently, but warrant revisiting:

> The Soviet side observed the critical inflation phase of the satellite optically and forwarded the data to us. They did not provide radar data, which would have been most desirable, but they had not committed themselves to doing so. The Soviets provided recordings and other data of their reception of the transmissions via ECHO from Jodrell Bank [United Kingdom]. On the other hand, the communications were carried out in only one direction instead of two, at less interesting frequencies than we would have liked and with some technical limitations at the ground terminals used. I do not want to over-emphasize any technical benefits from this project. It was, however a useful exercise in organizing a joint undertaking with the Soviets.[7]

Dryden's reflections on Echo-II reflect a general notion that collaboration—no matter how perfunctory—was in fact a feat of diplomacy. Unfortunately, the Echo-II experience was typical of most collaborative ventures with the Soviet Union, dating to the International Geophysical Year (IGY). Arnold Frutkin, before working with NASA in international relations, had served as deputy director of the US National Committee for the IGY and recalled that the Soviets would frequently attempt to initiate data exchanges and then cancel. As Soviets tended to be slower and more secretive, the Americans became increasingly suspicious. These frustrations surfaced in the press, indicating at least a limited public awareness of the many ups and downs of Soviet-American relations in space.

In February 1965 (13 months after the initial forced deadline for weather satellite exchanges), the *Washington Post* ran its piece, "U.S. May Terminate 'Cold Line.'" The *Post* detailed Dryden's report before Congress, in which he gave his colleagues in the Soviet Union a final ultimatum: unless satellite transmissions came across the Cold Line "in a reasonable time," the United States would terminate the link. NASA's deputy administrator continued, detailing Soviet promises that satellite data would be forthcoming in 1965, and perhaps most vexing, how his numerous letters to Anatoly Blagonravov (Soviet academician in the Soviet Academy of Sciences) regarding the Cold Line had gone unanswered. Though NASA hesitated to set an exact deadline, the article suggested "some officials feel that American patience could wear thin by June 30 [1965]." NASA was effectively kept on hold for another year, waiting until June 25, 1966, for the launch of the first announced Soviet meteorological satellite, Cosmos 122.

Historians have documented these and similar discourses, interpreting them at times as substantive offers for scientific and engineering cooperation and at other times as more politicized diplomatic posturing with complicated meanings.[8] NASA officials communicated their doubts and at times vociferous exasperation with the Soviets. NASA administrator Tom Paine reported before the Senate Committee on Aeronautics and Space Sciences that between 1965 and the autumn of 1970, NASA and the Soviet Academy of Sciences held no meetings regarding possible collaborative efforts, in spite of numerous proposals for cooperative activities from the United States.[9]

Paine had written the Soviets to invite proposals for experiments on US craft, to negotiate use of the laser reflector left on the moon from Apollo 11, to invite participation in the analysis of lunar material, to solicit Soviet attendance at the Conference on the Viking Mars mission, to consider coordination of planetary programs, and to mark his openness to further suggestions. Ten months later, the two parties succeeded in arranging a meeting.

For a decade, representatives of the Kennedy, Johnson, and Nixon administrations had expressed the desire for Soviet-American cooperation in space. Explains Walter McDougall, "Whether or not significant cooperation were achieved, the United States must be perceived as desiring it."[10] Thus, in a piquant twist of Cold War logic, Americans continued to offer joint work, but bore limited hope for projects more grand than the World Weather Watch and relatively limited exchanges of data.

NASA representatives pursued relations with other international partners. In their monograph analyzing the history of Soviet-American efforts at collaboration, Dodd Harvey and Linda Ciccoritti note that NASA "publicly established" plans for post-Apollo cooperation in space (see chapter 4). Central to this were "space goals 'internationalizing' the space enterprise *with or without the participation of the USSR*."[11] Frutkin observed that a substantial amount of COSPAR reports testified to America's cooperative associations. Without questioning the degree to which Soviet *researchers* shared the philosophies of the Soviet *state*, he contrasted US and Soviet policy: "Since the Soviet Union has so far given little more than lip service to such programs, virtually no references to cooperation with the Soviet Union are included."[12] Frutkin explained that "[t]he American space image abroad" was characterized by elements of openness, direct benefit to participants, generosity of research and results, a healthy drive toward technological and managerial preeminence, "and perhaps most important of all, the evidence of high national purpose."[13] He described the contrast between American openness and Soviet isolationism as "eloquent," and said that the American example was "clearly pushing the Soviet Union toward some more or less imitative effort."[14]

Frutkin, having participated in IGY administration, surely grasped the complex political environment his Soviet partners faced: travel restrictions, limits on the circulation of overseas publications, control over data, and the consistent prioritization of military over scientific pursuits. Years later, history would reveal the disappointment of even Sergei Korolev, whom Khrushchev personally restricted from participating in any international scientific symposia.[15] A similar (and ultimately more tragic) disappointment is documented in the memoirs of Iosif Shklovsky, a prominent Soviet heliophysicist. Shklovsky got his first taste of international science in the IGY and spent the remainder of his career fading in and out of the international scene—the ebbs and flows determined at least in part with his standing with the Soviet state. The 1958 Moscow Assembly of the International Astronomical Union was a great treat to the man who "was obviously thrilled to recognize individuals who he had known only by the proxies of their published papers."[16] While his publications circulated the world over through the course of his career, between 1958 and 1984, Shklovsky maintained sporadic contact with colleagues in the United States.

During this time, the solar physicist received many invitations to lecture and participate in scientific meetings abroad. In spite of being recognized worldwide as a leader in his field and his eagerness to travel, Shklovsky's "outspokenness about politics and human rights" jeopardized his requests to travel. But for rare International Astronomical Union (IAU) meetings and a couple scientific symposia, he remained homebound. Herbert Friedman, a colleague in the United States noted, years after the death of his friend: "[I]t was a bitter pill to swallow for a man who had such a burning desire to meet with his peers abroad."[17] By the time of the 1970 US National Academies of Science's annual exchange, Friedman was barely permitted to see his colleague at the Institute for Space Research, but never in private.

Whereas other fields of space research enjoyed an unprecedented thaw around 1972 (when bilateral arrangements were made for Soviet-American work in the Apollo-Soyuz Test Project and exchanging biodata from the Skylab and Salyut space stations), astrophysics experienced a setback. US researcher Herbert Friedman reported that in 1973 "many of the best Soviet astronomers" (including Shklovsky) were not permitted to attend the IAU in Australia. That same year Shklovsky was elected to the US National Academy of Sciences as a foreign associate, but, following a "courageous letter" in defense of Andrei Sakharov, he was banned from attending the 1976 IAU in France. This in spite of the fact that "he had been invited to deliver one of the most prestigious discourses of the occasion."[18]

NASA and the Shifting Political Climate, 1968–1972

Following this rocky period, between 1968 and 1972 Soviet-American relations encountered a point of departure at which the two maintained coordinated activities—be it even on one or two projects—until the present. In 1968, the World Weather Watch entered its operational phase (the World Weather Program), at which point both Soviet Meteor satellites and US TIROS satellites circled the earth providing continuous data to researchers and forecasters alike. In 1969 the United States cancelled its biosatellite program, making the Soviet offer for cooperation on the Bion biosatellites all the more attractive five years later. In the fall of 1970, Soviet academician Keldysh wrote NASA administrator Paine acknowledging that cooperation was to date "limited in character," leading eventually to the 1970–1971 agreement for an Apollo-Soyuz docking in orbit.[19]

The moon race, as it were, ended. NASA, which for many had come to be viewed as a single-issue agency, was now seeking new purpose in Spacelab, the Shuttle Program, hopes for additional planetary exploration, as well as sustained research and development in remote sensing. Congress reduced NASA's budget and priorities year after year, leading in part to the resignation of Administrator James Webb in October of 1968.[20] In March of 1969, Thomas Paine took over duties as NASA administrator, but remained in office a mere 19 months. James C. Fletcher followed as NASA administrator in April 1971, remaining through May of 1977.

During this time, initiatives for bilateral collaboration were in some regards a "bottom-up" phenomena. Historian Yuri Karash indicates that in late 1969 and early 1970, cosmonauts began making rare visits to the United States. At that time, Mikhail Millionshchikov, a vice president of the Soviet Academy of Sciences, spoke at the Second National Convocation on the Challenge of Building Peace in New York City, expressing the sentiment that the time was favorable for renewed talks in collaboration. In a remarkably short period of time, October 1970, leading officials from both US and Soviet space programs met in Moscow to discuss the possibility of joint ventures.[21]

In January 1971, NASA's acting administrator George Low and Arnold Frutkin met with Nixon's foreign policy advisor, Henry Kissinger. In their meeting, Low broached the possibility of formally inviting the Soviet Union to take part in a test mission involving an Apollo and Soyuz spacecraft. Kissinger assured Low, "As long as you stick to space, do anything you want to do. You are free to commit—in fact, I want you to tell your counterparts in Moscow that the President has sent you on this mission." (Kissinger's condition "as long as you stick to space" stemmed from the fact that astronauts had been quoted, indicating that bilateral negotiations at the national level ought to be as easy as those for space collaboration.)[22] With the Nixon administration's blessing, negotiations led eventually to the January 21, 1971, US-USSR Science and Applications Agreement.

These individuals signify shifting political climates—as both drivers and consequences of their times. The competition of the early Cold War gave way to détente, and a cautiously cooperative climate shaped the character of NASA-Soviet programs in the 1970s and early 1980s. This is not to say that the thaw in US-Soviet relations overwhelmed all other challenges to collaboration: what was possible in practice was determined at once by scientific direction, security restrictions, technical limitations, and fiscal realities. Thus a study of cooperative work in the fields of biosatellites, atmospheric science, and the Apollo-Soyuz Test Project will illustrate the structural flexibility inherent to NASA's principles and guidelines for international projects—how a wide variety of scientific and engineering communities managed to work under these adaptable guidelines, yielding scientifically and (to many) culturally meaningful returns.[23]

Under the 1971 US-USSR Science and Applications Agreement (renewed in 1974 and 1977), Soviet and US space researchers agreed to exchange lunar soil samples, share biomedical results from human spaceflight, and compare findings from Mars and Venus probes. In addition to this, they set up five joint working groups that supported the continuation of meteorological sounding rocket networks, coordinated maritime studies, joint vegetation surveys, and called for the flight of Soviet life sciences experiments on Skylab.[24] A number of these operations were, or led to, multilateral ventures.

Indeed, several multilateral endeavors overlapped with 1971, 1974, and 1977 arrangements between the United States and the Soviet Union. Realizing that it was in their best interest to invite the participation of other nations, policymakers on both sides of the Iron Curtain proposed cooperation in their research or at the very least opened a substantial amount of data to the public domain. Both Soviet and US lunar samples were distributed to a number of nations. Likewise researchers released results of biomedical and planetary research to international colleagues and continued to contribute standardized data and specialized observations to World Meteorological Organization data centers. In all cases the United States, it must be said, was far more forthcoming than the Soviet Union, in line with its far more positive commitment to international collaboration and openness.

In each of these fields, Presidential initiative per se appears to have played a limited role in sustaining cooperation. Indeed, several projects carried on in spite of executive policy intended to snub the opposing superpower. Following the initial thrust of the Kennedy administration, repeatedly pressuring Premier Khrushchev to work with NASA in space, bilateral collaboration operated for the most part under the inclination of NASA headquarters, NASA centers, the National Oceanic and Atmospheric Administration, and researchers located at various universities. Nevertheless, presidential administrations and the Congress together shaped NASA policy by setting budget priorities, demanding rigorous justification for innovative programs, or as in the case of the Carter and Reagan administrations, opting to not openly pursue joint objectives with the Soviets, but simply tolerating collaborative projects that were less prone to publicity.

LBJ and Webb: Seeking Balance for the 1970s

Johnson's leaving the White House in 1969 did not necessarily end a decade of unflagging executive support to NASA. In their determination to maintain the Apollo landing deadline of 1969, the Johnson administration wound up trimming or eliminating other scientifically meaningful projects from the NASA program. As numerous historians have noted, this placed NASA administrator James Webb in a complicated position, forced to prioritize among the Apollo timetable, post-Apollo projects, earth science, planetary exploration, and NASA's many other pursuits. James Webb fought bitterly for the funds to sustain robotic planetary exploration, fundamental research, the Nuclear Engine for Rocket Vehicle Application, all the while concerned for the minimum requirements for the Apollo mission.[25]

As development of Apollo spacecraft neared completion in the mid-1960s, operating budgets dwindled and initiatives cut back. Webb and his colleagues had anticipated flagging support and when negotiations commenced regarding post-Apollo priorities and funding they adopted a cautiously defensive posture. Former NASA chief historian Roger Launius observes that when the Johnson administration pressed Webb for post-Apollo objectives, "Webb was quite reluctant to commit NASA to specific goals and priorities in advance of any expression of political support."[26] In his 1965 "Summary Report: Future Programs Task Group" the administrator's only recommendation was that NASA plot out a "continued balanced program, steadily pursuing continued advancement in aeronautics, space sciences, manned space flight, and lunar and planetary exploration, adequately supported by a broad basic research and technology development program."[27] Webb emphasized that he saw no need to require an "overriding emphasis" in any of the aforementioned fields, nor did he believe that a new Apollo-style space race would secure the administration's future. NASA required a balanced (if self-contradictory) program, one that would meet demands for cost-effective administration, meanwhile maintaining a "pre-eminent role in aeronautics and space."[28]

By the mid-1960s and into the years following Apollo, lawmakers and the public alike frequently questioned the fiscal and political sustainability of speed-driven "crash" programs. Some critics voiced their doubts regarding the worth of space sprints such as Apollo or the rush to respond to Sputnik. Still others questioned the opportunity costs of space exploration as a whole—believing that the same funds that put men on the moon might somehow be reallocated to "urban blight," foreign aid, or be forfeited altogether to reduce tax expenditures.[29] In such a political environment, projects emphasizing the cost-benefit analysis of spin-off technologies or good stewardship of the earth's resources promised a logical counter to the harshest criticisms against "space spectaculars" both at home and abroad.

Multilateralism, Earth Resources, Life Sciences

Secretary of State Dean Rusk had already anticipated these criticisms in 1966, when he distributed a paper to the Space Council, pondering post-Apollo objectives and concerns in a climate of détente. Therein he identified a "Twofold International Objective" for the 1970s. Rusk first urged that the United States take action to "de-fuse" the space race between America and the Soviet Union. Doing so would not simply eliminate the hypothetical waste implicit in competition, but it would also thwart the sense of exclusivity and alienation imparted upon nonparticipants (i.e., Europe and the developing world). Second, he advised that for both the "technically unsophisticated as well as industrially advanced countries, the role of active participant offers a better route to awareness and understanding—*and responsible conduct*—than the role of passive beneficiary."[30] For Rusk, collaboration in space was never to take the form of "foreign aid."

White House officials harbored high hopes for remote sensing in particular, predicting that it would do more to establish the theme of using space as a resource for mankind. Earth resources surveying satellites, which we are now developing, should be of special help in this regard and open new routes to cooperation. By emphasizing such activities, we can not only help bridge the "have" versus "have not" gap but also begin the transition away from a race deeper and deeper into space toward a more (but not exclusively) earth-oriented program.[31]

In order to meet this objective the paper suggested educating and enlisting Western Europe and the developing world in space exploration. This alone would bridge the "technology gap" that loomed between the so-called space powers and others. The report continued, explaining that it was the United States' responsibility to enlighten budding or potential space powers: "It is even more difficult for technically unsophisticated countries to grasp the meaning of changes now in train. *Yet their reactions will be important if the international adjustment to these changes is to be responsive to our own interests.* Accordingly, we will need to use our programs still more effectively to broaden the base of cooperation."[32] The point bears repeating: "Broadening the base of cooperation" not only provided additional data to networks or instruments to satellites to satisfy the demands of globally oriented programs. For some, multilateral partnerships were viewed as a method to sway international sentiment, aiming to yield coalitions more responsive to superpower interests and build institutions of space research and development that exhibited values complementary to those of NASA.

Atmospheric Sciences: The GARP Years

Meteorology has a dual character as a public service and as a branch of the physical sciences which leads to its peculiar position as a part of both operational systems as well as basic research.
—1977 Report to House of Representatives[33]

The year 1968 brought big changes to the World Weather Program (WWP): Soviet and American satellites went operational that year and with that, member nations began to participate in the first of many regional observation experiments (the Global Atmospheric Research Program or the GARP). GARP experiments yielded basic data for atmospheric research, which was then applied to numerical models for computer forecasting. Morris Tepper, a key participant, claimed that "numerical prediction was made possible by the Global Atmospheric observations that the NASA program developed."[34]

In order to advance weather modeling, one must not only observe global phenomena on a daily basis, but also study with greater rigor seasonal occurrences or regional anomalies. Thus, WWP planners used the joint infrastructures of WWW and GARP to merge satellite data with conventional synoptic ground data. From the launch of the first experimental weather satellite in 1960 until both the United States and the USSR had operational systems up in 1967–1968, scientific understanding of meteorological phenomena became increasingly sophisticated, and models more detailed, due in part to data-sharing between the United States and the Soviet Union.[35]

Researchers took an increasingly systemic approach to meteorology and in so doing, plugged in an ever-growing number of variables that were lacking

Table 7.1   GARP regional experiments

Barbados Oceanographic and Meteorological Experiment (BOMEX)	Obtain observational data on exchange of energy, momentum, and water vapor between ocean and atmosphere	May–June 1969
Complete Atmospheric Energetics Experiment (CAENEX)	Study exchanges between the kinetic energy of the ocean and atmosphere	Summers 1970–1972
GARP Atlantic Tropical Experiment (GATE)	Analyze role of convective tropical systems in global circulation	June 15–September 30, 1974
Air Mass Transformation Experiment (AMTEX)	Study transformations of air moving from cold land over warm water	February 14–28, 1974, and February 14–28, 1975
Monsoon Experiment (MONEX)	Examine mechanics of monsoon circulation	January–February 1979 and May–June 1979 (both planned)
Joint Air-Sea Interaction (JASIN)	Analyze interaction between oceans and atmosphere	July–September 1978 (planned)
Polar Experiment (POLEX)	Examine role of polar regions in global energetics	January–February 1979 (planned)

in weather models. The atmospheric physics of the poles, ocean currents, and temperature ranges, as well as seasonal phenomena such as monsoons and hurricanes—each of these fields of knowledge demanded a more refined understanding of the Earth's atmosphere and oceans (table 7.1).

GARP's multilateral programs depended on an extensive mix of scientific instruments including sounding rockets, automatic weather stations, balloons, weather ships, and the newly developing weather satellites and computers. Such cooperation—often predicated on the agreement to merely observe the same phenomena from different vantage points and instruments—precipitated scientific advances that would otherwise have been impossible without a global assemblage of tracking stations and Soviet-American willingness to share satellite data. Nevertheless, American researchers and technologies dominated GARP research and also took unquestionable initiative in the formation of other spin-off international programs.[36]

By developing satellites and supporting networks, NASA officials bore considerable responsibilities to the NOAA and the WWP. As satellite systems engineer, NASA developed, procured, constructed, and insured Command and Data Acquisition stations. As government launcher, NASA selected and procured launch vehicles while maintaining launching sites. Even after launch, NASA tracked orbit through the entire useful life of satellites.

In times of malfunction, NASA staff monitored and commanded satellites, or simply made themselves available for consultation.[37] Together these responsibilities made NASA fundamental to the development of several overlapping fields of global atmospheric sciences. These included, but were not limited to meteorology, oceanography, and seasonal events such as hurricanes and monsoons. Working alongside NOAA, NASA helped construct the bureaucratic and technical infrastructure necessary for the development of global participation in—and benefit from—the WWP.

NASA's Morris Tepper, deputy director of Earth Observation Programs and director of Meteorology at NASA (Office of Space Science Applications), wrote to the executive director of the NOAA in 1972, enclosing a statement on NASA's maritime and meteorological programs in the coming decade. Tepper framed their relationship as one governed by NOAA's leadership. As "national meteorological representative" NOAA provided NASA with its specifications for all meteorological satellite observations. Embedded in this NASA-NOAA partnership lay numerous national and international demands, "requirements of a global nature." In the WWP, NOAA and NASA partnered with assorted national and international scientific organizations to produce what he described as "requirements on an international basis and areas of international cooperation."[38] Partnership with the Soviet State Committee on Hydrometeorology and Weather Control—in the form of satellite telemetry from their Meteor-1 and Meteor-2 satellite networks—remained the most important linkage to operation of this system.

Cooperation with the Soviets: Earth Resources, Weather, and the Environment

As envisioned by NASA-Hydromet-Weather Bureau planners in the early 1960s, a minimum of two polar-orbiting satellites provided by the United States and the Soviet Union would support GARP. Through the course of the 1970s NASA provided two generations of polar-orbiting satellites: the Tiros Operational System (TOS) and the Improved Tiros (ITiros). The Soviets made incremental upgrades to the Meteor system until 1975, when they introduced the Meteor-2 system.

American researchers and the press were quick to point out that Meteor satellites tended to have shorter functional and design lives than the US systems (six as opposed to twelve months).[39] One trade journal reported that although the Soviets were satisfied with the operation of Meteor satellites, the United States was not. It went on, explaining, "US dissatisfaction with Soviet meteorological data was expressed recently by NASA administrator James Webb, who noted that the Soviets were not living up to the agreement on exchange of data."[40] Although NASA officials described photos as "excellent," data tended to be delayed 24–48 hours (too late for use in weather prediction) and when they arrived, they were not gridded, adding to the difficulty of using them in a timely manner.[41] Were NASA representatives dissatisfied with Soviet technologies, management, or both?

In spite of lags in development and frequent lapses in data dating to the beginning of cooperation, the Weather Bureau (later Environmental Science Services Administration and finally, National Oceanic and Atmospheric Administration) continued to use what data was sent over from Moscow. Meantime, Meteor developers took credit for discovering several jet streams—many in the United States—as well as documenting hurricanes along the American east coast.[42] Soviet officials estimated that fishing and merchant ships saved 5–7 percent cruising time because they were able to map optimal routes according to weather forecasts.[43] In 1975, maritime savings amounted to approximately one million rubles ($1.35 million) annually and satellites had been crucial predictors of four hurricanes in the

Indian Ocean.[44] While it is difficult to evaluate the reliability of these Soviet claims, they do demonstrate the cultural and political significance attributed to the Soviet weather satellite program. In 1970, *Science* reported that both the United States and Soviet Union were still "heavily involved in GARP" with both nations orbiting Automatic Picture Transmission devices on their satellites.[45]

In 1975, the Soviet Hydromet committed to launching a geostationary satellite (along with the United States, Japan, and the European Space Agency) for 1977–1979 GARP experimentation, but were forced to withdraw the offer in 1978, due to technical difficulties.[46] Regardless, the polar-orbiting Meteor satellites continued to provide useful coverage of roughly two-thirds of the globe, including information on cloud cover, ice formation, radiation, weather fronts, and jet streams.

As evidenced by a broad array of periodicals—newspapers, trade journals, and the like—prestigious, cutting-edge projects such as GARP attracted considerable attention both in the public eye, as well as in professional fields. Morris Tepper, in recalling his work in developing NIMBUS satellites, explained, "By the time we're talking in this period, the Global Atmospherics Program was getting very hot. I was interested in providing the data to that Global Atmospherics Program that they could use in order to globalize the numerical prediction models."[47] Later, he suggested that the linking of the peaceful uses of outer space and meteorology was "an obvious thing and that's why GARP, which was to provide data to all the research groups all over the world, would be an excellent indication that NASA's providing data to the world in terms of peaceful uses of outer space."[48] Others feared that NASA's role in the day-to-day operation of programs such as GARP was at times underappreciated. Noted NASA's 1967 assistant administrator for policy analysis, "in the process of transferring space applications systems to user agencies the 'space identity' [i.e., NASA identity] is lost." Thus, he explained, "the general public does not fully appreciate this important continuing aspect of the NASA program. A clear recognition of our unique role in developing space systems for user agencies should be one of our important objectives."[49] This concern was particularly dire in the uncertain climate of the late 1960s, when NASA funding for even the Apollo program begged repeated justification.[50]

The Apollo-Soyuz Test Project, 1972–1975

In the summer of 1972, President Nixon and Soviet Premier Kosygin signed the Summit Agreement Concerning Cooperation in Outer Space for Peaceful Purposes, in which, among other fields, they agreed to engage in a joint training exercise and experimental docking of their spacecraft: the Soyuz and Apollo capsules.[51] This, the Apollo-Soyuz Test Project (ASTP), for a few years functioned as what Administrator Fletcher described as a "major visible space accomplishment," the likes of earlier Apollo missions and Skylab, or the upcoming Viking and Shuttle projects.[52]

With their administration a mere 14 years old, NASA staff had the foresight to consider documenting the history of ASTP even as Nixon and Kosygin were still in summit.[53] In April 1974, Edward Clinton Ezell and Linda Neuman Ezell went to work. Stalking the halls of joint meetings, sharing coffee with NASA staff, and chatting over photocopiers, the two wove personal interviews, "desk archives," technical data, and a flood of NASA's internal correspondence into a rich account of ASTP. With this monograph, Ezell and Ezell explain the operation of the joint working groups and the day-to-day engineering, the reflections of NASA and Russian engineers, as well as the activities of astronauts and cosmonauts in space.[54] They illustrated the complexity not only of executing the project, but explaining its programmatic justifications.

In the summer of 1973 the chairman of the House Committee on Science and Astronautics, Olin Teague, contacted NASA administrator James Fletcher regarding Apollo-Soyuz. Teague, well aware of NASA's difficulty in getting sustained support from their Soviet partners, wondered if it might be possible to add more scientific experiments to the payload, "making a justifiable, independent, scientific and technological contribution," even without the Soviet Academy of Sciences.[55] Recognizing that the Apollo-Soyuz mission was impossible without a Soyuz and its docking module, Teague went on in subsequent letters to explain that the "American public" must be well served by a productive US-only mission in the event of a Soviet pull-out. Teague offered his advice: "I believe that many of the alternate experiments identified by NASA are of sufficient importance to fully utilize the payload capacity of the vehicle." If not more science, Teague determined that an additional trip to Skylab, "would also seem prudent."[56]

With a budget of $10 million for experiments, and limits on both weight and volume, Fletcher agreed that the existing Apollo payloads may not fully justify a US-only flight. With an eye on the already tight budget, NASA was investigating "several possibilities" for increasing payload weight and volume.[57] Fletcher anticipated that his staff would complete the study in June and offered to brief Teague then.

Fletcher's notes indicate that this correspondence resulted in an October 2, 1973, Congressional hearing. Within two weeks, Teague wrote Fletcher, stating that the hearing had been "most productive," but reiterated concerns that NASA might do more to assure that alternative experiments or a visit to Skylab be used to justify the expense if the Soviets backed out.[58] Over the coming months, the two, with occasional interjections from (intended) Command Module pilot Jack Swigert discussed the possibility of adding a number of elements to the mission.[59] They considered earth observation experiments (in the fields of geology, hydrology, oceanography, weather, global tectonics, and atmospheric sciences) and even the possibility of including another camera, intended to help petroleum geologists who were at that time dealing with the oil crisis.[60]

Ezell and Ezell indicate that "George Low looked at the entire project from a political perspective" and, therefore, considered that a mission without Soviet participation was not practicable. However in the fall of 1973, Low explained that the $10 million budgeted for experiments *was* already enough to justify launch. In the October 2 testimony, Low had stated, "That is how the $10 million were arrived at. You asked the question, what would we do if the Russians for some reason were unable to fly with us, political, technical, or otherwise, and would the mission in itself with the $10 million worth of experiments…be worth flying." Low responded with candor, saying that that depended on how early NASA might be notified of a Soviet cancellation and more important, how much of the full $250-million budget had been spent.[61]

Furthermore, Low reasoned that if the Soviets and/or the American public heard too much of alternative missions without the Soviets they might begin to question the viability of the scheme altogether. "I think," Low stated, "that would be something that could be very easily misunderstood from the point of view of the other side if you started to plan what you are going to do if this mission doesn't happen."[62] In a booklet titled "Notes for Meeting Congressmen Teague and Dr. Fletcher," NASA staff explained (presumably to Teague) that ASTP was unlike previous Apollo missions that were based more upon scientific payloads. Contrasting Apollo 15 and ASTP, the booklet states,

> ASTP is primarily devoted to proving out the docking system. ASTP is carrying a good complement of scientific experiments but of more significance, I believe, ASTP is conducting experiments in space that could be the precursor of derivative applications in future manned spaceflight. Consequently, although the ASTP experiment payload is not exclusively devoted to science...it is a good viable package that has a potential of increasing the value of the use of space.[63]

Returning to Ezell and Ezell, "None of the alternatives seemed as desirable as the basic idea of a joint mission...It was a gamble, but the risk seemed to be a reasonable one."[64]

And it was a gamble. The Soviet partners tended to operate "shrouded in mystery," in Ezell and Ezell's own words. Paperwork moved slowly, but Fletcher, like Low, remained optimistic. A Senate Briefing Book dating from 1973 contrasted Soviet performance in the space sciences with ASTP, human spaceflight versus robotic, stating that Soviet performance had improved. Indeed, past experiences in the sciences did "not match the positive, businesslike approach Soviets have taken to ASTP nor the detailed information exchange in ASTP."[65] When representatives of the American press contacted Fletcher, disputing their exclusion from Soviet centers, Fletcher responded, pointing out that with the exception of "independent activities," the US press was welcomed to all ASTP proceedings. Additionally, news coverage of ASTP was by far the most liberal access ever to the Soviet space program. Never before had the Soviet public viewed such activities live. Fletcher added that this openness had been extended to the American news media as the "most comprehensive release ever" of real-time information related to a Soviet space mission."[66]

By the same token, historian Asif Sidiqi notes that with ASTP (or as it was called in the USSR, the Soviet-American Apollo-Soyuz Experimental Flight) an increasing number of key space complex officials became public figures. These disclosures, however, came alongside the standard fare of propaganda. Stakes were higher in the 1970s for the Soviets. In the public sphere, criticism mounted and due to a significant decline in the communist economy, citizens were less likely to be "vocally in favor" of the space program.[67]

ASTP, Fletcher, and NASA's "Balanced Program"

James Fletcher, too, was coping with a weakened national economy and likewise anticipated that the ASTP might function as a public relations windfall. In the years of ASTP planning, Fletcher's personal papers reveal a time of intense reflection on the operation and direction of NASA in the long run. ASTP held a crucial role in Fletcher's NASA and his vision for a long-term balanced space program. Communicating with President Nixon in 1973, Fletcher identified ASTP as one of several long-lead time "major visible *space accomplishments*" such as Skylab, Viking, and the Space Shuttle.[68] He suggested the programmatic complement to this would be a collection of short-lead time projects with "earlier practical return" such as remote sensing for earth resources, agricultural yield, forest preserves, hydrology, and minerals. Weather satellites, which "must be an international endeavor," may "in the long run have the biggest impact of any direct application satellite," he postulated. Regarding the environment and pollution studies, Fletcher observed a "growing interest both in this country and abroad for a move" in the direction of a global environmental monitoring system.[69]

Due to what Fletcher perceived as temporary budget shortages, he trusted that remote sensing and robotic exploration would sustain NASA (and the public's need for "a morale boost and an increased confidence in themselves") until the long-lead time Shuttle was operational and the federal budget had recovered. ASTP, the Shuttle, and Europe's Spacelab (see chapter 6) were crucial investments in the future of human exploration and Fletcher opined that "we should leave open the option of returning to the moon to establish permanent bases or to pursue further scientific investigations" or even a manned exploration of Mars.[70]

Thus, 1973 was something of a crossroads. Writing Roy Ash, director of the Office of Management and Budget (OMB), Fletcher indicated an understanding of the logic behind the mid-1960s budget cuts that accompanied the phasedown of heavy Apollo requirements. Yet, the trends that concerned Fletcher were temporary spending cuts turned permanent. NASA had made "major programmatic reductions" for FY1973 and 1974, but OMB and NASA were both aware that these cuts were made on the assumption that they were "temporary and that future budgets would again approach the 'constant budget' level" set in 1971 as $3.4 billion.[71] FY1975 "will become decisive," Fletcher predicted, explaining that at that point, it would be in NASA's best interest to forego the Shuttle, science, exploration, applications, or aeronautics, since cuts across the board were no longer tenable. At one point, using the term "balanced program" five times on one page, Fletcher asserted that there was a great deal of support for the current balanced program, but that "without this balance we would lose support for the remaining program in Congress, by the public, and by the scientific and user communities."[72]

In an economic climate that had cut the post-Apollo program short and postponed Shuttle development, ASTP functioned to help preserve the engineering know-how and managerial expertise of the Apollo program into the dawning Shuttle years. ASTP and Skylab might be taken as evidence of Apollo's sustained "vitality" in the US space program, a notion supported by Ezell and Ezell who asserted that in the closing days of ASTP, most staff transferred directly to the Shuttle program.[73]

The shared resources and expenditures implicit to international cooperation rendered it both diplomatically and fiscally attractive. NASA was entering a "new generation of space activity when we are called upon to *do much more with considerably less money*."[74] Whereas the Apollo program had cost $25 billion, the Shuttle was a mere $5.5 billion. "We are going to have to do more for less,"

Fletcher observed and again, ASTP was an important factor in years to come. To Fletcher, ASTP was

> an important step toward long-term, large-scale cooperation with the Soviet Union and other countries, such cooperation is, in my opinion, the *only likely hope in this century for large future steps in space*, such as establishing a base on the Moon or landing men on Mars. If we had to go it alone, my guess is that we would have to wait until the 21st century.[75]

However history and hindsight render a very different—and oftentimes far more critical—narrative of what ASTP has wrought. Through the course of ASTP, George Low and Soviet Academy of Sciences' academician Keldysh consulted one another on possible expansion of cooperation. They wrote of a joint Shuttle-Salyut mission (which used the last Apollo and last Soyuz craft that orbited the earth, and therefore posed no great risk of technology transfer) that would offer a much more meaningful and possibly sustained collaboration. They discussed a joint robotic mission to the Moon, retrieving soil samples from the far side. In 1977, the two nations signed an agreement for cooperation in human spaceflight, designating 1981 as a target year for a Shuttle-Salyut mission and establishing a joint task force studying the *possibility* of a joint space station.[76] Neither of these ever happened, only augmenting the accusations of some that the joint ASTP mission was, from an engineering and diplomatic standpoint, a dead end.

Even Walter McDougall, otherwise relentlessly pragmatic and eloquent in his assessment of the respective space programs, takes pause to observe of ASTP and contemporary manifestations of détente: "None of this did much to hobble Soviet technocracy," he groused. Rather, he asserts, the program "gave Soviet technicians the chance to traipse through US space facilities and study the hardware and flight operations first hand" (paralleling the visits of US engineers to Moscow and Star City). In conclusion, for McDougall, cooperation such as ASTP was nothing more than a "double boon" to the Soviets, appearing to restore their space program to an equal of the United States and "also provided access to American technology."[77]

In light of these plans that came to naught, a critic with an eye on human spaceflight alone (and not biosatellites or the rich field of remote sensing) might otherwise look to the years following ASTP as a "lapse" in cooperation *altogether*. While ASTP had debatable long-term positive influence on the American end (from the perspective of funding, follow-on projects, or perhaps even public relations) it does to some degree function as a foreshadowing of a warming and loosening of relations at personal and middle-managerial levels. These notions are explored in the pages that follow, under the 1982–1984 "lapse" in cooperation.

Biosatellites, 1974–1982

In 1974, Moscow's Institute of Biomedical Problems (MIBP) demonstrated a rare show of Soviet initiative in cooperation, inviting NASA's Ames Research Center (ARC) scientists to fly experiments on their 1975 Cosmos Biosatellite 782. Over the next several years, the Soviet Academy of Sciences launched these satellites at roughly two-year intervals, leading to a total of nine satellites carrying over one hundred US-led investigations into the effects of the space environment on biology and medicine. Principle investigators represented a number of research institutions and space programs from the United States, and East and West Europe.

In the interest of streamlining red tape and simplifying technological interfaces, US and Soviet space program officials maintained their policy regarding one another's hardware. This meant that American experiments operated on independent platforms. Sticking to the doctrine of clean interfaces, neither electricity nor data-recording were supplied by Soviet hardware.

Additionally, the two parties exchanged no funds through the course of experimentation: the MIBP would fund and command all activities associated with launch and reentry whereas the Americans would pay for all hardware and development costs for their biomedical experiments. These conditions changed in 1992–1993 following the organization of a Russian civil space program when Moscow asked NASA to cover half the expenses of launch—roughly $16 million (see chapter 8).

The Soviet Union began launching these satellites in 1973, identifying them by either their Cosmos nomenclature or their Bion number as laid out in table 7.2. Bion (a contraction of Biological Photon) satellite payloads were designed by the MIBP and carried an estimated design life of approximately 30 days in orbit. Being recoverable spacecraft, the Bions were a derivative of the Zenit reconnaissance satellite (used since 1961) and before that, the Vostok recoverable spacecraft (in use since 1960).

The Intercosmos space council invited participation of East European scientists from the beginning, but until November 1974 had not included the United States. At the fifth meeting of the Soviet-American Joint Working Group on Biomedicine, representatives of Moscow's Institute for Biomedical Problems shocked Ames researchers by inviting experiments from the United States. Following procedures drafted in the 1971 Agreement on Space Sciences and Applications (which was renewed in 1974 and 1977), Ames Research Center functioned as the manager of American participation.

Until the Cosmos 782 mission, cooperation between the two had been limited to sharing data, joint conferences, and publications. After Cosmos 782 landed in December 1975, it supplied data and specimens to researchers in Czechoslovakia, France, Hungary, Poland, Romania, the United States, and the Soviet Union.[78] In 1977 Cosmos 936 carried an impressive array of instruments sent from two more nations: Bulgaria and the German Democratic Republic.

Kristen Edwards points out that the Soviet Academy of Science's offer came at a fortuitous time for American biomedical researchers, in light of the fact that in 1969 NASA cancelled all biosatellite research.[79] Following the disappointment of an aborted program on Skylab, NASA bioscientists would otherwise have had to wait until the Shuttle was operational for space access.

Joint work on biosatellites stipulated that Americans design and build their instruments within the predesignated specifications of the Cosmos biosatellite, ship all materials, and when necessary train their Soviet counterparts in the use of such devices.[80] Edwards explains that the first invitation for US proposals placed NASA scientists on a tight schedule: experiment descriptions due to the USSR by December 1974, experiment hardware due by August 1975. In the

Table 7.2 Cosmos biosatellite flights

Cosmos/ Bion	Flight	US Payload	# US Experiments	Notes
3/782	1975; 20 days	25 rats, fruit flies, carrot tumor tissue, fish embryo	11	Fish egg experiment follow-on from ASTP
4/936	1977; 19 days	30 male rats, fruit flies	7	
5/1129	1979; 19 days	32 male rats, 5 female, quail eggs, carrot tumor tissue, cells	14	
6/1514	1983; 5 days	2 rhesus monkeys, 10 female rats, 30 male, quail embryos, carrot cell cultures	5	Planned and executed during lapse in agreement
7/1667	1985; 7 days	2 rhesus monkeys	1	Planned during lapse in agreement
8/1887	1987; 13 days	2 rhesus monkeys, 10 male rats	26	
9/2044	1989; 14 days	2 rhesus monkeys, 10 male rats	29	
10/2229	1992; 12 days	2 rhesus monkeys	7 US life science investigations	Last time Soviets shoulder cost of launch
11/	1996–1997	One rhesus died—US Cong/NASA cut funding	8 US life science investigations	Cong would have to approve 50% primate costs
12/ –	Never flew			Cong would have to approve 50% primate costs

Source: Assembled from information at: http://lis.arc.nasa.gov/lis/Programs/Cosmos/overview/Cosmos_Biosat.html; http://www.astronautix.com/details/cos21763.htm; Rodney Ballard and Karen Walker, "Flying US Science on the USSR Cosmos Biosatellites," ASGSB Bulletin 6, October 1992, 121–128; Kenneth Souza, Guy Etheridge, and Paul Callahan, eds., *Life into Space: Space Life Science Experiments Ames Research Center Kennedy Space Center 1991–1998*, NASA/SP-2000-534.

meantime, NASA life scientists engineered and constructed experiment hardware. Edwards points out that these machines were passive specimen modules fitted within containers measuring half a cubic foot. These "passive" modules functioned autonomously from the Soviet spacecraft, neither drawing electrical current nor making use of the Soviet data recording systems. Circumventing Soviet hardware and personnel hinged on matters of security and the fact that communications regarding Bion "were not always sufficiently open due to security concerns *in both countries*." Perhaps most important, the use of passive experiment modules eased anxieties regarding technology transfer—for Soviets and Americans alike.[81]

After payload elements were developed and tested, Soviet engineers took responsibility for system integration and testing of the overall spacecraft. As of 1992 (and likely before) persons in the USSR took charge of all animal training and biocompatibility testing. Soviet mission personnel took complete charge of launch activities. Following reentry, they coordinated the post-flight procedures between recovery sites and mission headquarters in Moscow.[82] Only when specimens were back in Moscow did NASA's ARC personnel take over their experiments.

Robotic and Human Spaceflight

Staff at Ames crafted several Cosmos biosatellite experiments specifically to complement projects on US human-rated spacecraft, beginning with the Apollo-Soyuz Test Project. James Connolly, chief of the Payload and Facilities Engineering Branch of the Life Sciences Division, functioned as ARC's project manager on Bion satellite experiments from 1986 through 1993. When interviewed, Connolly recalled some of the pros and cons of flying instruments on human or robotic craft. For one thing, "you have a lot more paperwork on a Shuttle mission," he explained, due to the safety considerations for astronauts. Bion satellites also had a quick turnaround: whereas the 3–4 year lead time on a Shuttle mission afforded advantages for more complex instrument and experiment development, the average Bion satellite permitted only 12–18 months preparation time (allowing for quicker turnover or faster revisions to studies). In the end, ARC staff found that they could use Soviet biosatellites as something of a test bed for Shuttle instruments. Connolly elaborated: "One advantage that we saw in the Cosmos program, as compared to the Shuttle, was that we could acquire technology components, do proof-of-concept development of a system, fly it, and then transition it into a Shuttle mission if the opportunity presented itself."[83] Looking to the future of biomedical cooperation, Connolly predicted that transferring experiments to the International Space Station would pose an entirely new set of demands on ARC equipment, having to function in space for long-duration flights of roughly 90 days (as opposed to the two- or three-week runs on Bion or the Shuttle). "On the Shuttle, we don't even consider changing out a filter. We have done some inflight refurbishment of water supplies and, of course, there were animal food change-outs that we dealt with in shorter flights." Perhaps these considerations contributed to his preference for robotic craft: "I'm in favor of as much automation as you can get," Connolly explained. Automated experiments allowed for greater consistency in operations and when sent on manned missions, require less attention from crews. Although automated missions accelerated the rate of experimentation and eliminated a considerable amount of red tape, biosatellites did have their costs.

For the most part, materials and organisms could only be viewed on Bions, not manipulated. This meant that in the event of a malfunction, it was nearly impossible for investigators to repair equipment. In spite of the scrupulous quality control and the necessity for high-reliability hardware to overcome such risks, the flight of nonhuman spaceflight experiments placed a significantly smaller burden on NASA budgets than did manned.[84] This relatively low-budget ceiling (paired with an equally low profile in the public eye) might well have made it possible for Bion cooperation to continue, even after NASA/Soviet Academy of Sciences 1977 Bilateral Agreement in the Peaceful Uses of Outer Space lapsed in 1982.[85]

Carter, China, and "Inducing Soviet Flexibility"

NASA and the Soviet Academy of Sciences signed the 1977 Bilateral Agreement in the Peaceful Uses of Outer Space as diplomatic relations were unraveling rapidly at the state level. In the late 1970s, President Jimmy Carter observed Soviet human rights violations against the Polish Solidarity movement with increasing frustration. This, coupled with involvement in conflicts in Ethiopia, Angola, Shaba, Yemen, Cambodia, Cuba, and Iran all reached a climax with the December 1979 invasion of Afghanistan. Cold War historian Odd Arne Westad characterizes Carter's response as that of "an activist president who was determined to make the Soviets pay a high price for their invasion of Afghanistan."[86] The Carter administration retaliated on a number of diplomatic fronts: recalling their ambassador, boycotting the Moscow Olympics, suspending the Senate consideration of SALT II, discontinuing various cultural and economic exchanges, restricting fishing rights in US waters, effecting an embargo on high-tech exports to the Soviet Union, and, most alarmingly, cancelling a 17-million-ton shipment of grain.[87]

At this time, President Carter flirted with capitalizing on Nixon's advancements in China to isolate and embarrass the Soviet Union as much as possible. Pondering cooperation across a broad spectrum of activities including space and nuclear energy, the Carter administration sought to reinforce diplomatic relations with the People's Republic of China. American technologies, together with scientific cooperation, were intended to "serve as a positive and constructive force in deepening US relations with the People's Republic, exerting influence on the PRC's future domestic and international orientation and, perhaps, moderating Soviet foreign policy conduct." In particular, scientific and technological exchanges stood to "place the USSR on notice that provocative Soviet behavior could stimulate increasingly intimate Sino-US ties with security overtones."[88]

In the fall of 1978, the president's Policy Review Committee met regarding science and technology programs with China. Acting on the president's instructions that they "move ahead" with student exchanges, technical aid in the field of energy, and space, the committee communicated a few suggestions. In particular, they noted that the Departments of State and Defense, the Central Intelligence Agency, and NASA each agreed that the United States could consider "allowing the PRC to procure" two 12-transponder C-band Westar Class satellites "from US industry under carefully designed controls that would limit undesirable technology transfer and unfavorable domestic and international reactions."[89] The satellite would be purchased and delivered in "turnkey" condition—that is to say, in geosynchronous orbit. Though no satellite hardware would enter the PRC, the committee did allow that US tracking-telemetry-control ground system technologies would have to be exported. As of negotiations in 1978, the Chinese would "pay all costs associated with activities which benefit them," and the United States would do likewise.

Up-to-date geosynchronous telecommunications satellites were, in Science Advisor Frank Press's opinion, the "definitive test of future US-PRC scientific and technological relationships." Carter also considered PRC interest in acquiring a Landsat ground station, capable of receiving multispectral data from the 1981 Landsat-D thematic mapper.[90] At the same time, the Department of Commerce began meeting with counterparts in the PRC discussing possible fields of scientific collaboration including metrology, meteorology, oceanography, fishery research and management, data center management and data interchange, and patents.[91] The committee acknowledged that these actions were calculated specifically to "help induce Soviet flexibility." Regarding the so-called Soviet-American Factor in Sino-American cooperation, the Committee reported:

> In [Soviet] propaganda they condemned the Frank Press visit and they can be expected to cast specific projects in the worst possible light. Yet, the prospects of expanded S&T contact may have helped induce Soviet flexibility. Clearly, they will be especially sensitive to any Sino-US collaboration which they see as enhancing the PRC's military capabilities vis-à-vis the Soviet Union.[92]

While this "turnkey" export of satellites to China never came to be, it does illustrate the lengths the Carter administration would consider.

The 1982–1984 "Lapse": Navigation and Rescue, Bion, and Atmospheric Science

The 1977 Bilateral Agreement on the Peaceful Sharing of Outer Space lapsed in 1982, not to be renewed until President Reagan signed a 1984 Joint Resolution of Congress, Public Law 98–562.[93] However, the end of détente did not dictate that the world's two leading space powers would resort to unilateral or bifurcated multilateral space policy in toto. In some respects, the two nations continued their tacit competition, such as maintaining leadership roles within their respective blocs of communications satellites. For the biosciences, atmospheric sciences, and navigation and rescue satellites, this period was more a time of "business as usual." Often the execution of these programs depended on the tenacity of a few key individuals.

It was at this time that the international search and rescue programs COSPAS (including the Soviet Union and several allies) and SARSAT (including the United States, Canada, and France) united. In addition to continuing research in space biology and medicine, NASA continued with planetary data exchanges on programs such as the exploration of Venus, solar-terrestrial physics, and the exchange of lunar samples and cartographic data.[94] In the meantime, researchers at the Fermi Institute sent their Dust Counting and Mass Analyzer (DUCMA) to Halley's Comet in 1983–1984.

A few nongovernmental initiatives took place among research institutions based in the United States and USSR. These included the execution of an agreement between the California Institute of Technology and Moscow State University. Dating back to the late 1970s, this agreement carried out joint work in gravitational physics—roughly 30–40 percent of which dealt with space-related fields such as the design of a gravitational wave detector.

In 1985 and 1995, the Office of Technology Assessment (OTA) released its reports *US-Soviet* and *US-Russian Cooperation in Space*. Therein the assistant director of the OTA, OTA's director of international security and the space program, and NASA's director of international relations labored to make sense of the rapidly changing field of international collaboration in space. Among a broad spectrum of policy concerns, the OTA took some time to reflect on the anomalous nature of collaboration following the 1982 lapse in bilateral cooperation. Rather than indicate a "lapse," the OTA reported a sort of premature *glasnost* setting in among researchers and policymakers. Life science in particular encountered an improvement in institutional relations. Before 1982,

> [t]here were significant difficulties in acquiring information on mission plans, and in obtaining accurate and complete scientific data. These problems varied in severity through time and across different fields. But workshop scientists believed that the situation was improving noticeably, with regard to both openness and data quality, when the intergovernmental agreement expired...At that time, US scientists were for the first time being taken into Soviet laboratories and shown instruments, performance data, etc.[95]

The authors continued, distinguishing between individuals and bureaucracy, "While recognizing more openness on the Soviet side," "scientists stress the still essentially closed nature of Soviet scientific and technical programs, and the difficulties Soviet scientists may have working through their own political bureaucracies."[96] The OTA report of 1995 supplies an intriguing list of cooperative activities over the so-called Soviet-American hiatus. Indeed, researchers in both nations exhibited an unmistakable determination to cooperate—with or without a state mandate.

Human relations played a significant role in the sustainability of biosatellites. In her history of the Cosmos Biosatellite Program, Kristin Edwards details how the low level of personnel turnover—in the NASA ARC as well as Soviet Institute for Biomedical Problems—fostered a level of personal trust and respect that spanned decades. In Moscow, Dr. Yevgeni A. Ilyin managed each Cosmos mission from its beginning in 1973 to its end. At Ames, internal promotions constituted the only major changes to personnel.[97]

NASA commenced with negotiations for two Bion missions over the course of 1982–1984. In 1983, Moscow's Institute for Biomedical Problems, partnered with space programs in the United States, Czechoslovakia, France, Hungary, Poland, Romania, and the German Democratic Republic, launched Cosmos 1514. Researchers held the 1983 Cosmos mission in especially high esteem since this would be the first such satellite to carry primates—two rhesus monkeys.

Immediately thereafter, negotiations commenced on the next mission, Cosmos 1667, which was slated to carry on biomedical research and experiments regarding extraterrestrial radiation. By the time the Soviets had launched 1667 in July of 1985, US-Soviet relations had not only warmed, but the Reagan administration had even advanced the idea of simulated space rescue mission between the Shuttle and the Salyut-7 Space Station.

Multinational Cosmos Biosatellite programs continued, essentially in the same vein through 1992's Bion 2229. This, the eighth consecutive mission to carry US instruments, was regarded as the most integrated set up for technical collaboration to date. Notes ARC's history of bioscience: "Russian and American scientists and engineers worked together more closely on Cosmos 2229 than on any previous space mission. NASA developed several flight hardware units for the mission, trained Russian engineers and technicians to operate the hardware, and in collaboration...developed postflight procedures."[98] These measurements included body temperature, electrical activity of the heart, and electrical currents generated in active muscles during space flight.[99] With principal investigators hailing from Mt. Sinai School of Medicine, University of Texas Medical Branch, ARC, University of California at Los Angeles, University of Louisville, University of California, and Davis, this mission still exhibited a drop in participation from the last satellite, which had more than 85 NASA-sponsored researchers from 19 states and 3 foreign countries.[100]

In practice, there were few marked differences between Cosmos 2229 and its successor, the Bion 11. Most notable is the fact that flight hardware on the satellite was the "most highly integrated combination of NASA and Russian systems. These supported research in musculoskeletal, neurovestibular, and regulatory experiments and necessitated a great deal of joint engineering in post-flight ground-based hardware."[101] However these two missions bridge an historically significant shift in NASA-Soviet principles and guidelines for cooperation.

Even before the launch of Cosmos 2229, Americans realized that they would be paying half the expenses of flight on the next satellite. These funds (roughly $16 million), in one respect, covered a portion of the operating expenses necessary for Bion satellites, but also functioned to infuse much-needed capital into the collapsing Soviet space infrastructure. A variety of factors led to the Institute of Biomedical Problems' request that NASA fund half the Bion 11 and 12 missions. In order to explore the circumstances that precipitated this transition in the principles of collaboration, we must first understand Soviet state infrastructural changes that took place in the preceding decade.[102]

The year 1992: Rethinking the Clean Interface and New Objectives

NASA funding is very important to the Russian space program.
—US Congress, *U.S. Soviet Cooperation in Space, 1985*[103]

The possible eradication of NASA's clean interface mode of cooperation with Russia raised a number of difficult quandaries for program officials regarding the relationships among private enterprise, the state, and the tenets of free market capitalism. Between 1990 and 1992, even as debates raged in newspaper editorial columns and on Capitol Hill on whether or not Americans ought to collaborate (more) with the Russians in space, policymakers questioned with whom, precisely, they ought to be negotiating. Often more than one bureau claimed ownership of hardware or intellectual property. NASA officials had difficulty deciphering who precisely was in charge, what Soviet priorities were, and even which assets were up for sale. One report, released in October 1991, illustrates the tenuous situation:

> In his diminished leadership role, Mikhail Gorbachev has had little to say about the future of the Soviet space program...A reorganization has begun involving the major Soviet space design bureaus and installations, some of which will be transferred to new private industries. Yet to be sorted out is the degree of influence and authority key personnel within the reconstituted bureaus, agencies, and industries will have.[104]

Table 7.3 NASA Russian-related activities: Summary of agency programs and costs with the Russian Republic ($ in millions—provided to Congress March 1995)

	FY1995	FY1996	FY1997	FY1998	FY1999
Russian Space Agency Contract	100.00	100.00	100.00		
Mir missions	141.7	102.7	54.3	16.3	.6
Space station-related developments	20.0	20.0	10.0	0.0	0.0
Space science	14.4	10.1	9.2	12.3	6.2
Earth science	3.7	3.1	3.3	3.0	3.0
Space access	2.7				
Aeronautics	11.7	3.0			
Tracking and data	2.3	1.9	2.0	2.1	2.1
Total [761.7]	*296.5*	*240.8*	*178.8*	*33.7*	*11.9*

Source: US Congress, Office of Technology Assessment, US-Russian Cooperation in Space OTA-ISS-618 (Washington, DC: US Government Printing Office, April, 1995).

As indicated here, it was not merely the floundering Soviet state that the American government sought to regulate—it was the engineers who might defect, scientists who may market technical knowledge, or industrialists who might withdraw from weapons compliance. Table 7.3 illustrates the range of fields supported by collaborative work. [105]

Chapter 8 describes how the pursuits of national and collective security figured prominently in a number of US federal agencies and departments. Deliberations regarding the purchase or lease of Russian aerospace equipment (much but not all of it factoring in to the ISS) took place in a variety of US state bodies including NASA, the National Space Council, the CIA, and the Departments of State, Commerce, Defense, and Transportation. Each exercised responsibility over its own corner of national—and international—security.

While the International Space Station figures most prominently among these projects between FY1993 and 1997 the Bion 11 and 12 spaceflights accounted for $16 million.[106] Meanwhile, the space sciences writ large accounted roughly 14.5 percent of all program costs, as detailed earlier. The 1995 Office of Technology Assessment frankly assessed the situation. In his foreword to this report, the director Roger Herdman notes that "much of the motivation for the expansion of cooperation with Russia lies beyond programmatic considerations." In particular, their report points out that continued cooperation, including large payments for Russian space goods and services, might help stabilize Russia's economy and provide incentive for some of Russia's technological elite to stay in Russia.

Often representatives of *Glavkosmos* used the justification of sunk costs to rationalize continued investment. (*Glavkosmos* was the Ministry of General Machine Building's Main Directorate for the Development and Use of Space Technology for the National Economy and Science Research, known as the commercial arm of the Soviet space program.)[107] Thus they carried on the hope that the sale of various elements and subassemblies already developed by the

Soviet space programs might provide foreign currency to the withering program. NASA officials likewise highlighted the thrift of collaboration, but with some important differences. In explaining these expenditures, OTA officials likened Bion to the ISS docking mechanism "and other minor goods and services" that "involve the use of unique Russian capabilities by NASA at a low cost compared with the cost of developing them indigenously." Some believed themselves to be buying or selling products; others believed themselves to be initiating a long-term commitment, a process. These individuals sought to build relationships as buyers and sellers, scientific collaborators, or in preserving formerly Soviet resources.

Many individuals expressed a desire to not simply denude the Soviet space infrastructure of all its useful persons and ideas, but to preserve the organizations and institutional memory within. Regardless of whether or not one views this monumental shift in NASA policy—the decision to pay the Russian space program for hardware and services—as an investment in the Russian space program or bargain for the United States, the OTA leaves us with one final thought-provoking observation. "[N]o other executive branch agency is transferring funds to Russia at anything approaching this rate. US government funds obligated for assistance to Russia through September 30, 1994, total something over $3B, but over a third of that total is for in-kind goods (food shipments, principally in FY 1993)."[108] While foreign policy, environmental, and national security considerations had always played roles in the principles and guidelines of joint projects, the next chapter describes how they were expressed in very different manners. In years past, national security concerns centered primarily upon fears of technology transfer. Following this, notions of "national security" came to be characterized as "international security" as the United States attempted to control the flows of former Soviet researchers and engineers to potential belligerent nations.

Chapter 8

Russian-American Cooperation in Space: Privatization, Remuneration, and Collective Security

As the Soviet Union awkwardly dismantled itself in the early 1990s, NASA policymakers labored to adjust their existing research and exploration initiatives to what was shaping up to be a new world. Having ostensibly won the Cold War, state officials now and again paused to consider the chances of a more enlightened coupling of capitalism and democracy. For some, waning tensions begged an unrestricted reassessment of government, cutting back on half a century's build-up of armaments, infrastructure, and spending. Vice President Al Gore oversaw the streamlining of American bureaucracy before taking the reins of the Gore-Chernomyrdin Commission (for economic and technical cooperation between the United States and Russia). For *both* former Cold War superpowers this cohort sought balanced budgets, smaller smarter government, and improved regulatory practices.

At the close of the twentieth century, new economic and security regimes took shape, carrying the promise of reduced tax expenditures, increased capital flows in the global economy, and the likely inclusion of the Newly Independent States in formerly "Western" multilateral security structures. Tightening budgets reflected a new skepticism of public spending on large R&D and scientific projects such as the Superconducting Supercollider (cancelled by Congress in 1993), the Strategic Defense Initiative (cancelled in 1994), and the Space Station Freedom (which later became the ISS in 1994). At the same time scholars began to seek links between Japanese commercial success and the shrinking percentage of profits being reinvested in American industrial research and development.

This political environment characterized by demobilization, fiscal belt-tightening, and bureaucratic reform combined to produce the curious situation in which the world's leading space powers collaborated for more than two decades, meeting some needs through innovation and others by coasting on the surpluses of Cold War science, engineering, and productive capacity.

In 1991 NASA sent an ozone mapping spectrometer into orbit aboard a Ukrainian *Tsyklon* rocket, originally designed as a Soviet Intercontinental Ballistic Missile.

In 1992 Rockwell, prime contractor for the Shuttle Transport System and Energia, Russian Scientific Production Association (NPO), began retrofitting a docking device intended for the Soviet Buran. It would be used for the American Shuttle's visit to Russian space station Mir.

In 1993 NASA's Space Station Freedom Office considered the possibility of purchasing a Soyuz capsule for use as a space station "life boat." It was later integrated into the International Space Station (ISS) as a Crew Transfer Vehicle.

The year 1994 brought the consolidation of the Western alliance's Space Station Freedom (SSF) and the Union of Soviet Socialist Republics' Mir II plans into the International Space Station.

What follows is an overview of the history and historic significance of Russian-American collaboration in space in the 1990s. The first half contextualizes the two nations' collaboration, considering its intended role in the post–Cold War reordering of international trade, demobilization, and environmental activities. It considers less the micropolitics of how and why NASA retooled preexisting projects and initiatives for collaboration and more how NASA's history dovetails with American foreign policy as it *was intended* to bring order and stability to the former Soviet Union.

The latter half of the chapter focuses less on international activities and more on US interest groups. It is again an overview, illustrating the complex of interests shaping space policy: would importing finished products from Russia come at the expense of American industrial prowess? Should national space program cooperation and liberalized trade be considered an effective preventative against weapons proliferation? To what degree would NASA (and Congress) be willing to reshape their preexisting national policies in the interest of international cooperation? The Gore-Chernomyrdin Commission bundled these oftentimes-conflicting interests when seeking to embed formerly Eastern structures of trade, science, and international relations in the West. Under these agreements, NASA officials labored to craft and often renegotiated agreements with the fast-degrading, but still very proud heirs of the Soviet space program.

Part I: International Interests

Throughout the 1990s, the two nations retrofitted and reengineered launch vehicles, spacecraft, and their support systems for the Shuttle-Mir Program, the International Space Station (ISS), cooperation in remote sensing, as well as the commercial launch of communications satellites. The adaptive reuse of these Cold War artifacts reflected new priorities for the US and Russian governments: the acceptance (or criticisms) and use of these technologies were shaped by concerns for trade liberalization, nuclear disarmament, and Cold War budget constraints. American fears over idle productive capacity and the lingering threat of postwar unemployment at home were coupled with the threat of Asian industrial ascendance in the 1990s.

In marshaling resources for cooperation, proponents of the ISS chose not to approach cooperation as a definitive set of one-time deals or off-the-shelf purchases. Instead, they suggested that the collaborative use and development of space technologies fostered relationships within and among governmental complexes—each in their own way coming to grips with the end of the Cold

War. The international policies of President Bill Clinton and Vice President Al Gore (in some ways an extension of the Bush administration) illustrate that Russian-American cooperation in space was but one element of many fields of postwar cooperation—each intended to foster enduring ties of trade, finance, technical development, and environmental stewardship.

That said, it was not inevitable that the United States and Russia join space programs a scant four years after the fall of the Soviet Union.

For Sale: Cold War Hardware

In the late 1980s, Soviet policymakers identified a new use for space infrastructures. Over time and in certain circles, the Russian space program came to be interpreted not simply as a collection of state assets providing public and defense services, but also as a collection of products that might be sold on the international market. As state finances plunged, the sale and lease of space assets promised cash from abroad.

Soviet space program officials had begun flirting with the notion of selling or leasing equipment as early as 1985. That year the Ministry of General Machine Building (MOM) created the Main Directorate for the Development and Use of Space Technology for the National Economy and Science Research—*Glavkosmos*. This, the "commercial arm of the Soviet space program," emerged from the Soviet industrial complex geared specifically to placing Soviet space technologies on the international market.[1]

Shortly thereafter, at a 1987 symposium for roughly four hundred foreigners, the world market perused many of the same goods and services that were offered for sale again in the 1990s. There, *Glavkosmos* offered microgravity space for rent in the then one-year-old Mir space station, space on returnable capsules, rocket launches on the Proton, entire communication satellites, along with communication satellite transponders. One American was particularly struck by the *Soyuzkart* mapping agency. Remarking on the quality of aerial and space imagery available for sale, he recalled, "[W]e bought what I think was the first print they sold, paying about $800 for a print of an area in Oregon with five-foot resolution—better than anything Landsat or Spot has for sale."[2]

However, this early engagement between Soviet sellers and prospective buyers revealed a limited understanding of late-twentieth-century market mechanisms. The director of Massachusetts Institute of Technology's Space Engineering Research Center (Edward Crawley), and a colleague in MIT's Soviet Space Policy Institute (Jim Rymarcsuk) observed that "the USSR appears to have a limited conception of the financial and decision-making of US firms. The business planning process (including market assessment, capitalization, product development, and marketing) is new to the USSR."[3] The authors noted the tendency of *Glavkosmos* to insist upon the immediate sale of hardware as opposed to entering long-term development agreements that were common to US contracting relations. Driven by a need for hard currency, Soviet marketing resembled "US practices of old" in which the supplier need only assure final functionality of a part, but did not invite user input in design or production.[4]

In addition to this, the marketing of this surplus equipment and space facilities allegedly faltered under a number of American federal controls protecting domestic industry from foreign competition or prohibiting the transfer of defense-related technologies to other nations. Among these were the regulations laid out in the Arms Export Control Act, intended to block the transfer of items falling under the Munitions Control List to communist countries.[5] This Western Bloc embargo dated to 1949, when seven nations signed on, including the United States, Belgium, France, Italy, the Netherlands, Luxembourg, and the United Kingdom, forming the Coordinating Committee on Multilateral Export Controls (CoCom).[6]

As of the spring of 1992, nearly all space-related hardware was included in the US Munitions List and regulated by the Arms Export Control Act and its International Traffic in Arms Regulations (ITAR) (see chapter 14). Additionally, Congress exercised control over the export of space commodities destined for third parties intending to launch American space hardware on Soviet launch vehicles. These protectionist measures were intended to benefit both US national security and the nascent private launch industry.[7]

Thus it came as a shock to many when in 1992, the Bush administration eased into negotiations for a handful of hardware purchases. In the interest of upgrading Strategic Defense Initiative (SDI) systems, the Space Defense Initiative Office considered the purchase of a Topaz 2 nuclear power system and $6 million of plutonium 238, a nonweapons grade isotope commonly used in NASA deep space probes as well as some Defense Department applications.[8] More important, but less publicized, SDI administrators were planning to purchase electric thrusters for station-keeping on a projected 40–60 Brilliant Eyes satellites.[9]

At the same time US firms GE Astrospace and Space Systems/Loral were considering the purchase of thrusters from the Russians. GE Astrospace intended to use four such thrusters (costing $200,000–300,000) for station-keeping on AT&T satellites. Space Systems/Loral considered higher-performance thrusters.

US-Russia Summits and Early Negotiations for Trade, 1991–1992

The year 1991 ushered in a flurry of activity, calibrated to the rapid disarmament of both Soviet and American Cold War era weapons arsenals. At the July 1991 Bush-Gorbachev summit, the two signed the START I treaty, agreeing to cut their weapons base by roughly two-thirds. While meeting, they also signed an Agreement Concerning Cooperation in the Exploration and Use of Outer Space for Peaceful Purposes. As with previous agreements (including 1987, 1977, 1972, 1971, and to a lesser extent, 1962), the agreement charged joint working groups (JWGs) to negotiate cooperation in a number of fields including space biology and medicine, solar system exploration, space astronomy and astrophysics, solar-terrestrial physics, and earth sciences.

As detailed in chapter 7, cooperation between the United States and Soviet Union tapered off considerably in the 1980s, but by no means stopped. Several instruments were being built and flown on host satellites: between 1987 and 1997 a total of 70 NASA life science experiments flew on three Soviet/Russian biosatellites.[10] In August 1991, NASA sent its Total Ozone Mapping Spectrometer into orbit aboard a Russian Cyclone (Tsyklon) rocket.[11] That same year, using a research ship, a plane, and a ground station, Soviet scientists made observations of chemical releases in the Caribbean for the American Combined Release and Radiation Effects Satellite program.

Four themes in the Agreement for Cooperation captured the limelight. For one, this agreement called for preliminary cooperation on the Shuttle-Mir missions. Second, in a revolutionary change in trade limitations, Russia would be permitted to submit a bid for launching one US-built Inmarsat 3 communications satellite (a tremendously important shift in international trade relations, permitting the launch of a US instrument on a Soviet lifter). Third, the United States would contribute a hard lander to Russia's Mars 94 spacecraft. And finally, the two nations would explore the possibility of using the Soyuz-TM as a "lifeboat" for the space station in medical or technical emergencies (also known as an Assured Crew Return Vehicle or ACRV).[12] NASA and the White House viewed these projects as just the beginning of a long-term relationship in space exploration.

In July 1992, one month after the Bush-Yeltsin summit and just three months after taking office as NASA administrator, Daniel Goldin joined a delegation of military, industrial, and scientific leaders heading to the former Soviet Union. Led by NASA and the National Space Council, this interagency delegation included Brian Dailey (new staff director of the National Space Council), Martin Faga (assistant secretary of the Air Force for space), and representatives of the National Security Council, State Department, and Central Intelligence Agency. Following the failure of the Space Exploration Initiative in 1989, the Bush administration was interested in implementing significant changes in NASA functionality. To this end they recruited Daniel Goldin from TRW Space Technology Group, where he had considerable success running space programs using minimal managerial structures and streamlined engineering practices. Goldin's mandate was to implement a top-down change in NASA practices and procedures (in parallel to the space sciences success with "Faster, Better, Cheaper").[13]

Following a visit to European Space Agency facilities in Germany, the delegation went on to a number of institutions in Russia and Ukraine. These included sites geared for both human and robotic space activities including the Russian Flight Control Center (TsUp), NPO Energia, NPO Energomash, Khrunichev, KB Salyut, Babakin Research and Test Center, Lavotchkin, NPO Zvezda, and the Yuriy Gagarin Cosmonaut Training Center (Star City). NASA and NSC representatives met with the directors of the institutes visited. They also met with the heads of the Russian Committee for Hydrometeorology & Monitoring the Environment, the Institute of Biomedical Problems (IBMP—responsible for the Bion satellites detailed in chapter 7), the Ministry of Industry, the Department of Aviation Industry, the Central Aerohydrodynamics Institute (TsAGI), and the Institute of Aviation Motors.

The Goldin-Dailey delegation departed on July 17. On July 18 a second American delegation arrived. It was headed by the director of the Office of Space Commerce (in the Department of Commerce). In the days to come, senior management and engineers from 17 leading aerospace firms visited more than 40 locations in Russia. They met with representatives of design bureaus, scientific production associations, research institutes, and production enterprises to discuss possible joint ventures. Administrator Goldin explained the significance of the second delegation—the US government aimed to cultivate "genuine" partnerships among Russian and US firms. Hoping to thwart accusations that aerospace firms were unnecessary middlemen in state collaboration, he went on to explain that this Department of Commerce trip was necessary for US aerospace companies to assess Russian technologies and that the Russians were planning a similar trip in reciprocation in the hope of finding routes for Russian-American business partnerships.[14]

Transnational projects occupied a minefield of political and economic considerations, making it a shrewd decision for the government and industry executives to travel separately. Might Russian launch vehicles infringe on the budding US commercial launch market? If NASA purchased surplus hardware from the Russians, would savings in taxpayer dollars outweigh the "cost" of engineering and production work lost? Might profits from civil space and aviation joint ventures help US firms weather cuts in defense spending? That being said, was the government merely offering Lockheed, Boeing, Rockwell, and the like new "subsidies" intended to help weather recent defense cuts?

Instead, proponents suggested that the adaptive reuse of Soviet surplus equipment such as the Topaz reactor, electric rocket thrusters, and the docking module originally intended for Buran-Mir missions provided *value-added* work to US industries. At the same time this hardware cut research and development overhead for NASA and the Department of Defense. Dan Goldin's observations, though understandably crafted to appeal to the appropriate audience, reflect the tricky nature of Russian-American business dealings. *Aviation Week and Space Technology* explained that "[a]dministration officials are eager to involve US companies in the cooperative process, but they do not discount the possibility that government funds will flow directly to Russia to purchase space hardware." Addressing the specter of post–Cold War unemployment, Goldin intimated that he desired to be particularly careful that the US civil space program did not add to the woes of the aerospace industry as defense spending dropped precipitously.[15] Rather than (in his own words) simply "ship money to Russia and get back a product," the new NASA administrator suggested that these new international deals could wind up a win-win situation: with taxpayers saving money and US firms acting as prime contractors on retrofit projects concerning Russian machines. "Do we want to make work for Americans, or do we want Americans to do value-added work?" he asked.[16]

In the weeks that the Goldin-Dailey delegation and the Office of Space Commerce industrial commission toured Russia and Ukraine, they observed a network of communities in painful transition. What remained of the Soviet scientific research base writ large were roughly 400,000 public and semiprivate institutions. Nested within complex hierarchies, some were at technical universities, others specialist institutions, and many situated within entire "closed" cities of technical specialization.[17] These included organizations such as the Institute of Microelectronics and High Purity Materials, the Research Institute of Robotics and Engineering Cybernetics, the Moscow Institute of Electronic Technology, and the St. Petersburg Aerospace Instrumentation University, all in need of funding and all in need of administrative direction. Some institutes, such as the Committee for Hydrometeorology, dated to the Cold War era. The Russian Academy of Sciences was reestablished, its origins dating to Peter the Great.

Russian Space Science and Technology

As the Western commercial sector began to tentatively explore former Soviet technologies for sale, assess Russian firms, and evaluate the institutional change necessary to make trade happen, representatives of the space sciences, too, engaged in their share of match-making. Whereas most Scientific Production Associations (NPOs, to the Americans) had been privatized through the course of market reform, many scientific institutes and agencies weathered the transition from state-directed socialism to neoliberal capitalism as government entities. Here are just a few of the key players.

Russian Space Agency (RSA), Rosaviakosmos/Roskosmos (RKA)[18]: Created in April 1992, the RSA functioned as a replacement for many Russian organizations including Glavkosmos, Interkosmos, Intersputnik, and the Ministry of General Machine Building's civil space policy functions.[19] This agency functioned as a rough counterpart to NASA, both a coordinator of space programs and procurer of technical systems. Russian Federation president Boris Yeltsin selected Yuri N. Koptev, a former senior official of the Ministry of General Machine Building (MOM), as agency head.

Managing a newly formed agency, Koptev's administration faced high expectations for performance as well as reform. Academician Roald Sagdeev observed: "[I]t took precisely 35 years to realize that the nation needs a unified organization to run its space program, not in the interests of the military or of the arms race, but in the interests of human kind, international cooperation, science and commerce." In 2001, experts observed that, "[t]he Russian space sector has come a long way." "If you look back ten years the space sector was totally within the military establishment, the so-called military industrial complex, this was, actually, a tremendously successful conversion; it is not complete, but still impressive."[20]

Russian Academy of Sciences (RAS): This institution was reborn at the end of 1991, becoming the default successor of the fast-dissolving USSR Academy of Sciences. As such, the RAS inherited many of Russia's key research institutions and space science organizations including the Space Research Institute (IKI), the Vernadsky Institute of Geochemistry and Analytical Chemistry, and the Keldysh Institute of Applied Mathematics. Until the early 1990s, the academy was NASA's primary partner in bilateral space agreements.

RAS: Space Research Institute (IKI): Directed by academician Albert A. Galeyev (and before that Roald Sagdeev mentioned earlier) and boasting a staff of more than a 1,000, this organization performed research in the fields of planetary physics, space plasma physics, astrophysics, space materials technology, optical studies, and physical studies. Since its founding in 1965, IKI staff prepared space research programs, designed, tested, and operated scientific instruments (including spacecraft), and engaged in extensive international cooperation.

RAS: Vernadsky Institute of Geochemistry & Analytical Chemistry: This institute focused principally on space research concerning geological analysis and mapping. The Vernadsky Institute's best-known work in planetary studies was the Venera Missions to Venus in the 1970s and early 1980s and the Mars 94 program. As a result, NASA staff were considering subcontracting with Vernadsky Institute through Arizona State University, seeking analyses of landing sites on Mars and weathering processes on Venus.

RAS: Intercosmos Council: For decades this council had coordinated international space science projects. However Administrator Goldin's July 1992 Briefing Book indicates that this institution's influence seemed to be waning with the dissolution of the Eastern bloc and rise of the RSA.

As of 1991, the most recent proposal developed by the Academy of Sciences for the development of space industries ("Program 2005") had been around for two-and-a-half years, unacknowledged by the Coordinating Committee for Space Research. This general lack of direction only fueled the demands of scientists and engineers for a more centralized and active command.[21] Though it may be impossible to generalize who welcomed privatization or realignment from statist to democratic operation, throughout this period each government entity and newly minted firm operated under at least one constraint: a precipitous lack of funding. The Russian Federation's revenue deficit, along with a maddening pace of inflation, a desperately weakened tax base, and an inefficient (if not corrupted) supply of financial support, left each organization in dire straits. Officials needed more (and more stable) currency; institutions required steady work in order to remain intact.

Policymakers, industrialists, and the American public alike shared in this awareness. The years 1991 and 1992 brought a flood of coverage in US newspapers, trade publications, and scientific journals, detailing the plight of Russian science and engineering. Workers went without pay while engineers took on supplementary work as taxicab drivers and auto mechanics. The Baikonur launch facilities weathered slowly while many production facilities sat idle.

Anatoliy Petrushin, deputy director for finance at Progress Plant TsSKB (Research and Production Rocket Space Center), explained that in an effort to avoid layoffs, his launch vehicle production facility had begun ersatz diversification. "For example," he pointed out, "we have set up a shop producing disposable syringes. And although only around 300 people work there, one half of the profit earned by the plant last year came from syringe production. Could this situation be more absurd?!" A disheartened Petrushin predicted the end of launch vehicle production for his plant: "Privatization will lead to just one thing: the replacement of space production by something that is short-term and ultra-profitable. The sophisticated equipment will then go out of commission and the plant will go under the hammer."[22]

More disturbing, the American Institute of Aeronautics and Astronautics reported that rocket and space industries in the Commonwealth of Independent States were "simultaneously seeking to exchange space competence for hard currency" and "attempting to convert their manufacturing capabilities to production of kitchen equipment."[23] A year earlier the *Washington Post* took readers to Ukraine's Yuzhny Machine Building Factory where trolley buses and airplane parts had supplanted ICBM production.[24] Yuzhny might best be identified as the facility at which Nikita Khrushchev boasted that the Soviets would "make missiles like sausages."[25] Indeed, what was an unemployed missile engineer to do? Several speculated that weapons scientists in more desperate straits were likely to sell their expertise to developing programs in Iran, North Korea, Afghanistan, or the like.

Even as debates carried on in newspaper editorials and on Capitol Hill as to whether Americans ought to collaborate with the Russians in space, policymakers questioned with whom *precisely* they ought even be negotiating. Often more than one bureau claimed ownership of hardware or intellectual property. NASA officials had difficulty deciphering who precisely was in charge, what Soviet priorities were, and even which assets were up for sale. A report prepared for the New Initiatives Office at Johnson Space Center illustrated the degree of uncertainty, if not confusion:

> In his diminished leadership role, Mikhail Gorbachev has had little to say about the future of the Soviet space program...A reorganization has begun involving the major Soviet space design bureaus and installations, some of which will be transferred to new private industries. Yet to be sorted out is the degree of influence and authority key personnel within the reconstituted bureaus, agencies, and industries will have.[26]

Stakeholders read the situation in different ways. Some called for US government and scientific organizations to send aid to the once thriving scientific and space infrastructures. Financial support might not only help individuals and their families, but perhaps dissuade weapons engineers from defecting, keep scientists from marketing their technical knowledge to "rogue nations," or keep industrialists in line with weapons compliance regulations. Noting that US aid was "but a drop in the bucket compared to the scope of the problems confronting former Soviet science," others begged that there be even the smallest, simplest demonstrations of support. One National Science Board official pointed out the benefits of offering "in kind" assistance in the form of journals, surplus personal computers, technically obsolete lab equipment, or the archiving of research data.[27]

Meantime, assorted critics voiced dismay. Some interpreted the former Soviet Union as an impotent and unstable giant that was best left to its own demise—be it the civil space program, military-industrial complex, or scientific research base. Others remarked on the confusion and limited liquidity that accompanied this rapid and haphazard privatization. Loren Graham, historian of Russian science, acknowledged the Russian state's problems with authoritarianism and corruption, cautioning that "[i]f money goes directly into the hands of directors, it might be slowing the process of reform," ultimately, "enforcing the authoritarian character of the Soviet science establishment that we've criticized in the past."[28] At the same time, some speaking in the interest of national defense questioned the ability of the administration to assure its voters that their tax dollars would benefit *civil* space and not be plowed back into military complexes.

Be that as it may, a coalition was in the making that viewed the weakened economic system as an opportunity to reshape Russian institutions—Americans might provide leadership in postcommunist market reform, ease the conversion to a free market, introduce Russian firms to global business, or produce profitable joint ventures among otherwise downsizing defense firms.

Party lines were not necessarily dependable predictors of behavior. Representative James Sensenbrenner (R-Wisc.) declared that such trade offered "a way to complement each other's civilian space programs in a way that allows mankind to advance, yet provides jobs for both the Russian and the American people."[29] Later he pointed out that cooperation with Russia would provide a way to "nudge" Russia down a "democratic path and [ensure that it] does not lapse into totalitarianism."[30] Senator Al Gore (D-Tenn), chair of the Senate subcommittee that authorized NASA funding, opined that "[t]he taxpayers would like to save some money if we can buy off the shelf some important components that [the Russians] have developed in their space program."[31] In 1993, Dana Rohrabacher—a Republican from California—went so far as to contemplate the hypothetical replacement of the Space Transport System with the Energia Rocket system. Such cooperation with the Russians, he postulated, "will not cost American jobs and will not cost taxpayers for us to work with these new friends and to help cement democracy in what was the Soviet Union."[32]

In the 1992 presidential race, Republicans chuckled over the similarities between the Clinton-Gore platform and George H. W. Bush-Dan Quayle's. "It's a hoot," commented one Bush administration official, who went on to describe Clinton's space statement as being little more than a carbon copy of Bush's. Like Bush policy, Clinton's position paper supported Mission to Planet Earth, directed NASA to give higher priority to innovation in the civil aircraft industry, prodded the administration to establish a permanent presence on the moon, send humans to Mars, maintain strong cooperative ties on the Space Station while, at the same time, using robotic exploration whenever feasible. Tellingly, Clinton criticized Bush's policy for only two shortcomings: failing to set clear enough priorities for NASA (which led to NASA being "saddled" with more missions than it could possibly achieve) and favoring military space spending over civil.[33]

The Bush and Clinton years are bridged by a broad collection of (borrowing from arms control language) "confidence-building measures" between NASA's Goldin administration and Koptev's Russian Space Agency. This cooperation led directly to some government-supported joint ventures, while easing the way for private sector activity in the months and years to come.

Shuttle-Mir Planning

Chapter 7 has demonstrated how from the 1960s onward, Soviet-American relations in the life sciences remained cordial, punctuated by exchanges of data and research findings, along with dozens of experiments flown by US researchers on Soviet Bion satellites. In the years immediately preceding the collapse of the Soviet Union, life science researchers on both sides of the iron curtain continued to sustain low-budget, but scientifically meaningful cooperation. As of 1991, NASA had already shipped and installed special X-ray equipment for measuring bone density before and after extended Mir flights. As detailed in the last chapter, from 1975 through 1992, NASA's Ames Research Center had been contributing experiments to Soviet biosatellites. In June 1991 a materials experiment "the size of two tuna cans" traveled to Mir aboard a robotic Progress M-8 cargo craft. This was a cooperative project between NASA and the Soviet Union's Institute for Biomedical Problems (the same institutions responsible for biosatellite cooperation that bridged the Carter-Reagan gap in cooperation). Soviets lacked data on solar radiation levels outside the Mir and Americans were collecting information necessary for long-voyage engineering and, more immediately, Space Station Freedom design. Other advantages included the fact that the radiation experiment required no electricity from Mir and that it occupied minimal cargo room on its returning Soyuz capsule (Soviet representatives happily took this opportunity to point out that Mir maintained a "backlog" of manufactured materials waiting to be returned to earth).[34]

Roughly three weeks before the failed August Coup against Gorbachev, President George H. W. Bush proposed a new twist. In a series of initiatives developed by the National Space Council and Vice President Dan Quayle, Bush suggested the exchange of an astronaut with a cosmonaut. Might it be possible for an American to visit Mir, if the Americans accepted a cosmonaut guest on the Shuttle?

At this early phase, NASA maintained life sciences as the primary research interest. The Soviets would provide data already gathered on long-duration flight research; both would share medical equipment for flight and participate in efforts to standardize scientific instruments and lab analysis.[35] The exchange of crew held great symbolic value, foreshadowing a possible decline in secrecy of the then Soviet state. It would entail cross-training at the respective partner facilities, as well as calling for new telecommunications links between human spaceflight centers. Whereas the Americans had only flown up to 84 days in orbit, their experiments tended to be carried out on more sophisticated equipment and performed in-flight. The Soviets, on the other hand, could boast Mir missions of a year's length, but conducted most of their physiological research pre- and postflight and still had no freezer aboard Mir for storing blood and urine samples.[36]

Some warned of disadvantages. Frank Sulzman, chief of NASA's life support branch, pointed out what critics might find less appealing. For one thing, some may fear the undue transfer of American biotechnology to Soviet counterparts, thereby enhancing their lead in long-duration flights. One official, preferring to remain nameless, speculated that the cash-strapped Russians may charge the Americans money for "the means of minimizing the effects of weightlessness on the body," which in the short-term include nausea, fluid redistribution in the head and legs, and disorientation.[37]

In June of 1992, NASA administrator Dan Goldin (appointed by President Bush in April 1992) explained that the Americans and now *Russian* partners were advancing to the "next crucial step in expanding cooperative space activities."[38] Now, in addition to the flight of a cosmonaut on the Shuttle and an astronaut on Mir, the parties had agreed to negotiate two more international flights: an in-orbit rendezvous of craft (meaning the Shuttle would circle, but not dock with the Mir) and the eventual docking of the two craft a few months later. With the second exercise, astronaut Norm Thagard would transfer from Mir to the Shuttle for his return flight (which took place in the summer of 1995). Table 8.1 lists the Phase 1 Shuttle-Mir Flights.[39]

In the fall of 1992, negotiations commenced between Rockwell International (since 1972, the prime contractor on the Shuttle orbiter) and NPO Energia for the use of a Russian-designed Mir-Shuttle docking module.[40] In the meantime, as NASA staff settled into cooperating with the Russian Space Agency *Roskosmos* (itself only five months old) and Rockwell began work with Energia, the American press discussed the likelihood that American firms and NASA might take any number of courses: purchase Mir outright, invite Russian participation on the Space Station Freedom, or commence with plans for an international human

Table 8.1 Phase I Shuttle flights

Shuttle Flight	Duration	Primary Objective	Details
STS-60 Discovery	3/2/1994–11/2/1994	Experimentation with SPACEHAB-2, attempt to grow semiconductor film materials for use in advanced electronics	First flight of cosmonaut on Shuttle (Sergei Krikalev)
STS-63 Discovery	3/2/1995–11/2/1995	Experimentation using SPACEHAB-3, deployment and retrieval of SPARTAN-204 satellite, Shuttle and Mir rendezvous	First female Shuttle pilot (Eileen Collins)
STS-71 Atlantis	27/6/1995–7/7/1995	First Shuttle-Mir docking (S/MM-01)	Spacelab-Mir combined science and logistical transfer mission
STS-74 Atlantis	12/1/1995–20/11/1997	S/MM-02	Delivered docking module and two solar arrays (one built by Russia and one by the United States
STS-76 Atlantis	22/3/1996–31/3/1996	S/MM-03, research and transfer of supplies using SPACEHAB	Linda Godwin and Michael Clifford conduct first US EVA around two mated spacecraft (MEEP experiments)
STS-79 Atlantis	16/9/1996–26/9/1996	S/MM-04, experimentation using SPACEHAB-05	Shannon Lucid departed Mir for earth after setting US record for time in space (188 days)
STS-81 Atlantis	12/1/1997–22/1/1997	S/MM-05, experimentation using SPACEHAB double module	Largest transfer of logistics between two spacecraft (approx 6,000 pounds to Mir and 2,400 pounds to Atlantis)
STS-84 Atlantis	15/5/1997–24/5/1997	S/MM-06, SPACEHAB double module	One-year anniversary of US continuous presence in space
STS-86 Atlantis	25/9/1997–6/10/1997	S/MM-07, SPACEHAB double module	Fourth exchange of US astronauts, first joint US-Russian EVA during Shuttle flight
STS-89 Endeavour	22/1/1998–31/1/1998	S/MM-08, SPACEHAB double module supplied Mir with more than 8,000 pounds of scientific equipment, logistical hardware, and water	Fifth and last crew exchange
STS-91 Discovery	2/6/1998–12/6/1998	S/MM-09, SPACEHAB experimentation	Last Mir docking mission, first time high-energy particle magnetic spectrometer placed in orbit

Source: Judy Rumerman, NASA Historical Data Book Volume VII: NASA Launch Systems, Space Transportation, Human Spaceflight, and Space Science 1989–1998 (Washington, DC: NASA History Division Office of External Relations, 2009), NASA SP-2009–4012.

mission to Mars. But this was all speculation. At the time, Russian ties to the Space Station Freedom (by 1992 a disheartening eight years in the making—see chapter 13) were limited to a study contract, exploring whether or not the Soyuz might be employed as an ACRV "life boat" on the space station.

At the beginning of the Shuttle-Mir Missions, the Mir Space Station consisted of four modules, launched incrementally.[41]

Mir Base Blok (also: FGB Universal Blok Salyut or FGB Universal and Adaptable Space Apparatus (SA)): This module, derived from the military space station Almaz, had been used to provide power, station-keeping reboost, tugging, and docking to a number Russian missions—human and robotic alike. A report provided to NASA by the Khrunichev State Research and Production Facility highlighted the adaptability, variability, and compatibility of the FGBs, explaining that they were identical, predesigned systems with the same engines, tanks, control units, thermal systems, and so on. Russian engineers achieved variability among FGB craft by moving engines, adding or subtracting tanks, or changing electrical power ratings. Thus, the FGB blok was compatible with all Salyut, Mir, and eventually Russian ISS modules and had provided power to at least seven robotic Kosmos missions as well as Mir's Kvant-2, Kristal, Spektr, and Piroda modules.

Kvant-1: This blok was launched in 1987, carrying instruments for scientific experimentation as well as six gyrodynes and a Salyut 5-B digital computer for station orientation.

Kvant-2: Launched in 1989, this module included an extravehicular activity (EVA) airlock, solar arrays, and additional life support equipment.

Kristal: Docked in 1990, carrying scientific equipment, retractable solar arrays, and an androgynous docking mechanism.

Spektr: A derivative of the FGB apparatus, Spektr had originally been designed for Soviet military experiments, but had never been launched due to a lack of funding. "Rescued" by US-Russian cooperation, this module was launched in May 1995. Americans and Russians used it for earth observation and atmospheric study.

Piroda: Supplied additional remote sensing capability, along with hardware for materials processing, meteorological and ionospheric research. Priroda also carried equipment for US, French, and German research.

Reforming Soviet Infrastructure: The Gore-Chernomyrdin Commission's Many Initiatives

[Y]ou have to see this [space station] not as a tinker toy, not as particular project, but as an infrastructure and as new kind of infrastructure.
—Clinton administration official[42]

With the Clinton administration, plans for the Shuttle-Mir flights adopted an important new meaning as "confidence building measures" between the United States and Soviet Union. Rather than being the end product, Shuttle-Mir became a means to more intensive cooperation in space that culminated in what eventually came to be known as the International Space Station Program. Thus, after August 1993, the Shuttle-Mir flight planning came to be retroactively defined as Phase I of the ISS. The Shuttle-Mir and ISS projects were bound in part by a comprehensive $400-million contract between NASA and the Russian Space Agency as well as by administrative jurisdiction—both projects operated under the International Space Station Program Office.

To appreciate the greater significance of NASA's collaborative work with the Russian Space Agency in Shuttle-Mir, and later the ISS, space exploration must be recognized as but one element within a clearly defined regime of the policy objectives of the Clinton administration. These fields fell under the jurisdiction of the 1993–1998 US-Russian Commission on Economic and Technological Cooperation (also known as the Gore-Chernomyrdin Commission or GCC). Through agreements reached by Vice President Al Gore and Prime Minister Viktor Chernomyrdin, the White House aimed to reshape Russian bureaucratic and free market relations for the protection of American investments, long-term political stability of Russia, and the control of weapons knowledge and hardware.

These objectives are evident in three fields. (1) They refined fiscal, administrative, and insurance procedures to make international trade safer for investors. (2) They set up bureaucratic mechanisms in the field of defense conversion and demobilization intended to aid Russia in the retooling of military production facilities for consumer goods and producer durables. (3) The commission introduced environmental measures enlisting Russian resources and personnel in the Mission to Planet Earth Joint Working Group (MTPE/JWG), the Earth Sciences JWG, and by founding a Russian Environmental Task Force. The point bears repeating: these working groups and task forces provided opportunities for collaboration in space as well as "non-space" activities.

Led by the Russian Academy of Science and NASA, the JWGs assembled entities that had since the early 1960s been swapping data and working in collaborative research projects. In addition to the RAS and NASA, these included the National Oceanic and Atmospheric Administration (NOAA, formerly the Weather Bureau), the Russian Federal Service for Hydrometeorology and Environmental Monitoring (also known as ROSGIDROMET, and formerly the Soviet HYDROMET), and the new authority on civil space, the Russian Space Agency. As of 1994, these agencies were engaged in approximately 22 activities. The most notable included the world's only orbiting ozone spectrometer, correlative measurement of the ozone layer, climatology studies, studies of the productivity of Russia's Boreal Forest, health, fire risk, and context in the global carbon cycle, American watershed research by satellite, vulcanological studies of Russia's Kamchatka peninsula, tectonics studies, a study of gravity and magnetics in Tibet and China, and ocean studies by satellite. One sign of the times: the agencies included joint work in Internet connectivity between NASA and Moscow's Institute of Space Research (IKI).[43]

In a gesture coupling demobilization and environmentalism, the Russians agreed to assemble an Environmental Task Force (ETF), fashioned after the example set by the Americans. This task force worked to combine geophysical research needs with data and images available only in classified systems and databases. Both the United States and Russia charged their ETF panels with "assessing the potential application of classified intelligence and defense systems and data to environmental studies." Classified data and information holdings were then reviewed to see if they were relevant to environmental researchers.

Eventually, the United States and Russia would swap old reconnaissance images, but as of 1994, the partners agreed to operate autonomously. Indeed, in the 1994 draft terms, the Americans explicitly noted that this cooperation was by no means an exercise intended to open Russian classified data to the West.[44]

Another project joining demobilization and environmental policy was that concerning the Total Ozone Mapping Spectrometers (TOMS). Flown on NASA craft since 1978, these instruments had monitored ozone concentrations and, in particular, annual depletion over the southern hemisphere and the Antarctic ozone hole.[45]

The first TOMS instrument to fly on a meteor was launched in 1991 from the Russian launch facility Plesetsk. Carried into orbit aboard a Cyclone launcher (designed as an ICBM during the Cold War), the TOMS was key to Soviet-American implementation of the Vienna convention on the protection of the ozone.

Unlike the earlier Bion satellites, TOMS instrument packages were not insular passive passengers. Instead, they demanded compatibility of electrical power supply, control, synchronization, data transmission, mechanical, and thermal utilities.[46] NASA engineers refurbished a Nimbus-7 TOMS Engineering Model, retrofitting it with an Interface Adapter Model (making it possible to "plug in" to the Russian Meteor-3).[47] One unanticipated advantage was that the Meteor TOMS was able to record the effects of the Philipino volcano Mt. Pinatubo (which had erupted two months before the TOMS launch). For a full two years, the Meteor-3 TOMS indicated that ozone had been affected by the scattering effects of the stratospheric sulfate aerosol layer from the volcanic eruption.[48]

In addition to the policies here, the two nations agreed in principle to a joint conference to "help Russian environmental scientists establish their data needs and begin to match those needs to Russian sources of relevant information." This conference would explore Russia's highest priorities in the environment, equipping researchers for studies in radioactive pollution, air and water quality, methods for dealing with industrial/ecological disasters, the effects of defense conversion, soil degradation, and forest management/deforestation.[49]

The two nations entered into a joint technology development project exploring alternative energy sources that linked environmental initiatives with private innovation. The vice president and the prime minister instituted an Environmental Equipment Commodity Import Program, providing $125 million in grants for the export of US-manufactured equipment to Russia, seeking to improve energy efficiency in production, transport, and use.

In trade and investment, the White House helped reshape Russian tax and tariff structures to better protect American investors. Additionally, the US Export-Import Bank, the Russian Ministry of Finance, and the Central Bank of Russia entered into a Project Incentive Agreement offering financial support for "project risk transactions" in all sectors of the economy. The two nations agreed to a new protocol for income taxation, intended to stimulate American investment in Russia. They implemented a memorandum of understanding for an American Business Center. Backed by $12 million, this program was intended to help US businesses invest capital, transfer technologies, and provide business-related training to Russians. The agreement provided $110 million in financing and insurance against transnational business deals for the Overseas Private Investments Corporation (as of 1993 centering on mineral companies and truck manufacturers). Similar plans abounded for "model" American gas stations, guidance in materials and product quality control, all intended to aid the transfer of US business models and practice.

In the fields of the environment, energy, and the complicated task of keeping educated nuclear industry workers employed, Russia agreed to review the safety of its older nuclear reactors, enhance their integrity, and participate in studies for research and development in the field of nuclear power generation. At the same time the two nations set up a legal framework, protecting US firms from liability when supplying safety assistance to Russian nuclear power plants. In addition to this, they planned an Oil and Gas Technology Center Announcement to facilitate the exchange and use of technologies between the two nations, hoping to improve the recovery of oil and gas and reduce production costs in Russia. Both sides believed that facilitating Russia's transition to a market economy still required that the US government adopt a degree of liability on behalf of American investors, Russian businessmen, and the ailing Russian state. Where tax revenues were not at stake, credibility was.

As the two countries methodically dismantled nuclear weapons arsenals under SALT-II, they drafted agreements on the principles and methods of defense conversion and the diversification of former defense industries. In addition to protocol for converting defense firms to civil production, the two parties set aside $20 million in Nunn-Lugar funds to help Russian industries retool for producing modular housing.[50]

Space exploration and research occupied a fourth field of collaboration, bridging the environment, trade, and science writ large. As noted earlier, most projects being pursued at this time (including the Shuttle-Mir, Phobos lander, Total Ozone Mapping Spectrometor-Cyclone, Konus, and WIND experiment) predated the Clinton administration but were in many regards appropriated into the defense conversion regimes of the Clinton White House. Table 8.2 illustrates the range of projects pursued and relative costs.[51]

Human spaceflight programs figure most prominently among these projects, though between FY1993 and 1997 the Bion 11 and 12 spaceflights accounted for $16 million.[52] Meanwhile, the space sciences accounted for roughly 14.5 percent of all program costs, as detailed later. At the five-year anniversary of the GCC, Prime Minister Chernomyrdin reported that overall commodities turnover between the two nations had doubled in the past five years and that American investment accounted for one-third of foreign investment in the Russian Federation.[53]

In 1995, the Office of Technology Assessment evaluated the situation. In his foreword, Director Roger Herdman noted that "much of the motivation for the expansion of cooperation with Russia lies beyond programmatic considerations."[54] In particular, the report pointed out that continued cooperation, including large payments for Russian space goods and services, might help stabilize Russia's economy and provide an incentive for some of Russia's technological elite to stay at home, so contributing to the nonproliferation of weapons of mass destruction. Whether intended only to stabilize the internal structures, or to control the flow of weapons knowledge *outside* the former Soviet Union, the need to maintain vibrant research programs in Russia were "essential program justifications" for cooperation, linking the survival of scientific communities with collective security.[55]

Table 8.2   NASA-Russian activities: summary of agency programs and costs with the Russian Republic ($ in millions—provided to Congress March 1995)

	FY1995	FY1996	FY1997	FY1998	FY1999
Russian Space Agency Contract	100.00	100.00	100.00		
Mir missions	141.7	102.7	54.3	16.3	.6
Space station-related developments	20.0	20.0	10.0	0.0	0.0
Space science	14.4	10.1	9.2	12.3	6.2
Earth science	3.7	3.1	3.3	3.0	3.0
Space access	2.7				
Aeronautics	11.7	3.0			
Tracking and data	2.3	1.9	2.0	2.1	2.1
Total [761.7]	*296.5*	*240.8*	*178.8*	*33.7*	*11.9*

Source: US Congress, Office of Technology Assessment, US-Russian Cooperation in Space OTA-ISS-618 (Washington, DC: US Government Printing Office, April, 1995), 56. These figures include the initial $400M agreement for Shuttle-Mir and ISS cooperation, plus cooperation in other fields and increases to the initial contract detailed below. See table 8.3 in this chapter.

As early as 1992 collaborators had begun to take a new host of factors in international relations into account. Similar to the fields of trade, nuclear energy, and environmental regulation, space exploration and research became levers of reform. In 1998 Boris Yeltsin explained that the principle role of the Gore-Chernomyrdin Commission projects had been to "create a solid economic foundation for the system of relations between Russia moving along towards market reforms and the United States." Yeltsin concluded that they had succeeded, observing, "We are working very closely together in a number of key directions—the development of science, technology, health care, environmental protection, the peaceful use of space, and reduction of the nuclear threat."[56] Cooperation with the Russians supported growing bureaucratic, commercial, and intellectual infrastructures between the world's two leading space programs. Thus, Russian-American cooperation on the International Space Station mapped on to administrative reform in the Russian space complex as well as NASA and its contractors. In 1993 and 1994 NASA narrowly managed to save the Space Station Freedom program from the White House and congress by streamlining management, cutting spending, and linking cooperation in space to post–Cold War regimes of international security—believing that space cooperation would keep Russian science workers employed, but also linking the promises of commerce and ISS cooperation to treaties such as the Missile Technology Control Regime.

Part II: National Motivations

Moving from international policy to national, the remainder of this chapter illustrates the variety and complexity of US national interests coupled to the Gore-Chernomyrdin agreements. It introduces the reader to the perceived doldrums the Space Station Freedom had fallen into, the financial savings at first anticipated by ISS reorganization, criticisms and concerns as voiced by Congressional representatives wary of various elements of ISS collaboration. The subsection "On Being More Equal" illustrates alternative trajectories that the ISS partnership may have taken when Russian partners (and Energia in particular) raised questions of national autonomy. The final sections address the linkages between US national security and Russian defense industries, including other motivations for trade liberalization.

Space Station Freedom and Perceptions of NASA's inefficiency

After nearly a decade of development and $9 billion in tax expenditures, NASA had no hardware, nor a singular plan to show for the Space Station Freedom project. On March 9, 1993, the newly elected president Clinton ordered NASA to begin a "rapid and far-reaching redesign of the Station," with the intention of "significantly reducing development, operations, and utilization costs."[57] Clinton wanted to reduce the planned cost from $14.4 billion to $9 billion and directed NASA to submit options to a redesign committee.

In the spring and summer of 1993 Charles Vest, vice presidential appointee and MIT president led a committee assessing three new possible space station configurations, all of which still averaged $10 billion over the Clinton administration's prospective costs of $5 billion, $7 billion, or $9 billion. Option A was estimated to cost $17 billion and required 16 Shuttle flights for assembly. Option B was larger than Space Station Freedom, required 20 Shuttle flights, and cost $19.7 billion. Option C cost $15.5 billion, was the least like Space Station Freedom, and required 8 Shuttle flights to place in orbit one US module and seven internationally contributed modules.[58]

While weighing Options A, B, and C for station redesign, the Vest Committee considered the ramifications of cooperating with Russia in space station construction. It eventually endorsed the notion of consolidating design plans and hardware from Mir-1 (still in orbit), Mir-2 (still on the drawing board), and Space Station Freedom, in spite of the fact that it would demand a higher inclination orbit—moving the space station from a 28-degree orbit to one that extended 51.6 degrees from the equator (and therefore necessitate expensive upgrades to the Shuttle).

NASA staff took the Vest Committee recommendations and ran with them. One year later, a number of former committee members and NASA staff alike agreed that they had successfully implemented a "single core NASA management team to optimize efficiency, accountability, expertise and cost effectiveness." Changes included setting up a single host center, identifying a single prime contractor, following the new Integrated Product Team approach to concurrent engineering, and refining Program Office-line organization.[59]

Thus, within a brief period of time NASA administrators, staff, and contractors weathered several interconnected changes. They completely overhauled SSF management, "co-locating" Boeing and NASA in one International Space Station Program Office. At the same time, NASA prepared itself for the possibility of cooperating with Russia, reviewing Russian space technologies and their possible contributions to the space station.

Why were all these changes necessary? Critics of NASA management including Dan Goldin himself believed that in order for an initiative as expensive and complicated as the space station to survive, it must operate more smoothly and inexpensively. One Clinton official demanded in 1993 that NASA would have to go through organizational reengineering similar to most major companies of the time, observing, "[I]ts decision structure is cluttered, it's circular, it's labyrinthine."[60]

It is important to note that these changes were implemented on the assumption that Russia would be integrated into the new space station, either as a contractor or partner, and that NASA made these initial decisions independent of the original Space Station Freedom partners. The (then hope of) political and financial benefits of Russian cooperation paired with drastic changes in NASA management to build a new coalition of supporters that was just barely strong enough to defend the International Space Station from a hostile Congress: the project survived by just one vote in the House in summer 1993.[61] Russian-American cooperation on the space station was finalized later that year.

Administrator Daniel Goldin used the ISS's redesign as evidence of greater changes taking place in NASA. Pointing out that his staff had reduced the SSF's projected annual operating costs from $3.5 billion to the International Space Station's $2.1 billion, Goldin explained, "The problem we had was we had 4 prime contractors and 4 NASA Centers. Now, that's an oxymoron in itself—4 prime contractors."[62] Not only was management hopelessly decentralized, but the four NASA Centers tended to compete for jobs, dollars, and autonomy. Observed Goldin, "And each prime contractor reported to a center Director and every so often, Center Directors would get together...And NASA Johnson didn't trust NASA Marshall. They did the pressurized modules, and NASA Lewis did the power system. NASA Kennedy did the launch integration. But who was responsible? Each Center Director was responsible for their budget."[63] Centralizing management accompanied drastic budget cuts at NASA (estimated at 30 percent).[64] At the same time, SSF's former Tier 1 subcontractors trimmed staff and budgets. McDonnell Douglas downsized from 1,800 to 1,000, Rocketdyne from 1,000 to 800, and Boeing from 1,230 to 1,100.[65]

Rarely did Goldin miss an opportunity to tout the estimated $2 billion savings that resulted from cooperation with the Russians. "We get a space station that has almost double the power," he raved.

We go from 60 kilowatts to 110. We get a space station over a year sooner. And we get a space station that costs America $2 billion less. We get a [space station] that has dual access from Cape Kennedy and Baikonur, which gives us tremendous flexibility. We get a tremendous knowledge base from the Russians, who have had astronauts in space since 1986 almost continuously.

Goldin continued, stating, "They have helped us solve some reliability problems already. So we have a more robust station earlier for less money," plus, he added, "we have a coming together of the scientific community in Russia with America."[66]

Russian Contributions and the $2 billion "Savings"

The Goldin administration at NASA anticipated that using Russian (read: Soviet) technologies would reduce the cost of getting a space station in orbit.[67] Listed here is a brief list and description of such elements. Note that several are identified by more than one name. This is due to changes in bureaucratic nomenclature as well as inconsistencies in translation, but also because Russian design philosophies value adaptability—meaning that one module or vehicle may be adapted to many new models, receiving a new name with each incarnation.

FGB Module 45,000 pounds (also: Zarya, Functional Cargo Blok, FGB Tug, Mir-2): This module traces its lineage to the Salyut space station (first orbited in 1971) and its basic design to Mir-1 and plans for a Mir-2. Much of the appeal of Russian cooperation on the ISS was that the FGB was near completion and could be launched much sooner than any comparable SSF module under US development. This, the first ISS module in orbit, had guidance, control, navigation, multiple docking ports, propellant stowage, and propulsion capabilities, meaning that it could provide station-keeping reboost power (to monitor and control the altitude of the ISS's orbit). The FGB also had Environmental Control Life Support Systems (ECLSS) to supply oxygen, remove carbon dioxide, monitor for airborne contaminants, store oxygen and nitrogen, and circulate air. The FGB was intended to provide guidance and control for the first five months in orbit until the Service Module was to be launched. It would also "keep alive" power to US labs until NASA's power module arrived.[67] It was launched in November 1998.

Service Module 64,000 pounds (also: Zvezda): Like the FGB, this module is based on the Mir service module and, as such, provides redundant systems of life support. As of 1997, NASA's administration anticipated that these two modules together would provide control, reboost, and life support for continuous habitation of up to three crewmembers until a US habitation module was to be launched in 2002.[68]

Science Power Platform 46,000 pounds (also: Power Mast): This mast featured 13 kilowatt power generation capability and was augmented by US-provided solar blankets. The pressurized mast section houses gyrodynes.

Progress (Resupply Missions): First launched in 1978, Progress was an automated spacecraft derived from the Soyuz spacecraft. NASA and RSA planners anticipated over one hundred automated Progress propellant resupply missions (six–nine launches per year) throughout the life of the ISS program, saving valuable space on Shuttle flights and reducing the number of Shuttle launches necessary for ISS upkeep.

Soyuz Crew Vehicle 16,000 pounds (SSF: Assured Crew Return Vehicle, later ISS: Crew Transfer Vehicle): First launched in 1966, Soviet space program officials used this vehicle model to resupply Salyut space stations. RSA and NASA officials planned for ten alternating Soyuz crew rescue vehicles to be made available to the ISS through 2002 when a US crew rescue vehicle was to replace it.

Launch Services: for ISS assembly including two Proton launches (first launched in 1965), approximately 40 Soyuz launches (first launched in 1966), and another estimated 70 Soyuz launches in the postassembly phase.

Initially, each of these launch vehicles and spacecraft were to be provided as "contributions" of the RSA—fully funded and delivered in orbit by the Russians.

It was in this way that NASA's space station redesigners justified Russian cooperation as saving a total of $2 billion from NASA's total budget.

Criticisms and Concerns

A broad spectrum of criticisms and concerns accompanied Russian inclusion on the ISS. Some voiced dismay over the move from a 28-degree orbit inclination to one of 51.6 degrees, questioning the expense of a Shuttle retrofit for an Advanced Solid Rocket Motor (ASRM). Others voiced concerns that the Americans might make themselves dependent upon another nation for ISS access or assembly, still others insisted on maintaining a leading edge in aerospace technologies above all else.

Congressman Sensenbrenner worried that NASA might use these changes to eke more money from appropriations, stating: "[N]ow, I think that increasing the inclination of the orbit…is going to cost money, and perhaps lots of money, because the higher the orbit, the more thrust is necessary with the rockets." He continued asking, "Does this money come out of the existing NASA budget? And if so which programs will be cut? Or does the Administration plan to request a supplemental appropriation so that the lift capacity of the shuttle and the other Western rockets would be able to comport with sending a space station into a significantly higher orbit?"[69] Sensenbrenner noted the fact that not only was NASA placing Russia on the critical path, they were paying the RSA directly for services—"something never considered with long-time allies."[70]

Noting that the House of Representatives had already voted down the ASRM, Sensenbrenner voiced his concern over the flow of American dollars to Russia: "As the Cold War ends, a chilling irony remains. Even though some say America won the Cold War, it is clear from looking at space policy that the spoils of victory are going to Russia."[71]

Dana Rohrabacher, another member of the House, disagreed with Sensenbrenner over his concerns with the new ISS orbit. He pressed for total dependence on Russian launch vehicles. The real problem, he observed, was people who were overcommitted to the Space Shuttle (which he described as the most overpriced transport system in the history of man). Rohrabacher suggested that the use of the Energia Rocket system would "actually bring down the cost to the taxpayers." Cooperation with the Russians, he predicted, "will not cost American jobs and [it] will not cost taxpayers for us to work with these new friends and to help cement democracy in what was the Soviet Union." He cautioned instead against spending "hundreds of millions of dollars more on an antiquated space shuttle system."[72]

Congressman Bacchus, on the other hand, supported cooperation in concept as good economic, foreign, and domestic policy. He added that, all the same, "my very strong view is that we must continue to focus on protecting American jobs, American technology, and an independent American space program even as we strive for cooperation with the Russians in space." The United States must remain the senior partner with the Soviets, he cautioned, and it should not place the Russian Federation on the critical path. He suggested instead that cooperation with Russia be based on the same approach as that invoked with the other foreign partners: "I would like to see us design something in which we could plug in the Russians if they are around to be plugged in."[73]

Under the auspices of the 1994–1998 Gore-Chernomyrdin Commission talks, NASA broke historic precedent, doing business with Russia in a substantially different manner. By taking on the Russian Space Agency as a subcontractor for the International Space Station, officials made a notable exception to the "no exchange of funds" tenet of NASA's international cooperation. In the August 1993 Economic and Technological Agreement on US-Russian Space Station Cooperation, NASA promised to compensate the Russian Space Agency $305 million in exchange for US astronauts' training and time aboard the Russian Mir space station. This money was disbursed to the RSA in FY1995–1997—a crucial time for the Russians, who by then were not only maintaining and improving the aging Mir, but also developing their contributions to the ISS. US officials presumed that much of the $305 million would be plowed back in Mir hardware for safety improvements, general maintenance, and retrofitting for Shuttle docking.

In June 1994, NASA released a joint statement on space station cooperation. It explained that "[a] definitized Contract Agreement was signed between the NASA and RSA for up to $400m of goods and services to be provided during Shuttle-Mir operations and during the early international Space Station assembly phase."[74] Thus, between 1993 and 1994 both US and Russian representatives realized that funds dispersed in and through Shuttle-Mir were intended to ease the financial burden of delivering space station equipment and services. While the $305 million was intended to support Mir systems upgrades, to help fund the docking module for Shuttle-Mir, and to help cover the added expenses of training and expanded management, the additional $95 million was considered a direct contribution toward the expense of "Phase II" activities, in particular, early development of ISS components. These included but were not limited to design costs for the joint airlock, service module, FGB Energy blok, power mast, Soyuz/ACRV.[75] (See table 8.2 for a detailed listing of Shuttle-Mir, ISS, and other collaborative projects covered by this contract.)

On Being More Equal

The first half of 1994 proved a rocky period in which both the RSA and NPO Energia tested the authority of NASA over the Russian space program. Through the course of negotiations—and renegotiations—NASA used the SSF structures dictating US leadership to legitimate authority over the ISS.

NASA reported that in this period the RSA (1) attempted to coerce NASA into fully funding all Phase II contributions, in spite of agreements to the contrary (outlined earlier in the $305 million/$95 million split); (2) expected to command and control their FGB cargo module and then, after the arrival of other segments, enjoy "joint control"; (3) wanted to be recognized as coequals with the United States: the Russians refused to sign an interim agreement on the ISS hinting that the Space Station Freedom power relations were inapplicable to them; (4) refused to sign up to the barter system, which tried to minimize the exchange of funds among partners; (5) denied the notion of a unified international crew, expecting to pilot "their" modules as they saw fit and be compensated for the transport of all crew to and from the station on their vehicles.[76]

In the ensuing negotiations, NASA officials were emphatic the Russians had been invited to participate in a *preexisting* managerial structure in which

"NASA has always taken the lead role in the Space Station program and had final authority to resolve conflict." As had been the case in the original Space Station Freedom plans and Space Station Alpha, the facility would operate as a single integrated vehicle, commanded and controlled by the United States, which had by far invested the most energy and resources into the venture.[77]

As a result, by electing to join the former Space Station Freedom partners, Russian officials not only committed to providing specific modules to the station, heavy lift capability, and auxiliary command and control centers, they also placed their technologies and workforce within preexisting structures of authority, designating NASA the "lead partner" on the International Space Station. Yet Russian Space Agency officials demonstrated obvious reluctance in submitting to American authority.

At a meeting on June 16, 1994, NASA and the RSA addressed a number of concerns centering on interpretations of what constituted "Russian territory" and the jurisdiction of Russian law over Americans. Initially, the RSA had intended to operate its ISS modules independent of the rest of the craft, staffed by cosmonauts using Russian as the operating language. It took considerable work for the Americans to convince their partners that enlistment in ISS presumed that it would function as a unified and integrated craft as SSF plans had dictated. Communication regarding safety and critical operations would be in English. This included labels, displays, placards, onboard flowcharts, schematics, and printed procedures.

Policymakers were equally concerned with legal jurisdiction back on earth. US officials took a keen interest in the allocation and use of American dollars, since funding for space cooperation was intended to aid the recovery of research and manufacturing (read: "nonprogrammatic concerns"). The fact that Russian contributions to the space station were being bankrolled increasingly by NASA led to a situation in which Americans sought a degree of authority over relations between the Russian government and industry. Initially, this was troublesome.

In 1994 NASA officials expressed concern over the awarding of American dollars to Russian subcontractors. Expressing a desire to preserve/contain the Russian R&D infrastructure, one official reported that the RSA had refused to farm out work to institutions that NASA had deemed "key subcontractors" in the Russian research community. Instead, the RSA maintained that they held absolute authority over subcontract allocation.[78] Moreover, the RSA refused to report back to NASA on subcontracting procedures. What NASA requested were characterized as "*minimal* information on research subcontractors" and even those reports were in a simplified and reduced format of "just a few pages."

This situation was troubling to NASA representatives, considering the fact that such information was needed to ensure that the Russian research community was being "properly supported by NASA funds" at a particularly precarious point in time. According to a 1994 briefing book, the Russian Space Agency was "generously paid" for such line items as subcontractor reports. Some officials went so far as to speculate that this evasion of responsibility was an expression of NPO Energia's influence over the RSA.[79] Whereas it was acceptable for NASA to provide advance notice of inspection prior to arriving at manufacturing facilities, the RSA's notion that "Russian law will apply to all aspects of the contract performance within Russia" was simply untenable. What resulted appears to have been a three-way competition for authority among NASA, the RSA, and NPO Energia. If the RSA insisted on being an equal to NASA, complete with final decision-making authority over its own subcontractors, then NPO Energia might exercise a higher degree of authority over itself and subcontractors. Administrator Goldin's briefing book explained:

> Since the signing of the 15 Dec Accord we have experienced a consistent effort by the Russians to alter the principles of the 1 Nov Addendum. The Russians consider themselves an equal partner and they wish to alter the IGA, MOU and JMP to reflect this concept. They expect to be paid for any services that are not needed for their 'core segment' which is basically the MIR II. They do not accept the concept of [the ISS being] a single integrated vehicle orchestrated by NASA.[80]

This tension among NASA, the RSA, and Energia was exacerbated by change orders to contracts, demanding extra funds from NASA for goods and services NASA believed were already settled. Internally, NASA officials characterized this as an "unacceptable" move on the part of NPO Energia that was "trying to control dollars" over which the space agencies ought to have had jurisdiction. NASA suspected that the Russian Space Agency had more or less been put up to requesting redundant contracts for research program support. Similarly, the two were charging "exorbitant" fees for cosmonaut time on American projects, even charging "multiple times in and out of the central contract."[81] Table 8.3 reflects what the Russian space program attempted to charge, not necessarily what the United States agreed to pay.[82]

Perhaps, too, high expectations for autonomy stemmed from conditions in the (still unfolding) Shuttle-Mir agreements. As guests of the Russian-built and operated Mir space station, astronauts and NASA officials reported that they could agree to RSA authority on the operation of the Mir. Indeed, "no one in NASA would want to challenge that RSA authority." However, ISS agreements

Table 8.3   What the Russians have added

Project	Description	Cost Added ($m)
Extra progress flights	Providing spare parts to Mir space station	40
Progress launches	Two launches to support US astronauts	80
Russian science	Payment for support to Russian science community	36
Mir crew payments	Payment for Russian crew time on joint Mir research	78
CTV DDT&E and other CTV work		112
Spektr solar arrays		4
Value-added tax*	20% tax	80
US astronaut consumables	(NASA believed already covered in initial contract)	33
Translation and transport		16
	TOTAL	*479*

Notes: * Meeting with RSA: "Congress will never agree to tax burden which causes loss of one fourth buying power."
Source: Meeting with RSA, June 16, 1994, Box 44, Folder RN 73851, Goldin Papers. This table reflects what the Russian space program attempted to charge, not necessarily what the United States agreed to pay.

dictated that the Russian modules on the ISS were a very different matter and there "we cannot accept that Russian law will apply co-equal to US law on the ISS."[83] In spite of these agreements, Russian Space Agency officials viewed their autonomy on Mir as a precedent for Russian-built modules on the ISS.

Indeed, ISS planners still operated under a number of uncertainties through the mid- and late 1990s. Between 1993 and 1997, Russian capabilities of meeting deliverables and deadlines slipped steadily. In 1994 NASA decided to purchase the FGB module outright in order to assure the RSA's receipt of funds as well as timely completion of the project. In 1995 the United States agreed to extend Shuttle-Mir operations in order to funnel more funds into the ailing aerospace infrastructure to help cover expenses in logistics support.[84] In 1996 the Russians acknowledged that the Service module would be eight months late, due to funding shortages suffered by the RSA, leading NASA to pursue backup plans, such as funding the Naval Research Laboratory's Interim Control Module (a 1980s project designed by the NRL's Naval Center for Space Technology).[85]

This spectrum of projects entailed a number of challenging obligations for Russians. Though a much-welcomed windfall, the money was not by any means to be seen as foreign "aid." NASA and the White House officials agreed that funds being sent overseas would help preserve Russia's aerospace infrastructure, but US law demanded concrete products and definable services in exchange—a docking mechanism, metallurgical data, technical training, and the like. During negotiations in November 1993, the Russians stated explicitly that they were prepared to adopt the expense of responsibilities beyond the $95 million "as a matter of national pride."[86] However, as time passed, missed deadlines, shortage of funds, and general noncompliance on the part of Russia began to complicate matters. To the consternation of many, the United States began to shoulder an increasing share of the financial burden.

Regulating People, Regulating Technologies

As we take these steps together to renew our strength at home, we cannot turn away from our obligation to renew our leadership abroad.
This is a promising moment...Russia's strategic nuclear missiles soon will no longer be pointed at the United States nor will we point ours at them. Instead of building weapons in space, Russian scientists will help us to build the international space station.
—President Clinton, 1994 State of the Union Address[87]

Critics and proponents alike recognized that the Gore-Chernomyrdin Commission agreements were intended to liberalize trade structures, introduce new regimes of environmental monitoring and protection, and preserve the institutions and infrastructure in Russia and the Commonwealth of Independent States in exchange for compliance with American standards of demobilization and technology control regimes. Policymakers anticipated that capital—in the form of increased trade flows among American and Russian firms, as well as direct payments from NASA to the Russian government, would not only preserve existing infrastructures in Russia, but *contain* dual-use technologies and know-how.

Many doubted that taxpayer dollars could reasonably be expected to divert flows of weapons knowledge. One source remarked: "[O]fficials involved in controlling the spread of weapons see the [ISS] plan as a way to give Russian industrialists incentives to adhere to Western nonproliferation rules. The two Russian companies with the biggest stake in a joint space station, *Energia* and *Krunichev* also build military spacecraft and missile parts."[88]

Referring to whether or not the Russians had a right to sell liquid booster technologies to India, another observed, "The space-station deal, for example, was both a reward to Russia's aerospace industry for not selling sensitive rocket technology to India and a chance for the US to enlist Russian scientists about [*sic*] the effort to control the future spread of dangerous weapons." The authors explained, "Washington's decision to deal in the Russians on the orbiting space station is the cornerstone of an ambitious...strategy for binding Russia to the US and Western style reforms by building links with its military, scientific, and industrial elites."[89]

Thus, this exception to the "clean interface" mode of cooperation raised a number of difficult quandaries for program officials regarding the relationships of private enterprise, the state, science, and the tenets of free market capitalism. The Gore-Chernomyrdin talks provided Clinton administration officials with an opportunity to shape policy in Russia in a number of fields. It was not merely the floundering Soviet state that the American government sought to regulate—it was the engineers who may defect, scientists who may market technical knowledge, or industrialists who may withdraw from weapons compliance. In 1993 Central Intelligence Agency director James Woolsey observed that delays in pay, deteriorating work conditions, and uncertain futures were "apparently spurring Russian specialists to seek emigration despite official restrictions on such travel."[90] Such fears led to a number of public and private relief efforts, intended to preserve and contain the former Soviet military-industrial complex.

NASA officials displayed a similar philosophy when dealing with Ukraine, linking participation on the International Space Station with compliance to Missile Technology Control Regimes. In May 1994, Administrator Goldin met with Ukrainian deputy prime minister and director general of the National Space Agency of Ukraine (NSAU) Vladimir Gorbulin. In his premeeting briefing, he was informed that, in March, Gorbulin had "pressed the issue of Ukrainian participation in the Space Station." The brief continued, pointing out that Russia "has indicated its desire to employ the Ukrainian Zenit [launch vehicle] to support the Station." However, "these launchers are being coordinated *directly between Ukraine and Russia.*" The report stated that, in an apparent effort to secure a more direct NASA partnership, Gorbulin also discussed the use of Ukrainian guidance, control and navigation for the FGB, as well as other ISS components.[91]

This was not the first time Ukraine had courted NASA. In June 1993 Deputy Prime Minister Shmarov had met with a number of NASA officials, wanting in part to use the former strategic missiles SS-24 and SS-19 as well as the Zenit launch vehicle and AN-225 aircraft for "national economic purposes." Covering all his bases, Shmarov also informed NASA that Ukraine had produced 65–75 percent of the earth sensing satellites flown by the Soviet Union and that as of the summer of 1993, the company had been broadening its work with satellites in the international arena. It had plans to work with Intelsat, Inmarsat, Eutelsat, and

COSPAS-SARSAT. In addition to this, the country supported joint programs in space geodesy and global climate change research. Ultimately, the report advised: "Mr Shmarov may want to develop a role for Ukraine in the...Space Station relationship. If he broaches the subject, you should be non-committal and reply that we have no objection if Ukraine also talks to Russia."[92]

Would NASA dismiss Shmarov and Gorbulin entirely? Not likely. Due to National Security Council's Rose Gottemoeller's "particular interest" in Shmarov and the "delicate" nature of negotiations surrounding nuclear warhead dismantlement, Goldin was advised to bide his time and yet "NOT" encourage the possibility of direct NASA-Ukraine coordination in space.[93] The report concluded by stating that, although Ukraine had "significant launch capabilities, including the Zenit and Cyclone launchers....until Ukraine becomes a signatory to the MTCR and other international treaties, the US Government does not wish to pursue this."[94]

Industry and International Relations

US and Russian cooperation in the Space Station entails not only government to government cooperation but also industry to industry agreements. The bottom line is that while government agreements will formalize cooperation, the actual building of the station will be accomplished primarily by private industry.

—NASA administrator Dan Goldin[95]

The case of Ukraine is instructive. Trade restrictions might function as one of many American bureaucratic mechanisms channeling the flows of US resources or they might lessen the negative impact of foreign competition on American firms. However, the advantages of American protectionism diminished with the increase in joint ventures between Russia and the United States. At the same time, US aerospace firms began to vertically integrate: launch providers merged with satellite builders. Initially, as of 1992 one policy analyst noted the division of the aerospace industry into two powerful blocs: General Dynamics, Martin Marietta, McDonnell Douglas, and Rockwell international demanded strong protectionist policies against Chinese and Russian boosters. Hughes, Loral, and General Electric Aerospace, however, lobbied for access to the less-expensive foreign launchers.[96]

The years 1993 through 1995 brought the merger/acquisition of several key firms: Martin-Marietta acquired General Electric Aerospace, then General Dynamics. In 1995 Lockheed (which in turn had been collaborating with Energia and Khrunichev) merged with the Martin consortium forming International Launch Services. Thus, the Lockheed-Martin group pressed for the total elimination of Proton launch quotas while Boeing (and its new subsidiaries McDonnell Douglas and Rockwell) entered into business arrangements with Ukraine's Sea Launch, marketing the Zenit.[97]

Globalization is by no means a new phenomenon for the aerospace industry, which for decades has seen joint ventures in aviation research, development, and production.[98] However the trade liberalization of the 1990s brought US and former Soviet complexes together for the first time. Hughes, Lockheed, Martin Marietta, and General Electric had been key figures in Cold War era reconnaissance, military communications, and early warnings satellites.[99] With Europe's market share rising steadily and defense spending dropping precipitously, the industrial lobby, proponents of defense preparedness, and congressmen became increasingly concerned. As of 1969, US firms held an astounding 91 percent of the world market share. In 1993 this figure had dropped to 67 percent.[100] What follows gives nuance to the significance of US-Russian partnerships.

In 1993, the United States permitted Russian firms for the first time to launch American telecommunications satellites into geosynchronous orbit, providing they sold their launch services at a cost comparable to Western prices. In 1998, Ukraine and Russia entered into Technical Safeguard Agreements designed to protect American satellite and missile technology and allow US industry to launch satellites from foreign locations. Between 1997 and 2006, Proton launchers captured a market share equal to the Atlas (11–12 percent and 10–12 percent, respectively), but it must be recalled that the Proton was by way of joint ventures, now also an American product.

It is indisputable that Russia's rise on the world market is due, at least in part, to Russian-American joint ventures that brought about a convergence of Western management, marketing, and perhaps most important, customers. These factors were evident in the logic and execution of the Gore-Chernomyrdin Commission for Economic and Technical Cooperation. However the American aerospace industry stood to gain as well—not so much by opening new markets, as finding new business partners. These included the commercial space launch ventures of Lockheed-Khrunichev-Energia (ILS) and the Energia-Boeing-Yuzhnoe venture, Sea Launch.[101] Additionally, Pratt-Whitney, Rockwell, and Aerojet initiated business deals with the former Soviet Space complex, while the Russian manufacturers of the Cosmos, Cyclone, and Rokot launch vehicles each found international partners to launch their vehicles. Analysts speculated that Europe's market share would drop from roughly 50 percent in 1996 to 25 percent in 2006.[102]

Thus, the United States helped shape the formation of a privatized aerospace industry in the former Soviet Union. The US government opened itself and American firms to Russian space industries, but—as mentioned earlier—in exchange, the United States demanded the formation of a civil space agency as well as agreements concerning compliance in the demilitarization of former weapons facilities. It is at best doubtful that their optimistic wishes for weapons control were successful. Nonetheless, the United States attempted to woo the remnants of the Soviet Union into military and economic compliance by offering a combination of trade and fiscal incentives. With it came more than $760 million (as detailed in table 8.2) to buttress their faltering aerospace infrastructure

In the long run, these government dollars were but a drop in the bucket—or more aptly a foot in the door—compared with the profit intake of private industry.[103] As of 1998, Western customers were paying more than $880 million a year for space services. This accounted for roughly 70–80 percent of the Russian space program's operating costs.[104] In 1997 alone, Energia Corporation claimed over $350 million in commercial earnings, roughly half the total foreign sales for the entire space industry. While cooperative space work did not release the largest sum of money to the Russian space program, it did provide a politically palatable environment for reforming state infrastructures to favor trade on the global market.

Concluding Remarks

We know that right now your options at home are limited and outlaw regimes and terrorists may try to exploit your situation and influence you to build new weapons of war. [the physicists and engineers scribbled in tiny notebooks] But I think we should talk about a brain gain solution, and that is a solution of putting you to the work of peace, to accelerate reform and build democracy here, to help your people live better lives for decades to come.

—James A. Baker III,
US secretary of state to Soviet Nuclear Weapons Lab employees,
February 1992[105]

This chapter, by illustrating the broad scope of technical cooperation in trade, environmental regulation, scientific research, and space policy has demonstrated how the new conditions of cooperation placed both the Russian and American space programs in new positions of accountability (and vulnerability) to one another. Americans invested capital and credibility in exchange for regimes of surveillance of the aerospace industry, weapons trade, and the environment. At the same time, Russians agreed to become liable to American inspections, answerable to American contracts, and subject in limited degrees to American prescriptions for trade and business organization. Compliance was another matter.

In the 1990s, several (at times conflicting) post–Cold War objectives shaped the discourses and intercourses of space work. These included pressures for reduced budget expenditures, a new élan for streamlined budgets, desires to reduce nuclear arsenals, as well as a new science policy that often encouraged private industry to invest in its own R&D. The waning of the Cold War did not render space cooperation inevitable, neither did it necessitate amicable relations. Instead, Russian design philosophies of adaptability, variability, and compatibility combined with the abundance of Soviet era defense spending, providing NASA and American firms with a number of prospective bargains. The globalizing aerospace industry and 1990s trade liberalization both facilitated these transactions and benefited from them.

While Soviet-American competition in space no longer operated as quite the same driver to funding and political consensus as was characteristic of the 1960s, the people and artifacts of the Cold War continued to shape policy. Thus, for the Russians, idle productive capacity and surplus launch vehicles took on a new meaning in a new geopolitical environment.

For Americans, international scientific and technological collaboration in space were used in an attempt to promote American interests abroad with Missile Technology Control Regime (MTCR) regulations and later the Iran Nonproliferation Act (INA). Clinton officials anticipated that ISS contributions and US leadership would facilitate the emergence of a consensus for a *new* US-led Western Alliance—one that co-opted the former Soviet republics against a new block of adversaries: Iran, Iraq, North Korea, Afghanistan, and other "rogue states."

Between 1994 and 1998, the United States paid out approximately $800 million through ISS-related activities. The Congressional Reporting Service states that in 1996 "reports surfaced of Russian entities providing ballistic missile assistance to Iran, including training; testing and laser equipment; materials; guidance, rocket engine, and fuel technology; machine tools; and maintenance manuals (see CRS Report RL30551)." In 1998, George Tenet, director of Central Intelligence, testified to the Senate Intelligence Committee that Russian aid had, "brought Iran further along in ballistic missile development than previously estimated."[106]

These revelations set Congress at odds with the White House, kicking Section 6 of the INA into action, threatening to cut off funding associated with the ISS, and leaving NASA's largest program potentially dead in the water. Controversy ensued regarding what elements of ISS collaboration applied to the "crew safety" exception of the INA, allowing for a minimum continuation of funds to the program in the interest of US astronaut safety. These discussions became all the more heated following the orbiter *Columbia*'s tragic accident in 2003, when NASA became completely dependent upon Soviet transport and again when President George H. W. Bush's Vision for Space Exploration cancelled US plans for a Crew Return Vehicle, again, increasing dependence on the Soviets for access to the ISS.

Critics of the INA (including the CIA) questioned whether or not it was realistic to presume that the Russian Space Agency could be held accountable for proliferation activities that could take place among any number of firms, the Ministry of Defense, or the Ministry of Atomic Energy (which for all appearances had indeed committed proliferation "crimes" associated with Iran). INA compliance rested upon the apparently naïve presumption that a carrot offered to the Russian Space Agency might (influence) behavior of the Russian government writ large. The Russian citizens responded with a range of improvisations including acquiescence and alignment as well as extortion, illusion, and outright noncompliance.

Foreign policy and national security considerations have always played leading roles in the principles and guidelines of Soviet-American space projects. Yet from 1992 onward they were executed in very different manners. Before then, high-profile collaboration in space followed nonproliferation regimes such as the 1963 Comprehensive Nuclear Test-Ban Treaty (which made a joint lunar mission offer plausible) and the 1969 Strategic Arms Limitations Talks (which made the ASTP plausible). In both instances, abstinence from bilateral security regimes could thwart collaboration, but by no means was collaboration offered as an explicit incentive *for* enlistment in nonproliferation regimes.

Specifically because collaboration in space was linked to a multitude of other cultural, bureaucratic, and capitalistic linkages, enrollment in the ISS became a plausible reward ex post facto. Thus, into the 1990s, cooperation in space continued to function (to varying degrees) as one of America's tools for legitimating power, spreading democratic ideologies, reproducing cultures of regulation, and teaching the mores of liberalized trade. How successfully?

Given the near incomprehensibility and near catastrophic disorder of the former Soviet military industrial complex, is it surprising that weapons technologies did in fact leak out? Instead we might ask, parallel to the much-debated "achievements" of the Apollo-Soyuz Test Project, to what degree did ISS and its associated attempts at post-Soviet order prompt at least a minority of Soviet representatives to "show their hand"—delineating industrial capabilities, identifying the critical state of their R&D institutions, and ultimately, reappraising their own bureaucratic potency if only to increase their legibility to the West? While the entire exercise was a categorically unsuccessful *replication* of Western structures and ideals, it did present at least an *extension of* Western capitalist order into the post-Soviet world and, therefore, a useful glimpse into the logic of American international leadership as well.

Part IV

NASA and Emerging Space Powers

Ashok Maharaj

Chapter 9

An Overview of NASA-Japan Relations from Pencil Rockets to the International Space Station

A bird's eye view of Japan's space history since World War II reveals the gradual and difficult emergence of the country as a major space power that, with US assistance—but also to bypass US restrictions on the transfer of sensitive technology—fruitfully channeled its quest for independence into a robust national program that enabled it to collaborate successfully with its erstwhile mentor and other partners.[1] The United States, through NASA and private industrial corporations, supported Japan's fledgling program early on, but deep internal divisions in the country made it difficult to build a durable arrangement. What little cooperation existed between Japan and NASA during the early 1960s was limited to small space science experiments using sounding rockets and data collection from ground stations. A 1969 agreement to provide launcher technology to Japan, strongly promoted by the State Department, was a major stimulus to the ongoing rationalization of a national space program, though this came too late for Japan to participate actively in the post-Apollo program. In fact NASA's relations with Japan began to achieve significance only during the late 1970s and grew extensively in the later years to include a variety of cooperative space projects that benefited both countries, including human space flight and participation in the International Space Station.[2]

This contribution traces the broad outlines of these developments with particular emphasis on three of the most significant phases of US-Japanese collaboration in space: (1) the frustrations of the 1960s caused by internal rivalries and a strongly nationalist agenda in some sectors of the Japanese space science community that hampered international collaboration and that eventually crippled Japan's ability to participate meaningfully in the post-Apollo program; (2) the transformations precipitated by the 1969 agreement to provide Japan with Thor-Delta technology that not only provided the country with much of the hardware needed to reach the geostationary orbit but also, by restricting the scope of technology transfer, accelerated the country's independence and self-confidence in launcher development as the 1970s wore on (treated separately in the next chapter); and, finally (3) the contribution of Japan to the International Space Station in the 1980s.

Domestic Rivalry in the 1960s

The San Francisco Peace Treaty of September 1951 removed the prohibitions that had been imposed on Japan's development of atomic energy and aerospace research for peaceful purposes. Local elites, determined to modernize the country, seized the opportunity to pursue atomic and space science research for international prestige and scientific and economic benefits.

The early history of space in Japan is marked by the tension between opposing concepts of how to secure a position for the country in space. On the one hand there was the nationalist impulse of Hideo Itokawa who was determined to remain independent of foreign help and indeed of government "interference" in his research agenda. Itokawa advocated the pursuit of space sciences using sounding rockets and believed in the incremental development of solid propellant sounding rockets to launch scientific and application satellites. His views were diametrically opposed to those of Kanuro Kaneshige.[3] Kaneshige aimed to use space technologies for economic and commercial benefit and sought international cooperation for forwarding his country's space goals. He was open to international cooperation and sought assistance from other foreign countries, mainly the United States and Europe, to nurture the fledgling program through cooperative endeavors.

Born in 1912, Hideo Itokawa graduated from the Imperial University of Tokyo and was involved in designing aircraft at Nakajima Aircraft Company during World War II. Concerned about the decline of the Japanese aerospace industry after the war, he galvanized the scientific elite at select institutions and created a niche within Tokyo University—the Institute for Industrial Science (IIS)—for research in aeronautics and space sciences.

In 1954 Itokawa's group obtained a modest research grant to develop sounding rockets.[4] A Japanese committee was formed in the spring of 1955 to coordinate a rocket project to coincide with the International Geophysical Year (IGY). The momentum generated while preparing for the IGY led to the establishment of a team that promoted the development and launching of sounding rockets for the collection of scientific data.[5] In April 1955 the IIS exhibited to the public the first results of Japanese space research: a tiny rocket with tail fins called the Pencil. As the name implied this rocket was in the form of metal tubes measuring 23 centimeters in length and 1.8 centimeters in diameter and weighing around 200 grams. It was filled with solid propellants, similar to gunpowder.

The IIS was renamed the Institute of Space and Aeronautical Science (ISAS) in 1964. Building on the experience gained with the Pencil rocket experiments the Itokawa group gradually scaled up their research and development to build the Kappa, Lambda, and Mu series of sounding rockets. Restrictions imposed on postwar Japan limited the launch vehicles to diameters of 1.4 meters or less. Itokawa's personal view was that there was no need for Japan to develop rockets larger than the Mu because miniaturization would permit smaller payloads to do greater tasks.[6] Stressing the possibilities inherent in miniaturization, he said that the Lambda series rocket could orbit a 100-kilogram satellite by increasing the booster's diameter from 750 to 850 centimeters. Tokyo University could thus handle the application satellite program. He was not in favor of developing liquid fuels, though they had advantages for control purposes, and he dismissed suggestions that he collaborate with the National Aerospace Laboratory's (NAL) nascent liquid fuel program.[7]

ISAS collaborated reluctantly with NASA in some experiments and research. Itokawa argued that having the United States launch Japanese satellites would take more time and money, would be less flexible, and would prevent the growth of Japan's own technology. He also feared a loss of autonomy for his university-based group, believing that if he received assistance from abroad he would be accountable to the United States and to the Japanese government. As Emmerson and Reischauer put it, "As a matter of policy, the Japanese preferred to sacrifice short-term gains in speed and budgets in the interest of the made-in Japan principle. The technological experience and the pride and prestige of an exclusively Japanese effort were at that time more important than the speed of the space program."[8] Itokawa's quest for autonomy came at a price. Four failed attempts to launch a scientific satellite using rockets developed by ISAS led to considerable public criticism. It also bedeviled relations with NASA. Since ISAS was the dominant space group in Japan at that time, the United States took ISAS's negative stance more generally as indicative of a Japanese policy of noncooperation.[9]

As mentioned earlier, Itokawa's group was not the only one engaged in developing rocketry in the 1960s. The other was the National Space Development Center (NSDC) set up in July 1964. The NSDC and its governing body, the Science and Technology Agency (STA), were open to international collaboration and wanted to emulate the "leader-follower model," meaning "identifying the leader in technological capability and learning as much as possible from its accomplishments, then building on that learning to develop a strong indigenous technology base."[10] The NSDC took responsibility for developing liquid-fueled rockets (the so-called Q series) for launching applications satellites rather than the pure space sciences pursued by the academic team at ISAS.

The NSDC was established by the National Space Activities Council (NSAC) chaired by Kankuro Kaneshige, also of Tokyo University. The council's role was to coordinate the mushrooming space activities in Japan after the IGY. It had representatives from universities and key organizations like the Atomic Energy Commission, the Meteorological Agency, the Tokyo Astronomical Observatory, the Institute of Industrial Science, and the National Aeronautical Laboratory. It also acted as an advisory body to the prime minister Eisaku Sato and formed the central node for governing the scattered space activities. In February 1964 the NSAC presented a report to Sato stressing that cooperation among the various government ministries and agencies alone was not enough to attain success in the development of launching vehicles, the construction of satellites, rocket launchings, and research on related matters. It recommended the establishment of a central executive organ on space development to promote comprehensively and efficiently the development of techniques in various fields.[11] Thus was the NSDC born, both to foster international collaboration and to create an alternative technological path to that being pursued by Itokawa with a view to launching telecommunications satellites.

The attempt to centralize space-related activities in Japan was not only a response to domestic divisions; external factors also influenced the shape of the institutional structure. First, the desire to be a player/participant in the emerging field of space sciences and applications was motivated by seeing the advances made by the United States and the Soviet Union. Second, officials in Japan felt obliged to participate constructively in international negotiations over the Outer Space Treaty that was opened for signature in January 1967. And finally, the creation of a governmental body was invoked by government officials so as to position the country favorably in the negotiations over the definitive Intelsat agreements that got under way in 1969 and that would define the terms of access to a global telecommunications satellite system (see chapter 5). Though Japan was keen to be party to the Intelsat agreements, its officials were cautious not to accept any unfavorable conditions that would jeopardize their own technological capabilities in satellite development or the domestic and regional use of comsats.[12]

NASA-Japan Relations in the Early 1960s

Apart from the spectacular transmission of the 1964 Olympics held in Tokyo, space relations between NASA and Japan during the early 1960s remained very superficial and were limited to sharing data and flying small probes in sounding rockets.[13] Two factors hindered NASA's cooperation with Japan during the early years. First, there was the controversial domestic behavior by Itokawa who sought autonomy for his ISAS group, coupled with open hostility between the Japanese scientist and NASA authorities. Itokawa complained to the State Department that, unlike the United States Air Force, which was more cooperative, NASA always imposed some terms or conditions that made the cooperation unattractive. He particularly resented an incident that occurred around 1962–1963 when he visited the United States as a representative of the government of Japan to arrange for the use of Wallops Island facilities for launching Kappa rockets that had grown too large for the Akita range, only to find that his requests were flatly turned down by NASA.[14] Arnold Frutkin put the blame squarely on Itokawa's shoulders. "The team at ISAS was not open to international cooperation," he said in a recent interview. "We offered them collaboration exactly as we did to the Europeans and they were more laggard than the Russians in picking it up."[15] Frutkin also deeply resented Itokawa's interpretation of NASA's launch policy that the Japanese scientist published in a national newspaper, pointing out that "Itokawa wanted to develop a launch vehicle and was willing to completely misrepresent what we were willing to do and were not willing to do in order to get a free hand in Japan."[16] This mutual mistrust undermined any hope of productive cooperation.

The second factor subverting durable space collaboration with Japan was the diffused nature of space activities in the country. In 1962, Richard Barnes, chief of cooperative programs for the Office of International Programs remarked that

[t]here is clearly a substantial reservoir in Japan of the scientific and technological resources (manpower and facilities) needed to carry out a sophisticated space research and development program. These skills are, however, not concentrated in any one segment of the Japanese community. They are diffused among universities, government laboratories, military and industrial organizations. Also, the close working relationship which exists between the U.S. scientific community and the U.S. government is nonexistent in Japan...there is currently no Japanese "NASA" i.e. an organization with the assigned mission of promoting and coordinating space research.[17]

Barnes concluded that without strong central direction the internal divisions in the country would seriously impede both the construction of a coherent national program and substantial cooperation with the United States.

As pointed out earlier, the NSAC chaired by Kaneshige was well aware of these difficulties and was determined to remedy them. He was helped in that goal by the first Chinese nuclear test that led the American authorities to move proactively at the highest levels to collaborate with Japan in space, notwithstanding NASA's qualms.

The Effects of the Chinese Nuclear Test

On October 16, 1964, the government of the People's Republic of China (PRC) successfully tested a 22-kiloton atomic bomb at the Lop Nur site. The balance of power in Asia clearly tilted toward Beijing. The government of India began to reconsider its non-nuclear posture, and eventually tested its own bomb a decade later—an option denied to Japan by Article 9 of its postwar constitution that prohibited the development of nuclear weapons. Instead the press and officials in Tokyo emphasized the loss of prestige suffered at the expense of a third-world communist country, and suggested that a robust space program would be a valuable technological antidote that could save national pride.[18] Indeed, as one leading politician and space advocate put it in 1966, "[I]f Mainland China should succeed in launching a satellite ahead of Japan, the sense of hopelessness of the Japanese will be so great that no one will have the heart to see it. It is the national responsibility of the leaders of our country," Yasuhiro Nakasone went on, "to take the initiative so that this national confidence cannot be lost, even a little."[19]

The risks of nuclear proliferation to nonaligned countries led the State Department to plan for an appropriate response even before the explosion occurred at Lop Nur. The imminent Chinese test, the State Department suggested, provided "an opportunity to demonstrate U.S. cooperation in sharing of advanced technology with countries of Asia." Granted the strict limits on nuclear collaboration with Japan, an alternative like "full and active cooperation with the Japanese in such outer space endeavors as space communications and the launching of a Japanese space satellite" suggested themselves.[20] This was not going to be easy, however, as officials in the American Embassy in Tokyo pointed out after China had tested its bomb. The Japanese would not leap at the opportunity to collaborate with the United States since there was "a feeling among Japanese space officials that independent development of a successful space program is important to Japan's prestige, especially in view of the recent 'Chicom' [State Department abbreviation for the PRC] successes in the nuclear field."[21] Certainly the prime minister wanted to see a Japanese satellite aloft to counter the impact of the PRC's nuclear test, and to demonstrate Japan's advanced scientific and technological capability. What is more, "Assistance from the U.S. in tracking and communicating with such a satellite would be well received in Japan and would contribute to U.S.-Japan relations." However, the Embassy emphasized, granted Japanese sensibilities, "[t]he position of the U.S. was to remain one of cooperation and assistance, rather than guidance or domination, if the political objectives of the Japanese were to be met."[22] Too much engagement would obviously expose Tokyo to a propaganda onslaught from Beijing for being dependent on the United States.

In September 1965 President Lyndon B. Johnson suggested to NASA administrator James Webb that the American space program "should have more visibility abroad and should yield more return to American foreign policy objectives."[23] Assisting the Europeans and the Japanese with their space programs would help strengthen the alliance within the capitalist bloc and assure greater American involvement in those nations. By helping its allies, the United States could also impress them with its technological superiority vis-à-vis the Soviets. Following up on this, Vice-President Hubert Humphrey, who was also chairman of the Space Council, visited Japan late in December 1965. There he suggested that the two countries work together on a major project akin to the Helios mission that the American president had proposed to German chancellor Erhard just a few days before (chapter 2). "We in the U.S. have watched Japan's remarkable advance into this field with interest and admiration. We look to your country," Humphrey went on, "for a major contribution on the leadership role as the world crosses the threshold into the space age." Hence the value of cooperating on "major space projects which none of us can do alone."[24]

This high-level willingness to collaborate constructively with Japan was thwarted by Itokawa's determination to remain autonomous, and his contempt for his competitors. Itokawa made official statements beginning in 1964 calling for Japan to launch a satellite in 1966. He wanted his country to be the fourth nation to orbit a satellite after the United States, USSR, and France. However, he insisted that his group achieve the feat alone and without help from foreign countries, unlike Canadian and European scientists who had sought US assistance in launching their satellites. He chose the three-stage Mu series rocket to launch a Disturbed Ionosphere Patrol Satellite (DIPS) or an All-Wave Radio Noise Receiving Satellite. He viewed both the satellites as a distinctly Japanese contribution to space science and as an extension of the experiments with Japanese instruments sent up in NASA sounding rockets from Wallops Island. Responding to critics who argued that "lack of coordination might result in duplication of effort within Japan," he said he saw "no harm in duplication." He also dismissed all efforts by the NSAC to rationalize the program by discrediting the Council: "[T]here are no space scientists among the members of the NSAC," said Itokawa, "and its chairman Kaneshige could hardly be called a space scientist. His field was textile machinery."[25]

In exchanges with State Department officials in April and May 1966 Kaneshige confirmed that the internal strife that so struck Barnes and Frutkin was damaging the Japanese space program. The chairman of the NSAC remarked on the "lack of a good program," and said that "the fact that Japan has not yet succeeded in integrating its two space programs—the Itokawa program sponsored by the Ministry of Education and the Program of the Science and Technology Agency—[was] causing embarrassment." According to Kaneshige, the prime minister was making policy with regard to space, but "there [was] nobody now who can speak for the Japanese space program." He amplified this statement by claiming that no one (presumably other than the prime minister) was even authorized to request American tracking assistance in the event of the launching of a Japanese satellite. Kaneshige believed that Japan might ultimately establish some sort of national space agency, a "little NASA," though he felt that it was first essential to work out a sensible, long-range plan for space research.[26]

Kaneshige's gloom led him to pour cold water on every suggestion made by senior State Department official Herman Pollack for closer collaboration between the two countries, no matter how tentative. A memo summarizing an exchange between the two men, in which Pollack emphasized how much store he placed on collaborating with Japan, concluded that Kaneshige's replies "in general carried the impression that until Japan's internal problems with its space program are settled by the Japanese themselves, Japan would find it difficult to discuss with the U.S. the details regarding a useful program of international cooperation in space."[27]

To sum up, during the 1960s NASA collaborated sporadically, and with difficulty with Japan. Absent a coherent national space program and a single government-sponsored organization to serve as interlocutor, there was no reliable point of contact in Tokyo. Frutkin was emphatic that he would not deal with individuals unless they were empowered by their national authorities. Itokawa's strident nationalism and public misrepresentation of NASA's launcher policy ruled him out as a partner. Kaneshige's intentions were sound but he was not able to rein in his rival, a man who enjoyed wide public visibility and who contemptuously dismissed him as a meddling bureaucrat. Relations with the United States were further soured by provocative remarks by Itokawa that may have struck a popular chord at home but that only increased consternation in Washington. The Chinese nuclear test particularly irked the head of ISAS. In 1964 the State Department reported that Itokawa had said that "some Japanese scientists had been considering the possibility of publicizing Japan's potential to produce nuclear weapons if it so chose, as a means to counteract any claims about the superiority of Chinese Communist science in connection with its nuclear program."[28] ISAS's work on solid propellant research also raised eyebrows. It was noted, for example, that the Mu series of rockets, which were being developed by ISAS in collaboration with firms like Nissan and Mitsubishi, had the potential to evolve incrementally to an Intermediate Range Ballistic Missile (IRBM). Under these circumstances NASA could not but tread cautiously, above all in the domain of launchers, and notably since NSAM334 of July 1965 specifically prohibited technological assistance to foreign entities that might enable them to acquire independent access to the geostationary orbit for comsats (chapter 3).

That said, it is all the more remarkable that in 1969 the two governments signed an agreement to provide Thor Delta technology to Japan. The steps taken by the president and the State Department to draw closer to Tokyo after the Chinese nuclear test in 1964 planted the seeds of this agreement. Those initial contacts, however, were limited to discussions of what might be done at a general level to foster space collaboration between the countries. The narrowing down of the field to one major project required a determined push by senior officials in the State Department against the wishes of NASA and other arms of the administration. This episode is important enough to merit a study of its own, and is handled in depth in chapter 10.

Japan and Post-Apollo Talks

The 1969 agreement on the transfer of launcher technology to Japan catalyzed renewed efforts in the country to establish a centralized body responsible for space that was similar to NASA. Japan's National Aeronautics and Space Development Agency (NASDA) was established to that end. Though ISAS was sidelined in favor of NASDA, both these bodies along with a few other government agencies and private corporations steered the Japanese space program until an umbrella organization called the Japanese Aerospace Exploration Agency (JAXA) was formed in 2003.[29] Cognizant of the "growing pains" of building and establishing a space program in Japan, of the geopolitical realities during the Cold War, and of domestic politics in Japan, over the last 50 years NASA has identified selective niches within ISAS, NASDA, and later JAXA for scientific and technological collaborative endeavors.

NASDA was established as a public organization on October 1, 1969, with strong support from both the minister of science and technology and Prime Minister Eisaku Sato. It operated under the policy guidance of the STA who provided its budget, along with some government agencies. NASDA took over the functions of the National Space Development Center and of the Ionosphere Sounding Satellite Division of the Radio Research Laboratories of the Ministry of Posts and Telecommunications and included engineers and scientists from both academic and industrial circles.

The timing of the creation of NASDA reflected the trajectory space was taking toward the application needs of nation-states. The agency took the lead in the development of space application capabilities in Japan, including satellites for remote sensing, communications, and meteorological observation, the development of launch vehicles for those satellites and the development of facilities for production, testing, and tracking the satellites. It also benefited from a change in Washington's foreign policy initiatives in the 1970s that saw the waning of a "special dependency relationship" that had characterized US-Japan relations since the end of World War II. The opening of China during the Nixon administration and the "changing nature of the cold war—détente with Soviet Union, the evolution of a new world economy, and domestic forces transformed the Pacific alliance."[30] This was reflected in NASA administrator Tom Paine's invitation to Japan in March 1970 to participate in the post-Apollo program (see chapter 4).

While the Japanese space community was eager to participate in the post-Apollo program, it was unclear what they could contribute. Uncertainties over the evolving configuration of the post-Apollo program itself (chapters 4 and 5) were compounded by the reorganization of the national program, and the limited resources Japan had for space. Minister Nishida noted that the country could only make a useful contribution to post-Apollo if it had achieved something significant of its own, and was suitably advanced technologically: "real international cooperation" was otherwise impossible.[31] Notwithstanding these reservations a special committee was formed by the Space Activities Commission on July 1, 1970, to consider what contributions Japan could make. It sought clarity from NASA on its detailed plans, but to little avail given the fluid nature of the situation in the United States and Frutkin's determination that potential partners should bring their own suggestions to the table (see chapter 4). A top-level team visited NASA field centers and contractors in July 1971 and had extensive discussions with Arnold Frutkin at the NASA headquarters.[32] The lesson that was drawn was that Japan should first close the technological gap with other countries by developing space technologies indigenously. The Special Committee backed off from any major participation in the shuttle, recommending instead, in its final report filed in May 1974, that Japan prepare experiments to use the shuttle and Spacelab, doing its best to develop and supply the hardware itself.[33] It also recommended that when the next generation system for human spaceflight was developed it was in Japan's interest to extend its cooperation to the full development of a space laboratory and to sending a Japanese astronaut into space.[34] This came in handy when deliberations on participation in the space station came up in 1984.

The Hesitant 1970s

Although the ambitious post-Apollo initiative by Thomas Paine did not bear fruit, there were ongoing if sporadic talks between the two parties in the early 1970s.[35] A meeting between President Nixon and Prime Minister Tanaka in 1973 led to the creation of a high-level binational panel to explore avenues for cooperation. The panel identified "space science and applications as a promising area for expanded cooperation with Japan."[36] Specifically, NASA promoted the "utilization of the space shuttle/Spacelab system by Japanese scientists and facilitating Japanese funding construction of a Landsat ground station."[37] A team from NASA visited Japan in October 1976 "to promote opportunities for Japanese utilization of the space shuttle for both scientific experiments and launching commercial payloads. [...]."[38] Logsdon suggests that little progress may have been made due to Arnold Frutkin's known antipathy to working with Japan, perhaps because of his experience in the Pacific theater in World War II. In any event after President Jimmy Carter entered the White House in January 1977, a new team of NASA managers took over. Norman Terrell replaced Arnold Frutkin as the director of international affairs (Frutkin in fact left NASA soon thereafter). Terrell encouraged NASA administrator Robert Frosch and his deputy Alan Lovelace to take up an offer to visit Japan in July 1978, to stimulate a more concrete discussion of Japan's plans for STS (shuttle) use. He also suggested that the visit could provide "the opportunity to offer ideas for planning more of Japan's international cooperation with the United States."[39] This visit led to the establishment of a joint study group that first met in December 1978. Its US chairman was Anthony Calio, deputy associate administrator of the NASA office of Space Science and Applications; Shozo Shimosato of the Space Activities Commission (SAC) led the Japanese participants. By June 1979 they had identified 17 areas in which US-Japanese cooperation might be initiated relatively quickly, respecting Frutkin's now-classic principles (chapter 1). After a permanent Senior Standing Liaison Group meeting on a regular basis had taken up the baton, another round of cooperative agreements were signed that provided the basis for effective partnership in space science and applications between Japan and the United States.[40]

NASA, Japan, and the International Space Station

Sweeping adjectives abound when one reads about the construction of the International Space Station in scholarly journals, newspapers and trade publications.

True to its name it brings together a team of international players—mostly developed countries—to contribute components for assembling in orbit a platform for basic scientific research and for ambitious future exploratory missions (see chapter 14). Japan is one of the key partners in this international venture, and this collaboration remains to date the largest space effort between Japan and the United States. By deciding to participate in the space station in the early 1980s Japan gained the needed visibility as a space-faring nation. The 1980s also saw Japan's participation in many international scientific programs and joint science and technology collaborations with the United States.

Japan's main contribution to the Space Station comes in the form of an in-orbit floating laboratory called the Japanese Experimental Module (JEM) or Kibo meaning "hope." The first element was successfully added to the International Space Station in the spring of 2009, the complete package was assembled in fall 2009.[41] Kibo's main purpose was to create an ideal environment for the study of the earth's environment and perform microgravity experiments.[42] It will also house the world's largest wide angle X-ray camera for galactic studies. The module consists of two facilities: the pressurized module that simulates a condition similar to what we experience on Earth and an exposed facility for long-term experiments in outer space.

Kibo was not the only contribution foreseen for the ISS. JAXA also planned to build a Centrifuge Accommodations Model (CAM). CAM's core was a 2.5-meter-long centrifuge that would have provided controlled exposure of various biological specimens to a range of gravity levels from as little as 0.01 g to 2 g. The program was cancelled as part of NASA's response to President Bush's January 2004 Vision for Space Exploration.[43] That vision called for the development of a Crew Exploration Vehicle (Orion) to take astronauts to and from the moon and the ISS, and a Crew Launch Vehicle (Ares I). It also directed NASA to restrict research on the ISS to elements that supported the vision. To meet these requirements NASA reduced the number of launches to the ISS before September 2010 from 28 to 16, and dropped plans to launch Russia's Science Power Platform and Japan's CAM, whose flight model, along with the engineering model of the centrifuge rotor had already been manufactured.

The Japanese motivation to cooperate in the Space Station dates back to the early 1980s when Japan's space program was still in its growth phase—it was yet to build its own application satellite and launch vehicles. The earlier invitation by the United States to cooperate with the space shuttle was turned down by Japan because of deep concerns about its own technological capabilities and the financial commitment involved in a cooperative endeavor when it was struggling with its fledgling space program. In the words of John Logsdon, "Japan, forced to sit on the sidelines during Shuttle development, was determined not to be left out of the next major cooperative opportunity."[44]

Though the invitation to participate in the space station was made during President Reagan's State of the Union address on January 25, 1984, the negotiations and planning started much earlier. Significant meetings were organized in 1982 and 1983 to plan for the space station with potential partners that included Japan. In May 1981, a special Space Station Task Force was formed under the Space Activities Commission in Japan to coordinate station-related activities through interaction with other government, semigovernment, and private agencies. Though Japan was positive about participating in the Space Station, the financial commitment to develop their own indigenous H-II launch vehicle demanded negotiations with their own space team for simultaneously committing resources for both projects—the H-II and the space station.

As Japan wanted to be an international player in human space flight, it committed itself to contributing to the Space Station. It efficiently allocated its resources and did the preparatory work well in advance. The Kibo module remained steadily on course throughout the period from approval in 1989 to arrival at the Tsukuba space center in 1997, weathering the storms of the transformation of the station from President Reagan's Freedom to President Clinton's International Space Station that included Russia. For Japan, the ISS in general and its own module in particular offered the opportunity for a permanent participation in manned space flight and a platform where research could be carried out into manufacturing technologies in weightlessness and vacuum.[45]

Japan's participation in the space station was not welcomed by many scientists and policy analysts in the country. They saw it as a needless drain on resources when Japan should be concentrating on building a robust space program.[46] As John Logsdon put it, this led the government to recognize that "it could not both accept the U.S. offer and satisfy its other space objectives without increasing its financial commitment to space." Having decided to do so, a broad consensus was brokered between government and industry from 1982 to 1984 in favor of collaborating with the United States in exploring the potential of human space flight.[47] Seen in this light, the Space Station has both increased resources for the Japanese space effort and contributed to building that autonomy in space that the country has pursued for the last 60 years.

Chapter 10

NASA and the Politics of Delta Launch Vehicle Technology Transfer to Japan

As described in the previous chapter, Japan's quest for the development of an indigenous launch capability began with the pioneering efforts of Itokawa and his team at IIS in the 1950s and at ISAS in the 1960s. Their program to develop solid propellant vehicles (Kappa, Lambda, Mu) for launching mini satellites to low earth orbit was thwarted in the late 1960s by three consecutive technological failures along with ongoing internal problems. These setbacks left the field open for the rival solid- and liquid-fuel program being undertaken by the NSDC, which progressed from developing a three-stage Q rocket in the 1960s to an N series constructed with American help in the 1970s. This was established with an intergovernmental agreement in which Washington undertook "to provide to the Japanese Government or to Japanese industry under contract with the Japanese Government, unclassified technology and equipment [. . .] for the development of Japanese Q and N launch vehicles and communications and other satellites for peaceful purposes."[1] This chapter focuses on the circumstances leading up to this arrangement, which was strongly promoted by the State Department, and the difficulties that NASA faced in interpreting its scope, and in cooperating with its implementation. That experience, in turn, enabled Japan to develop a "home-grown" H series of rockets in the 1980s, the latest being H-IIA capable of placing application satellites weighing more than two tons in geosynchronous orbit.[2]

As mentioned in the previous chapter, Hideo Itokawa's determination to build solid-fuel rockets without foreign assistance caused some consternation in the United States. The State Department noted that the Japanese were not only offering these rockets for sale—which it "did not consider to be a significant source of proliferation of solid fuel technology"—but were also offering licensing arrangements for their production abroad, especially to Indonesia and Yugoslavia—which was of considerable concern.[3] The Arms Control and Disarmament Agency (ACDA) that was set up as an independent body by Congress is 1961, specifically to deal with all aspects of arms control, nonproliferation, and disarmament, emphasized the possibility that the rockets could morph into strategic nuclear-capable ballistic missiles within three years no matter what America did. The United States, it suggested, could counter this development by offering Japan liquid-fuel rocket technology. As the ACDA put it in

September 1966, the United States had the "ability to influence the course of Japan's rocket developments" by making "certain areas of space-rocket technology" that were less relevant to missiles "more attractive."[4] Such a move would also bolster Japan's prestige and would be in line with the kind of support the United States was offering to the European program in ELDO (see chapter 3).

The moves made by the State Department in the mid-1960s to engage Japan in closer collaboration with the United States in the wake of the 1964 Chinese nuclear test (see previous chapter) were part and parcel of a general effort to contain Tokyo's nuclear aspirations, if they should ever emerge. As a memo sent to the embassy in Tokyo put it, "[G]iven Japanese capability to develop—if it chose to change current policy—nuclear weapons delivery system unilaterally and without foreign assistance," space cooperation could serve US policy objectives "of both discouraging proliferation tendencies in Japan and encouraging continued Japanese focus on exclusively peaceful exploitation of space." The alternative, "denying to Japan certain unclassified technology relating to space exploitation," would, the State Department suggested, "encourage unilateral program and very nationalistic tendencies and suspicion of U.S. which could stimulate decision by the government of Japan over next decades to exercise its nuclear option."[5]

These views were part and parcel of an evolving quest for collaboration with Japan, notwithstanding Kaneshige's gloomy prognosis in summer 1966. An official visit by Prime Minister Sato in November 1967 provided the occasion for a collective reaffirmation of US policy by officials in NASA, the Department of Defense, the Office of Munitions Control, and the science team in the State Department. A white paper prepared in anticipation of the state visit expressed continuing concern regarding Japan's determination to pursue an independent peaceful space program. It reiterated the advantages to Tokyo of working with the United States: savings in time and money, increased prestige in Asia vis-à-vis the People's Republic of China. And it suggested that Sato's visit provided an appropriate occasion for the United States to once again express its "willingness to broaden space cooperation with Japan." There were, though, a couple of areas in which that cooperation would have to be qualified: the launching of comsats, which had to satisfy Intelsat's conditions (see chapter 5), and "assistance in the development of Japanese launch vehicles including guidance systems [that was] limited by our policy against the proliferation of nuclear weapon systems."[6] In these areas, requests for technological support would be handled on a case-by-case basis: there could be no blanket technology transfer agreement.

On November 15, 1967, President Johnson and Prime Minister Eisaku Sato agreed that the two countries should look more closely into the possibilities for space cooperation. Possible avenues for collaboration were then reviewed thoroughly in Washington. A policy statement outlining the nature of prospective cooperation with the Japanese was agreed by State, NASA, Defense, ACDA, and the White House. It was forwarded to the US ambassador in Tokyo, U. Alexis Johnson, on January 5, 1968, with authorization to inform the Japanese government of the readiness to negotiate a space agreement. The offer was conveyed to the prime minister shortly thereafter, and 18 months later, on July 31, 1969, an exchange of diplomatic notes confirmed the terms of a new US-Japanese collaborative space project. It explicitly narrowed the scope of collaboration to technology and equipment for the peaceful development of launch vehicles and communications and other satellites.[7]

This arrangement deviated significantly from the white paper drafted in inter-agency discussions before Sato's visit. Not only did it identify as core items for collaboration just those items that had been singled out as particularly sensitive—launchers and comsats—, but it also made no explicit mention that technology sharing in these two areas would be decided on a case-by-case basis, as NASA had insisted.

Privileging launchers was defended in general terms by the State Department as essential to curbing potential militaristic ambitions in Japan. As one official put it to the secretary of state, "[T]o deny cooperation in unclassified technology oth-erwise available to European partners would stimulate suspicion of U.S. motives, encourage nationalistic tendencies and could well contribute to an eventual deci-sion by the Japanese government to exercise its option to develop a military deliv-ery capability."[8] U. Alexis Johnson himself later reiterated the argument in his biography: as he put it there, "since space launchers always presented the possibil-ity of conversion to military rockets [...] we would be much smarter to be in bed with Japan from the outset rather than have it develop a new rocket of which we would be ignorant."[9] The deal would also "benefit U.S. business interests and help with our balance of payments."[10] Johnson stressed this aspect in response to the criticism that the United States was "gratuitously providing the Japanese with scientific and technological data of inestimable worth": US industries who were "interested in this matter estimate that the return to them and our balance of payments should amount to a total of approximately $350 million by 1975."[11] This was a not insignificant sum when "by the mid-1960s the trade balance was beginning to swing in Japan's direction as its manufactured goods, especially elec-tronics, began to make big inroads in the wide-open American market."[12] There was, then, more than one good reason for agreeing to share launcher technology with Japan: it was quite another matter, of course, to explicitly encourage Japan to develop with American help the most sensitive components that had been identi-fied as candidates for technology sharing, namely, launchers and comsats.

Needless to say, neither NASA nor the Department of Defense were happy with this arrangement: Frutkin in particular "resisted it as strongly as I could [...]."[13] The State Department was able to override their objections by arguing that it was in the overall foreign policy and national security interests to foster closer collabora-tion with Japan in these crucial technological sectors. They did have to make one concession to their opponents however: the US government undertook "to per-mit United States industry to provide to the Japanese Government or to Japanese industry under contract to the Japanese government" access to the unclassified technologies mentioned earlier.[14] By identifying US *industry* as the agent of the exchange Washington effectively construed the agreement as a commercial arrange-ment between the Japanese and the manufacturer of the Thor-Delta rocket.

There is no doubt that U. Alexis Johnson, ambassador to Japan from 1966 to 1969, and then undersecretary of state for political affairs from 1969 to 1973, was the driving force behind the agreement. Frutkin was explicit about this in interviews:

> See, the prime advocate of a generous hand to Japan on a vehicle was Alex Johnson. He worked awfully hard for that in all the positions he had, first as ambassador to Japan, then in State, and then this ultra special committee dealing with intel-ligence and so on. I felt it was wrong for policy to be pushed by a single person. He

was very much interested in the Japanese interest in launch vehicle technology and tried to encourage us to be more forthcoming to them. Now, in my opinion, he should have known a lot better, because he was a member of the little intelligence group, an interagency intelligence group, that would have known far better than I did that the U.S. was not interested in Japanese access to launch vehicle technology [at that time].[15]

Johnson himself was unambiguous about the importance he attached to the agreement. Writing to Robert Seamans, the secretary of the Air Force, in 1969, he emphasized that "[t]his is a project close to my heart on which I did the original spade work with Prime Minister Sato." He added that, "on balance, I think it is very much in our national interest to proceed with the project as rapidly as we can."[16]

U. Alexis Johnson made an immense contribution to the growth of US-Japanese relations during his tenure first as ambassador to Tokyo and then as undersecretary of state for political affairs. In both positions, he played a crucial role in facilitating the return of Okinawa to Japan and in space cooperation. In an interview with John Logsdon he stated that "he had always wanted to find a way to counter balance what he perceived as a pro-European bias in U.S. foreign policy by increasing U.S. interactions with Asia, and particularly Japan."[17] Johnson chaired the Space Council subcommittee on international cooperation between May and October 1966 and thus was quite familiar with the discussion regarding increased US-Japan space interactions. When Johnson was named US ambassador to Japan in October 1966, he carried with him the desire to use space cooperation as one way of strengthening the overall Japanese-US alliance. He was sensitive to Japanese "national pride" and its "technology capacity" to develop launch vehicles and satellites. Seeing the earlier overtures toward space cooperation from the United States to be ill-defined—"we had offered several times in general terms to cooperate with Japan but never spelled out what we meant"—he began working closely with Washington for a "specific proposal" to engage with the Japanese space effort.[18] During the spring and summer of 1969 in anticipation of the July 1969 exchange of notes, Johnson worked hard for such a specific agreement with the Japanese, something that would entice them and give substance to the more general proposals that had been in the air since 1965. Seeing no progress made by the staff of the undersecretary's committee, he emphatically stated that

[t]he present course can have no result other than the Japanese going it alone or forcing them into the arms of the Europeans. As you know, I deeply feel that this would be contrary to every interest that we have with the Japanese; also, it is obviously urgent that this matter be resolved before the Cabinet Committee meeting in Tokyo at the end of July.[19]

In the event it was resolved to his satisfaction, and thanks to his passionate and determined pursuit of his objective. Indeed as Johnson explained in an oral interview in 1975, he was a "great believer in getting things done, going to the core of the problem. It's been very, very rare in my career that I ever write a memorandum to anybody or do a 'think-piece' about something. [...] I like to sit down and write the telegram" giving instructions to the ambassador in the field.[20] This close identification with the project and his determination to "write the telegram" rather than draft policy papers surely helps to account for

Johnson's tenacity in bringing the US-Japan agreement to fruition on terms that strongly favored Tokyo, against strong opposition in other arms of the administration, and in NASA in particular.

Implementing the 1969 Agreement

When the 1969 Agreement was signed there were no less than 24 pending requests for the transfer of launcher technology from the United States to the Japanese fledgling N-program. Here is a typical example of one such case that was pending in June 1969. It indicates how difficult it was to decide what could reasonably be passed on to Japan, and the importance that NASA officials attached to the final terms of any agreement between the two nations[21]:

Case No Company

64–69 TRW

a. *Commodity*: Assistance in performing a "Sizing Study" of the Japanese N launch vehicle, including computer simulations.

b. *Comments*: NASA finds it difficult to evaluate the significance of this case. It recommends that a final decision should be left in abeyance until after the agreement. DOD, in an interim reply on the case, said it would not object to those parts of the assistance that are within the scope of the agreement. ACDA gave an unqualified no objection.

The agreement itself authorized American industry, with US government permission, to provide "unclassified technology and equipment [...] for the development of Japanese Q and N launch vehicles and communications and other satellites for peaceful applications." As regards launchers, an attachment specified that the agreement would hold "up to the level of the Thor-Delta vehicle systems, exclusive of reentry and related technology."[22]

This was fine as far as it went, but it did not specify just which Thor-Delta configuration was to serve as a benchmark. Successful implementation, in the view of a State-DOD-NASA team, thus required the formulation of "a package guideline that a) would enable the Japanese to reach their objective of placing a synchronous satellite into orbit, b) would not raise any security problems for the U.S. and c) more importantly would serve as a yardstick to measure specific cases, as to whether they are within the scope of the agreement and therefore approvable."[23]

The task of setting that yardstick was entrusted to a multibody group called the Technology Advisory Group (TAG). It was composed of representatives from the DOD, NASA, the State Department, and the OMC (Office of Munitions Control).[24] First chaired by Mr. Vincent Johnson, deputy associate administrator of NASA for Space Science and Applications, the TAG broadly acted as a control mechanism for limiting the technology that was transferred and made sure that the equipment that was offered to Japan provided the bare minimum configuration to place a satellite in geostationary orbit.[25]

The immediate task before TAG was to clarify the wording of the agreement that was signed in 1969. This baseline would be used by the OMC to evaluate the licenses for exporting technology and equipment. However, as Vincent Johnson

Table 10.1 Thor-Delta baseline configuration definition agreed by TAG

First Stage	Second Stage	Fairing	Spin Table	Third Stage
DSV-2L-1B	DSV–3E-4	SDV-3E-7	DSV-3E-17 (TE-364)	TW-364–3
TX 354–5				FW-4D
Adapter Section			DSV-3E-5 (FW-4)	*Attach Fitting*
DSV-3L-2				DSV-3E-6

Source: Vincent Johnson to John W. Sipes, October 30, 1970, RG 59, Box 2962, NARA.

put it, "[T]he task of generating an explicit, single faceted and easy to administer definition of the level of technology authorized and/or intended under the U.S./ Japanese Space agreement [was] not a simple one." He pointed to the agreement providing "reasonable latitude in interpretation" notably as regards the "level" of the Thor-Delta technology that could be shared.[26] Since the level could be interpreted differently depending on the specific set of conditions surrounding a particular situation, the TAG wanted to have an unambiguous baseline to use as a yardstick against which to evaluate specific requests for release. The TAG provided OMC with such a detailed Thor-Delta definition on October 30, 1970 (see table 10.1).

The TAG chose Thor-Delta 58 as the baseline launcher for collaborative purposes. This was the model that provided the first two stages of Thor-Delta 71, the vehicle in use when the US/Japan space cooperation Agreement was signed in July 1969, and it had a geosynchronous capability of 156 kilograms. Thor-Delta 58 was the least sophisticated launcher capable of achieving the geosynchronous target of 120–130 kilograms the Japanese had set for their first experimental test satellites. It had also been used in May 1969 to place an Intelsat III communications satellite weighing about 145 kilograms into geostationary orbit. To the above baseline TAG inserted this caveat:

> It should be clearly recognized that such a definition cannot be used as the sole criteria for approval or rejection of a given request. Many cases will arise where it is either impractical, undesirable, or not in our own best interest to provide the specific hardware and/or technology defined in the base line system. In these instances the judgment must be exercised as to the need, suitability and relationship to the general Thor-delta "level" or "class" of hardware and/or technology. In these instances, a rationale should be provided setting forth the reasons for departure from the base line system.[27]

Negotiations for 'N' Upgrades

The original goal for the N rocket program was for the launcher to be capable of placing a 130-kilogram satellite in geostationary orbit. Japan soon began to negotiate an upgrade to their N-1. In late 1971 and again in 1972, NASDA director Hideo Shima, responding to demands from private corporations for heavier application satellite payloads, "informally indicated they were interested in upgrading their launch capability from the initial 130 kg geosynchronous N-1 to 300 kg by the late 1970s and 500 kg by the mid-1980s." TAG was approached by the Office of Munitions Control in January 1972 to inquire if the model of Thor-Delta or other launch vehicles would help the Japanese in meeting their desired 300 kilograms to geosynchronous orbit goal. The TAG resorted to the baseline it had framed earlier and replied that the 300-kilogram limit could be achieved by a "change in configuration but not in the level of technology." However, for the 500-kilogram goal TAG was skeptical if the target could be met by reconfiguring the baseline and it suggested that a "technical approach to such a target was premature at that time." The OMC circulated the TAG memorandum to the concerned government agencies for information. Arnold Frutkin was emphatically opposed. He informed the State Department that "NASA would not want to concur in any escalation of Delta technology for the Japanese." He also took steps to "make sure that our people on TAG would not be involved in anything that would appear to be a recommendation for any increase; they could give only a technical assessment of the increase in performance which would be required if the USG decided to meet Japanese program needs."[28]

The Department of State, on behalf of the Japanese, raised the question of an N-1 upgrade again in October 1972. They asked for the provision of nine strap-ons to the first stage that could be achieved with the so-called Universal Boat Tail (UBT); an ablative cooling thrust chamber for the second-stage engine instead of regenerative cooling; an enlarged second-stage propellant tank; an eight-foot fairing to handle larger payloads; and a larger third-stage motor. The Japanese proposed to buy the hardware first, then move to "kit-type" assembly in Japan and finally to production under license. NASA replied to these Japanese requests in November 1972, limiting cooperation to hardware sales only: as one document put it, "*since the US would benefit little if at all from the sale of technology in this field, we recommended that the requested items be provided on a hardware-only basis*" (emphasis in the original). Frutkin emphasized that "the selling of hardware to Japan does not produce an independent launch capability for Japan beyond that available through launchings for Japan from the US." It was also in line with overall US national interests, and was favored by contractors such as McDonnell Douglas. In sum Frutkin recommended that

we should go along with hardware sales of the items requested by Japan since they did not represent more advanced technology (larger fairing and third stage motor), or would be exceedingly hard to reverse engineer (thrust chamber), or could be easily developed in Japan without U.S. assistance (UBT)...and since the income to industry from continuing hardware sales could be substantial.[29]

Repeated requests from Japan, and pressure from the State Department, frequently undermined NASA's preferred policy that, beginning in 1972, emphasized the sale of hardware rather than the licensing of technology. For example, NASA tried to resist Japan's efforts to secure approval for the licensing of the technology of the solid-fueled CASTOR II strap-on rocket that was manufactured by Thiokol and used on the Thor-Delta 58 that served as the baseline for the 1969 agreement. The agency tried to persuade Japan to purchase the CASTOR II rockets directly from Thiokol, rather than develop the capability to manufacture them in Japan. The State Department objected, suggesting that it had "considerable difficulty with the proposition that Japan must engage in debate in order to secure benefits to which it is entitled under an existing agreement."[30] Bowing to pressure, NASA informed the State Department that although it continued to prefer that Japan buy the rockets from Thiokol, it would no longer object to the licensing of the relevant technology, as long as there was added to the price of licensing "an additional reasonable recoupment fee to compensate the U.S. for the R&D costs incurred."[31]

After an extended interagency review the United States, by the end of 1973, had agreed to sell Japan hardware to allow for the upgrade of the N rocket to a geostationary capability of 250 kilograms. That done, in 1974 NASA dug in its heels. The agency would "fully support the original agreement which provides a synchronous orbit capability of either 130 or 150 kg, depending on how the baseline is interpreted." But NASA officials unambiguously stated that "any changes to the baseline vehicle which has the capability of improving the synchronous orbit payload capability becomes the subject of a new policy decision."[32]

TAG was disbanded in mid-1974, and subsequent consideration of Japanese requests for launch-related technology was handled through normal interagency procedures.[33] NASA became directly involved in the approval of anything related to the transfer of N1-SLV technology to Japan. When Tsuyoshi Amishima, deputy chairman of Japanese Space Activities Commission, visited NASA in April 1974 to further discuss raising the level of technical assistance, NASA's negative stance was clear. NASA knew that it had to assist Japan to satisfy the Department of State, but beyond those obligations it had "no real interest" in doing so. NASA had helped Japan on its N-1 launch vehicle, not out of conviction, but only to play a "purely" technical advisory role.[34] Now, as Deputy Administrator George Low put it,

> The fact remains that when the original agreement was reached with the Japanese the baseline Thor Delta (Vehicle No. 58) had a capability of 130 kg to synchronous orbit. Over the years this has been upgraded by various means to a 250 kg capability, whereas the United States has upgraded its vehicle to 315 kg. In other words, the Japanese have received the benefit of a high proportion of all of the upgrading activities. With these results in mind, I have no alternative but to require that either Dr. Fletcher or I approve *all* (emphasis in the original) future changes in U.S. activities having to do with the Japanese "N" vehicle.[35]

In particular, as far as NASA was concerned, any changes to the baseline vehicle that further improved its geostationary orbit capability would be the subject of a new policy decision taken by Fletcher or Low themselves.[36]

This response led to extended discussions between the United States and Japan for close on two years after the visit of the Japanese team in 1974. NASA's position was that the requested assistance for the upgrade of the N-1 vehicle was technically beyond the terms of the 1969 agreement and the agency did not wish to amend or supplement that earlier agreement. The United States proposed an exchange of letters between the Department of State and the Japanese Scientific and Technological Agency to establish sufficient understanding by both governments as to the level of technology and hardware assistance the United States could make available for the N-1 through normal export channels.[37] The United States agreed to provide Japan with the capability it desired but primarily through the export of already manufactured hardware systems and subsystems. Japan disagreed with this position, believing that the upgrades it was requesting were indeed covered by the 1969 agreement and that the licensing of the desired technologies should be allowed to proceed. The US line prevailed and in August 1976 the Space Activities Commission (SAC) announced that Japan would purchase from the United States the hardware required to upgrade the N-1 launcher. This was embodied in a 1976 exchange of diplomatic notes that constituted a new US-Japan space cooperation agreement. Because of the hardware-only limits set by the United States, all stages of what came to be known as the N-2 vehicle were of US design and manufacture (see table 10.2).

In 1971 the US assistant secretary of the East Asia Bureau, Marshall Green, gave his unequivocal support to space collaboration with Japan, defending it as a valuable nonproliferation strategy. As he put it,

> The key reasons why we originally supported the Space Cooperation Agreement with Japan were: to provide Japan with the means of satisfying requirements for national prestige that would not involve nuclear weapons; and, to get the U.S. in on the ground floor of the Japanese space program to assure an American orientation. Other related objectives were to demonstrate the value of cooperation with the U.S. to broaden our bilateral relationship in areas less contentious than security, to spur the sale of U.S technology and hardware, and to shape and influence Japanese policy in areas that hopefully will involve only the peaceful application of space technology.[38]

Green was emphatic that those same objectives were still valid in 1971, and that space cooperation provided an opportunity to bolster US-Japanese relationships that were suffering "temporary strains" in the political and economic fields.[39]

Table 10.2 US firms providing technical assistance, production license, or hardware for Japanese launch vehicles

	N-1	N-2	H-1
Vehicle integration	McDonnell Douglas	McDonnell Douglas	McDonnell Douglas
First stage			
Airframe	McDonnell Douglas	McDonnell Douglas	McDonnell Douglas
Main engine	Rockwell	Rockwell	Rockwell
Vernier engines	Rockwell	Rockwell	Rockwell
Strap-on boosters	Thiokol	Thiokol	Thiokol
Second stage			
Airframe	McDonnell Douglas	McDonnell Douglas	Japanese
Engine	Rockwell	Aerojet	Japanese
Reaction control system	TRW	Aerojet	TRW
Third stage			
Airframe	McDonnell Douglas	McDonnell Douglas	Japanese
Engine	Thiokol	Thiokol	Japanese
Fairing	McDonnell Douglas	McDonnell Douglas	Japanese
Guidance/control	Honeywell McDonnell Douglas	McDonnell Douglas	Japanese

Source: Steven Berner, *Japan's Space Program: A Fork in the Road?* (National Security Research Division, Rand Corporation, 2005).

NASA took a different tack. In the Intelsat agreements, the agency was particularly concerned about the commercial implications of helping Japan develop access to the geostationary orbit. It had little option but to yield to pressures from the State Department and Japan and went along, albeit reluctantly, with the 1969 agreement. Subsequent negotiations with Japan over sharing Thor-Delta technology took place in parallel with negotiations with Western Europe over its participation in the post-Apollo program, discussions that had been dominated by concerns of technology transfer. By 1974, after stretching the interpretation of initial agreement to accommodate requests for an escalation of payload capability to the geostationary orbit to 250 kilograms, and allowing the sale of hardware to achieve that end, the agency took control of the situation and cried halt. It had been an uphill struggle. As Frutkin put it in a memo to Low that year, working through the TAG, NASA had "consistently and successfully slowed down the outflow of production know-how and technology with potential commercial implications. We have tried especially to limit the number of Japanese assigned to U.S. plants (GE and Hughes)," he went on, "and also to provide that only results of tests, design, review etc. would be provided to the Japanese, rather than give them access to the process of exercising know-how. In all of this," Frutkin concluded with exasperation, "we are working up-hill against the fait-accompli of the State Department."[40]

The Japanese authorities, for their part, became increasingly resentful. As the head of the National Space Development Agency (NASDA), Hideo Shima, put it in 1976,

> The philosophy of the U.S.-Japan agreement was that the United States would help Japan until Japan would become a colleague...Up to now we have made efforts in line with this philosophy. Hereafter, however, the United States is saying that, although it will sell manufactured hardware related to large launch vehicles beyond the technical level of the U.S.-Japan agreement and will also provide launching services, it will not teach Japan how to manufacture hardware. Japan's position is, that is OK, and we will develop it for ourselves from now on, building on what we have learned.[41]

From Q to N to H, from Technological Dependence to Independence

While ISAS was working on the K, L, and M series of solid booster rockets in the early 1960s, NSDC, the precursor of NASDA, worked on solid- and liquid-fuel rockets designated as the JCR (Jet Controlled Rocket) and the LS-C series. Both the JCR and LS-C rockets were two-stage rockets, the first using solid fuels only, the second a combination of solid and liquid fuels. Both were later built into a three-stage Q rocket, which was in turn overtaken by the more powerful N (Nippon) launcher built with American help. Work on Q was not entirely wasted, however. Flight-testing of the JCR helped develop the control system of N-1, while a liquid stage from the LSC was later adopted as the second stage of N-1, which made its maiden flight in 1975.

The successful N-1 launch vehicle built after the 1969 agreement comprised three stages. The liquid first stage was adopted from the Thor-Delta Vehicle produced by McDonnell Douglas. The liquid fuel used was LOX and RJ-1 propellants. The engine for this first stage was produced in Japan under license with technological assistance from Rocketdyne. To give added thrust it had three strap-on boosters, Thiokol's Castor II-TX354–5, which were also produced under license in Japan. The second-stage engine was adopted from the Q rocket, as mentioned a moment ago, with some American assistance. It used nitrogen tetroxide and Aerozine 50. NASDA wanted to build this stage in Japan indigenously so as to retain some Japanese component and as a platform for building its own stages in future. Mitsubishi Heavy Industries (MHI) constructed the rocket engines for the second stage. The third-stage motor was imported from the United States.

During the 1970s N-1 launched six satellites into orbit including Kiku 2 (1977), Japan's first geostationary satellite that was built indigenously based on American technology. It was upgraded for launching heavier satellites up to 350 kilograms and was designated N-2. Following the N series the logical step toward launching heavier application satellites led to the development of the H series of rockets. Preliminary studies on the H began in the mid-1970s and two test flights were conducted in August 1986 and August 1987. The first stage, strap-on boosters, and fairing were manufactured under license and the rest—cryogenic second-stage, inertial guidance system and the third-stage solid motor—were developed indigenously. Thereafter a fully indigenous more advanced rocket called H-II was developed in the mid-1980s with the first test flight on February 4, 1994. Though this was a technological triumph for Japan it was not a commercial success. The launch cost was around $190 million, which was twice the cost of a launch with the European Ariane or American Atlas.[42]

To overcome the cost problem Japan initiated the H-IIA development program, with the primary goal of cutting launch costs in half by increasing the launch rate. While Japanese technological independence was a primary purpose of the original H-II program, the overriding commitment to low cost in the H-IIA program led to contracts with ATK Thiokol in Utah, who supplied solid rocket booster technology. Boeing and Man technologies of Germany were also selected to produce core stage tank domes.[43] Table 10.2 gives one some idea of the extensive presence of American firms in Japanese launcher development, and the gradual reliance on national industries to provide key components such as guidance and control.

Two Disputes over Geostationary Satellite Launches

In 1967 Japan's National Space Development Center strongly recommended that the country launch its own comsat by 1970 to ensure that it had some weight in shaping the negotiations on the definitive arrangements for Intelsat that got under way in 1969 and that lasted more than two years (see chapter 5). The alternative, as one document put it, was to have Japanese skies "dominated by the U.S. which as a member of INTELSAT (International Telecommunications Satellite Consortium) now has practical control of space communications networks."[44] This concern doubtless catalyzed U. Alexis Johnson's determined effort to accelerate American technological help for Japan's domestic launcher in 1969. In the event, the slow progress made in the negotiations to upgrade the N-1 led the Japanese authorities to seek alternative routes to the geostationary orbit for both a meteorological satellite and two communications satellites.

NASA was willing to consider two options: it could provide a reimbursable launch on a Delta 2914 from American soil or it could sell a Delta 2914 to Japan for launch there. The latter option was soon shelved. The agency was concerned about the transfer of launch operations know-how to a foreign country. A National Security directive (NSDM187 of August 30, 1972) specifically restricted the transfer of launch vehicles to other counties for communications satellites.[45] Finally the high cost of launching a Delta 2914 from the Japanese site at Tangeashima persuaded the authorities in Tokyo that it was preferable to request reimbursable launches from the United States for their first generation of geosynchronous satellites.

A reimbursement agreement between NASA and NASDA was signed in 1972 for three satellites. Himawari (sunflower, 325 kilograms) was a meteorological satellite built by Hughes Aircraft for Japan's NEC. Sakura (cherry flower, 350 kilograms) was a telecommunications satellite built by Ford Aerospace for MELCO. Yuri (lily, 350 kilograms) was a broadcast satellite built by the Space Division of General Electric for Toshiba. They were launched in quick succession between July 1977 and April 1978 from the Kennedy Space Center, though not before a major misunderstanding between the two partners had been resolved.

At the core of the dispute was the question of responsibility for the insertion of the satellite in the geostationary orbit. Early in 1974 NASA decided to offer geostationary orbit insertion services only for US government spacecraft launched on a reimbursable basis.[46] For other clients, NASA's responsibility extended only to the separation of the satellite from the launch vehicle at the point of insertion into geostationary transfer orbit. At that point an apogee kick motor integrated into the satellite, and provided along with it by the client, would move the satellite to its final desired position. Soon thereafter it emerged that the Japanese, for their part, were under the impression that the reimbursable launch contract with NASA included placing the satellite at the desired location on the geostationary orbit. On learning otherwise they took NASA's advice and asked for bids from five American firms that had provided software support and insertion into the geostationary orbit for foreign satellites (Hughes, Philco-Ford, General Electric, Systems Development Corp, and Comsat General). These came in at about $12–15 million per satellite, excluding hardware, a figure to be compared with the launch cost of $10 million per satellite.[47]

Early in September 1974, in the light of this information, and an imminent visit by NASA administrator Fletcher to Tokyo, the Japanese embassy asked NASA to reconsider its decision. It wanted the agency to provide a complete package after the spacecraft was delivered to the Kennedy Space Center, from checkout, installation in the launch vehicle, insertion into synchronous orbit, in-orbit check out, and, finally, movement of the spacecraft to its desired orbital position. It was only at that point that control over and responsibility for the spacecraft would be turned over to the Japanese.

NASA's associate administrator for tracking and data acquisition, Gerald Truszynski, explained what this commitment would mean to NASA. The agency would have to extend its span of responsibility considerably, and far beyond the normal provision of tracking and data acquisition support from its existing tracking stations. Providing a full range of services for three satellites launched in quick succession meant establishing a dedicated Spacecraft Project Office (probably at Goddard Space Flight Centre (GSFC)) to carry out the activities involved.

Operation of the control center and the development of the project-unique software would be major undertakings. Its personnel would not only have to be thoroughly familiar with the spacecraft design and characteristics but would probably also have to have access to the technical specifications to assure overall compatibility with the ground control systems. They would have to conduct mission analyses to determine optimum mission profiles. Also NASA would have to contract with the spacecraft manufacturers to provide the support at KSC before launch and in the control center during and after launch. In summary, the response from NASA clearly stated that to accept overall responsibility, it would have to divert significant civil service manpower for about 18 months or more. Further, it would result in a complex administrative structure since it was very probable that NASA would be essentially placed between the Japanese and their US spacecraft manufacturers. In sum Truszynski suggested that the best that NASA could do was to compute and supply definitive orbit data in real-time, and to track the spacecraft during transfer orbit. It could also lend a couple of people to each Japanese project to provide technical advice of various kinds, and could host some Japanese engineers to work in its mission control centers and other NASA locations to learn how the agency did the job.[48]

NASA's reluctance to satisfy Japan's demands was reinforced by input from US industry. Bud Wheelon of Hughes Corporation let NASA know that he would be happier if Fletcher did not strike a deal with Japan on orbit insertion during the administrator's forthcoming visit to Tokyo. As George Low explained to the NASA administrator,

> Apparently, each of the U.S. companies is in a major loss situation with respect to the satellite being built for the Japanese and had planned to use the orbit insertion business to "get well." In Bud's words, "if the government now steps into the orbit insertion business, we would in effect be subsidizing the Japanese at the expense of U.S. industry."[49]

The Japanese fought back. Their Japanese scientific counselor at the embassy in Washington, Hisako Uchida, pushed the orbit insertion case further by citing the example of the Italian Sirio satellite, where NASA offered to insert the satellite into the geosynchronous orbit. In reply to the query by Uchida, NASA again detailed its general policy associated with orbit insertion services. NASA's responsibility was limited to "insertion of the space craft into transfer orbit and all subsequent mission operations is lodged totally with the requesting agency or its contractors." NASA categorically "denied providing such services for any non-U.S. government spacecraft launched on a reimbursable basis and does not contemplate doing so in the future." NASA had to offer geosynchronous orbit injection support services for the Italians because of the formal commitment made to Italian National Research Council (CNR) in 1971. "In recognition of this commitment prior to adoption of the 1974 policy, NASA agreed in late December 1974 to honor its previous commitment and provide minimal geostationary injection support services for SIRIO only." In all other reimbursable non-US government cases injection into geostationary orbit "has been and will continue to be conducted from facilities other than NASA's." For example, the geostationary orbit injection of the Franco-German Symphonie satellite launched by NASA in December 1974 was conducted from ESOC (European Space Operations Center) in Germany.[50] With that the matter was apparently closed in NASA's favor.

As we have seen Japan's quest for launcher autonomy was intimately linked with its determination to gain access to the geostationary orbit for telecommunications satellites, both to enhance its influence in Intelsat and to strengthen its position in the global market for comsats. To secure the strength of national industry the Japanese authorities took a number of measures in the 1980s to close the home market to outside competition. NASDA channeled "all government satellite procurement to Japanese firms, prohibited the procurement of all kinds of satellites, and banned the procurements [abroad] of Japan's telecommunication giant, NTT, despite the lower price and superior quality of foreign satellites."[51] The result was that local content in comsats increased from 24 percent in 1977 to 80 percent in 1988, while local content in broadcast satellites grew from 14 to 83 percent in the same period.[52]

The US authorities, with widespread domestic support, objected strongly to the restrictions on foreign procurement by Japan in this sector. It not only excluded American firms from the Japanese market but also signaled Tokyo's determination to secure a leading position in the global telecommunications satellite market. Section 301 of the Omnibus Trade and Competitiveness Act, passed by Congress in August 1988, provided the United States with an instrument to lever open the Japanese market. The overall legislation had been in place for almost 15 years, and was a response to the change in the American balance of trade beginning in the late 1970s from a modest surplus to a massive deficit. Section 301 was tightened up in 1988 by introducing a so-called Super 301 amendment that was unusual in being "targeted against the behavior of governments in their home markets instead of focusing on the competition provided by imports in the United States" (e.g., by illegal dumping).[53] The US trade representative subsequently charged Japan as being engaged in unfair trading practices in three sectors: supercomputers, wood products, and telecommunications satellites, and threatened to impose trade sanctions against the country if it did not open these markets to US exports.

The Japanese were outraged as the United States was basically telling them to rein in their ambitions to be major competitors in the world market for comsats. Tokyo caved in all the same, canceling plans for the development of the fourth series of its communication satellite program. US producers such as Loral Space Systems, Hughes Space and Communications Group, and GE successfully won bids to supply satellites to Japanese firms, so pushing them out of the local market. As one representative from Hughes Space put it, the 1990 agreement opened a few more opportunities for the American company but, more importantly, prevented Japan from sheltering "an infant industry that might eventually become a world-class competitor."[54] By the 1990s Japan had its own launchers, and it had built up immense in-house capability in the manufacture of geostationary satellites. Its aspirations of becoming a world leader in the development and sale of space technology had not, however, been realized.

Chapter 11

An Overview of NASA-India Relations

NASA's cooperation with India began with the establishment of satellite tracking stations and space science. Cognizant of the contributions made by Indian scientists in the field of astronomy and meteorology, a scientific tradition that stretched back several decades, NASA outlined a cooperative program that focused on mutual exploration of the tropical space for scientific data. The cooperation started in the early 1960s with the loan of sounding rockets, launchers, and the training of Indian scientists and engineers at selected NASA facilities dedicated to astronomical and meteorological research. This initial collaboration gradually expanded and more advanced space application projects brought the two democratic countries, in spite of some misgivings, closer together in the common cause of using space sciences and technologies for developing and modernizing India. In the process NASA ended up coproducing a space program that articulated the sentiments of the postcolonial scientific and political elite of India. Conversely, the experience with India imparted a new meaning and architecture of what a space program should be in developing countries in Asia and Latin America.

NASA's relation with India is contextualized here in the framework of the United States' relations with India beginning in the early 1950s. The global Cold War and the ambiguities, desires, and tensions of a postcolonial nation-state vying for leadership among the newly decolonized states in the Afro-Asian region forms the essential backdrop to understanding the origins and trajectory of NASA-India relations. Using theoretical underpinnings from postcolonial, diplomatic, and science and technology studies, complemented with oral histories, this chapter weaves a narrative describing the motivations, justifications, and the trajectory of NASA's relations with India.

Two interconnected themes frame its organization. First, the history and discourse of modernization and development will be used to situate US-India foreign relations in the postwar period. In the wake of the Bandung conference (1955) leaders of newly decolonized states hoped to construct a third, "nonaligned" force in the international arena that was independent of the competing ideologies of progress that defined Cold War rivalry. Bandung also became a platform for developing nations to embrace the mantra of rapid modernization and self-reliance to leapfrog into modernity. This movement was not always welcomed by the United States, which remained at arm's length from India until its defeat in a border war with China (1962) and the Chinese nuclear test (1964). The Chinese threat was given a global dimension: the People's Republic of China (PRC) would become the model for newly liberated countries in the "Third World." To counter this threat the United States hoped both to accelerate India's emergence as a major regional power and to use its technological advantage to direct India's nuclear and missile ambitions into civilian space projects. US-India cooperation in space-based technologies was seen as a prestigious and useful alternative for the development needs of the country. The Indian scientific and political elite, aware of the evolving nonproliferation regime defined by the United States and the Soviet Union, sought to "indigenously" develop their own space technologies both for civilian and military purposes by creating new institutions domestically, and through the transnational traffic of experts, systems, and software. These themes are explored in what follows by tracing NASA's relations with India on four technological systems—tracking stations and sounding rockets, communication satellites, remote sensing, and launch vehicles.[1]

US-India Foreign Relations

One cannot understand postindependent India without reference to the United States. Scholars who have studied the history of Indo-US relations over the last five decades have almost exhausted the English vocabulary to describe the tensions that prevailed between the two largest democracies.[2] In the Cold War that ensued between the United States and the USSR soon after the independence of India and Pakistan from British rule in 1947, the United States favored an alliance with Pakistan owing to its strategic location, bordering the USSR, China, and the Middle Eastern countries. The ensuing partnership was intended to counter any communist expansion from China or the USSR into the South Asian region. While India espoused the policy of nonalignment, Pakistan sided with the United States, joined the Baghdad Pact and the Southeast Asia Treaty Organization (SEATO), and received extensive military supplies. This close alliance between the United Sates and Pakistan resulted in increased alienation between the United States and India and in the words of Dennis Kux, there was "a failure to understand each other's political, economic, and geo-strategic complexities," which ultimately "deepened these asymmetries."[3]

However, though the political relations between United States and India seemed "estranged" on the surface during most of the Cold War, it is rather intriguing to see, underneath this "cold peace," the extensive role the United States played through different government institutions and agencies to modernize India and to establish it as an alternative to the communist model adopted by the Soviet Union and, above all, China. As decolonization gathered momentum, the United States felt that it was imperative to stabilize and develop the country along capitalist and democratic ideals so as to win the hearts and minds of millions of people in the Afro-Asian region. This is evident through the massive economic aid India received from the United States during the first two decades of India's independence and the constant traffic of experts—from science and technology to cultural, linguistic, and economic fields—between the "metropolis" and "periphery."[4] Early nuclear cooperation, the origin and development of the Indian space program through NASA, artificial rainmaking experiments, oceanography studies, hybrid seeds and green revolution experiments through the Rockefeller Foundation—all of these technological projects during the 1950s and 1960s can only be seen as part of a sustained attempt by the United States to pull the Indian elite into the Western sphere of influence.

India's humiliating defeat in the border war with China in 1962 briefly brought the United States and India diplomatically together. The defeat by China was a "Sputnik shock" for the Indians that led to a rapid rise in defense budgets. Renewed importance was given to science and technology for defense purposes. John F. Kennedy's administration made use of this opportunity to promote India's democratic credentials. Kennedy's policy toward developing countries, India in particular, showed a striking difference compared to previous administrations. While Eisenhower's secretary of state John Foster Dulles divided countries into pro- and anticommunists, Kennedy and his advisers were sensitive to the needs of new postcolonial states and gave room for the expression of independent foreign polices by different countries in the developing world. They also believed that economic stability would bring prosperity and political stability that in turn would be a bulwark against expanding communism. However, ongoing distrust of India's neutrality colored Kennedy's perception of the country and restricted the scale of his innovative approaches to improved bilateral relations.

Viewed through this geopolitical contextual grid, NASA's significant cooperative endeavors were not uniform but ebbed and flowed and were constantly shaped by this larger bilateral foreign policy framework. Significant punctuation points that altered, for better or for worse, NASA's relations with India were: India's border war with China—1962; the Chinese nuclear test—1964; the Indo-Pakistan War—1971; India's first Peaceful Nuclear Explosion (PNE)—1974; the start of the Integrated Guided Missile Development program, IGMDP, in India—1983, after the successful orbiting of India's satellite Rohini through an indigenously built Satellite Launch Vehicle (SLV-3) in 1980; the impact of the Missile Technology Control Regime, MTCR—1987; the Pokhran II nuclear weapons tests—1998; and the closer diplomatic relations that ensued after the 9/11 terrorist attacks on the United States.

This study of NASA-India relations is divided into two chapters. The first is a chronological narrative spanning five decades, beginning with space sciences initiated by NASA in the early 1960s and ending with a scientific moon mission called Chandrayaan I (Moon craft) in 2008. Built and orchestrated by India, Chandrayaan I, carried two NASA-built instruments on its maiden voyage; it was a proud moment for both parties to see the maturation of a space program that NASA helped to found with the Indian scientific elite in the early 1960s. Chapter 12 describes a joint application satellite project called the Satellite Instructional Television Experiment (SITE) in 1975–1976, often quoted by NASA officials as a prime example of the agency's international collaboration. SITE led to a follow-on project in which US business corporations sold communication satellites—the INSAT series—and launches to India.[5]

Vikram Sarabhai and Homi J. Bhabha

The origin of the Indian space story often begins with the visionary scientist and technocrat Vikram Sarabhai, who is credited as being the father of India's space program. But the early phase owes as much to the pioneering efforts of Homi Bhabha who was instrumental in the establishment of scientific institutions for the growth of science and technology in modern India.[6] By the late 1950s, when initiatives were taken to pursue space research at a systematic level in India using scientific balloons and miniature rockets, Bhabha had already established the scientific ethos and the rationale for the pursuit of cutting-edge technologies in the nuclear and space domains, and had launched a major nuclear program. The duo, however, were not new to the field of space sciences; they had established their niche in cosmic ray physics even before they thought about huge scientific projects for the emerging nation state.[7] In particular, Sarabhai was aware of the research opportunities, made possible by space technology, to study the upper atmosphere. Together they initiated early interactions with NASA for possible cooperative space projects.

Vikram Ambala Sarabhai was born in Ahmedabad, India, on August 12, 1919, to a wealthy family of industrialists.[8] After finishing his early education in Ahmedabad he moved to St. John's College, Cambridge, United Kingdom, in 1937 to read for his Physics and Mathematics Tripos. He returned to India after the outbreak of World War II and continued his professional training as a research scholar for several years under Nobel Laureate C. V. Raman at the Indian Institute of Science (IISC), Bangalore. It was at IISC that he nurtured a friendship with Homi Bhabha who was also doing research on cosmic rays, and who established a research unit for this purpose at IISC.[9]

Sarabhai returned to Cambridge after the war in 1945 and his extensive fieldwork carried out in Bangalore and Apharwat in the Kashmir region of the Himalayas enabled him to receive his PhD in 1947 for his doctoral thesis, *Cosmic Ray Investigation in Tropical Latitudes*. He returned to a new India that became independent in 1947 after centuries of British colonial rule. Jawaharlal Nehru became the first prime minister. Trained in Harrow and Cambridge, Nehru believed science to be a panacea for the innumerable problems faced by humanity. As he put it,

> [I]t is science *alone* that can solve the problem of hunger and poverty, of insanitation and illiteracy, of superstition and deadening custom and tradition, of vast resources running to waste, of a rich country inhabited by starving people. Who indeed can afford to ignore science today? At every turn we have to seek its aid. The future belongs to science and to those who make friends with science.[10]

There was an air of optimism and vibrancy, and the trio—Nehru, Bhabha, and Sarabhai—with patronage from the government and private industrial enterprises like the Tata family and Sarabhai's own family business conglomerate, would become potent actors to wield science and technology to cement the new nation-state.[11]

Vikram Sarabhai started a laboratory for the study of cosmic rays in Ahmedabad, which was later institutionalized by the setting up of the Physical Research Laboratory (PRL) in 1954.[12] Thanks to the business enterprise his family had established, he became in Thomas Hughes's phrase, a heterogeneous engineer with a multitude of portfolios.[13] Popular writings portray him as an institution builder, a visionary, a pragmatist, and a Gandhian. Sarabhai never wavered in his view that the development of a nation was intimately linked with the understanding and application of science and technology by its people. He selectively appropriated the modernization theories emanating from premier institutions like Massachusetts Institute of Technology (MIT) and Harvard University in the United States and created a rationale for a space program in a poor developing country, characterizing it as "a space program for rural development." Thus for him, "Pursuit of cosmic rays and space research does not require an apology in a developing nation provided the activities are within a total scheme of priorities in the allocation of national resources." He stressed that many of those who were engaged in pure science were also "involved in the organization and conduct of education, of planning and of industrial development in fields such as electronics and chemicals." He himself "was actively interested in the application of science for the improvement of agricultural productivity and in the implications of science to society and problems of security."[14] For Sarabhai it was under these conditions that the indigenous development of advanced science and technology, including in space, was an imperative, not a luxury for a huge and relatively poor country like India. Indeed he believed that "what applies to the economy of India applies to the economy of most of the countries in the Indian Ocean region." This is also why, while advocating for space, he was vocal against the superficial craving for prestige as justification for a developing nation to embark on a space program.

Vikram Sarabhai was also known in international circles as an apostle for world peace and disarmament. He was a member of Continuing Committee of Pugwash, and participated in Pugwash Conferences on Science and World Affairs and set up the Indian Pugwash Society. In the late 1960s Sarabhai became the scientific chairman of the UN Committee on Peaceful Uses of Outer Space, and so became a vector, as it were, to publicize the benefits of space for the newly formed developing nation-states. He was also interested in integrating India with the global community. His familiarity and high esteem at international meetings was recognized and he was also elected president of the fourteenth general Conference of the International Atomic Energy Agency (IAEA) in 1970. The sense of isolation that India experienced under colonial rule served as a major impetus for building a domestic program, to engage with scientific institutions around the world as an equal partner, and to create avenues for technology and idea sharing.

US officials were captivated by the unusual combination of industrialist and physicist in one man. His participation in international conferences, his accomplishments in basic scientific research, his visiting faculty status at MIT, the list of committees, commissions, and boards that he was a member of impressed the American embassy officials in India who wrote long reports to the US State Department of his activities. These qualities played a crucial part when NASA sought to build collaborative partnerships with the developing regions—here was a member of a native elite who could be a suitable ambassador for NASA's peaceful programs in developing countries, and not just in the region but across the world.

Seeing the emerging field of space science and technology, and the promise it offered, Homi Bhabha, the secretary of Department of Atomic Energy (DAE), also carved a niche within DAE in August 1961 for a rudimentary space science and research cell. Before the formation of this cell the Indian National Committee for the IGY had been asked to serve provisionally as the National Space Committee for India adhering to COSPAR. The initial cooperative relations between NASA and India were primarily negotiated by the DAE through Bhabha and Sarabhai. Prudence was the watchword at this time since, unlike Pakistan or Japan, there was no security arrangement in place with India for the protection of sensitive or classified information. The State Department was also aware of the presence of Soviet scientists and technicians, including a number of Soviet airmen, in the country, and took note of the close cooperation that India sought with the USSR covering peaceful uses of atomic energy, operation of nuclear power stations, and the production and processing of uranium. Bhabha was particularly worrying: the State Department described him as one who expressed communist sentiments and who was also involved in "communist front activities." He was seen as a potential technocrat who could "utilize contacts with both sides involved in the East-West struggle in order to achieve the most advantageous opportunities to advance his objectives."

Sarabhai died in 1971. The direction the Indian space program took under Vikram Sarabhai and the close cooperation he enjoyed with NASA, both technical and managerial, for more than a decade shaped the future trajectory of the country's space ambitions.

NASA and the Origins of the Indian Space (Science) Program

The United States' relations with India in the civilian aspects of space dates back to 1957 when the Uttar Pradesh State Observatory at Nainital, situated in Northern India, began the optical tracking of satellites in collaboration with the Smithsonian Astrophysical Observatory (SAO).[15] This was initiated in the framework of the Indian IGY program. The technical equipment provided was the highly specialized Baker-Nunn satellite tracking camera and a quartz clock. It was one of twelve in the world that filled an important gap between Iran and Japan in the global network of tracking stations. Through these stations, the approximate positions of satellites (both Soviet and American) were obtained.

Following the launch of Sputnik by the Soviet Union in 1957 the United States, through the newly formed NASA, made several overtures to emerging "third world" countries, inviting them to participate in the space program by experimenting with sounding rockets. Some countries, seeing the prestige associated with modern space technologies, immediately responded to the offers made by NASA to establish sounding rocket bases and develop nascent space programs at home. Working on space sciences offered the newly decolonized states and developing countries the promise of a march toward modernity—the native elite viewed experimenting with rockets as a source of pride, prestige, and a visibility among nation-states. However, very few countries that accepted the offers (tracking stations and sounding rocket facilities) actually sustained and built their own space programs for socioeconomic and strategic needs.

India's first encounter with NASA came in the form of tracking stations. These became the channel through which the agency began to extend its reach to include other nations in a worldwide data acquisition system for satellites launched by the United States. By 1963, 28 such stations in 16 countries were established.[16] They not only functioned as scientific instruments for disseminating data for the United States but also served as conduits for host countries to begin their own space programs. Milton C. Rewinkel, the US consul general, remarked that "[i]t is a matter of some pride to us, too, that by making America's space knowledge experience and facilities available to foreign scientists, the United States has enabled several other countries to initiate their own space program and develop their own space technology."[17]

The initial motivation for NASA to cooperate in a sounding rocket program with India was the perceived benefit of getting scientific data on the tropical atmosphere. These ambitions neatly merged with India's long scientific tradition of studying cosmic rays and the sun's ultraviolet rays. This work had been started by physicists such as Megnad Saha, who was later followed by scientists such as K. R. Ramanathan,[18] Raman Pisharoty, Homi Bhabha, Vikram Sarabhai, and others.[19] The early space science experiments using balloons and miniature rockets during the 1950s and 1960s were gradually nurtured into a space program by Sarabhai. The implementation of his ambitions was possible thanks to NASA's help, gifted scientists, the Cold War, India's geographic location close to the magnetic equator, and the political will of the Indian leaders.

The first recorded mention of Vikram Sarabhai expressing an interest in NASA's international cooperative programs was in the spring of 1961, while he was enrolled as a visiting professor at MIT. Following his previous discussions with world-renowned physicists such as Bruno Rossi at MIT, James Van Allen at Iowa, and J. A. Simpson and P. Mayer at Chicago, Sarabhai told NASA of India's plans to start a space science research program at select facilities: the Physical Research Laboratory (PRL), Ahmedabad; the Tata Institute of Fundamental Research (TIFR), Bombay; and the Tata Institute of Nuclear Physics (TINP), Calcutta. He also described his plans to recruit trained Indian physicists in European countries and the United States.

During the meeting with NASA officials Sarabhai explored possible cooperative endeavors that could be mutually beneficial to both NASA and India, including magnetic fields, solar radio astronomy, geomagnetism, atmospheric studies from 30 to 150 kilometers, trapped particles in radiation belts and electro jet studies. In furthering these fields of research he discussed the possibility of a cooperative sounding rocket program between India and NASA and also a telemetry receiving facility at the PRL, Ahmedabad. It was also in this meeting that Sarabhai learned about the work of atmospheric scientist Lawrence Cahill of the University of New Hampshire. Cahill would later visit India to conduct a number of sounding rocket experiments. This included launching an experiment to study the equatorial electro-jet by flying a magnetometer to an altitude of approximately 200 kilometers.[20] Encouraged by this account, in July Frutkin sent a memorandum to Sarabhai proposing a working arrangement with his PRL to record data from the Explorer Number XI Gamma Ray astronomy satellite using telemetry-receiving equipment loaned from the United States. This arrived on September 6, 1961, and was the first instrument from NASA to enter India.[21]

Frutkin hoped that this ad hoc arrangement could stimulate a more durable and centrally coordinated collaborative program between NASA and a government-sponsored Indian space research committee that Sarabhai spoke of.[22] Homi Bhabha, who combined nuclear matters with space science and technology topics during his periodic visits to the United States, confirmed that such a committee was being formed when he visited NASA Headquarters between

November 9 and 15, 1961. He stated that the committee would be responsible for selecting appropriate programs for India, and for training young people in the field of space sciences and technology. It would also send representatives to participate in meetings organized by COSPAR. Bhabha suggested that the "committee" would become the principal point of contact with NASA.

Frutkin's reply to Bhabha after his visit suggested possible areas of cooperation. He saw the establishment of a sounding rocket range close to the geomagnetic equator to be "most desirable" for launching scientific payloads prepared by PRL and TIFR for detecting high-energy neutrons emitted from the sun during periods of significant solar activity. Second, he suggested the launching of Indian sodium vapor payloads to investigate various properties of the upper atmosphere near the geomagnetic equator and the possible launchings of rockets during the International Quiet Sun Year (IQSY) as part of a proposed large-scale effort to make meteorological and ionospheric soundings on a synoptic basis. Third, he stressed the importance of participation in low-altitude meteorological rocket observations in conjunction with the International Indian Ocean Expedition (IIOE). After stating the possible avenues of cooperative endeavors, Frutkin drafted a memorandum of understanding (MOU), between India and NASA outlining the broad areas of mutual program interest and indicating the general guidelines for the conduct of the program.[23]

The body mentioned to Frutkin by Sarabhai and Bhabha, the Indian National Committee for Space Research (INCOSPAR), met for the first time on February 22, 1962. It was formed within the Department of Atomic Energy (DAE) under the chairmanship of Sarabhai and was composed of eminent scientists who were instructed to manage all aspects of space research in the country.[24] The establishment of this institution brought organization and coordination to isolated space activities that were carried out in different regions across the country. It dealt with both national and international affairs, until a separate Indian Space Research Organization (ISRO) was formed in 1969. In 1972 ISRO was separated from the DAE and was constituted under the newly created Department of Space (DOS). INCOSPAR, however, did not cease to exist; it was reconstituted under the Indian National Academy of Science (INAS) and retained responsibility for the promotion of international cooperation in space research and exploration and peaceful uses of outer space, and liaison with the UN Committee on Space Research (COSPAR).

A memorandum of understanding was signed between NASA and the DAE on October 11, 1962.[25] It provided for collaborative research on the upper atmosphere using sounding rockets. Under the agreement, NASA provided nine Nike Apache launchers, a trailer-mounted telemetry receiving station, a trailer-mounted DOVAP tracking system, a trailer-mounted MPS-19 radar with 016 computer and 70 KVA diesel generators, a Judi-Dart launcher insert, K-24 cameras for vapor cloud photography, and tracking and telemetry equipment and ground instrumentation on a loan basis.[26] These were to be used for joint scientific experiments to explore the equatorial electro-jet[27] and upper atmosphere[28] winds from the geomagnetic equator. Considering that India was pursuing a policy of nonalignment at the height of Cold War rivalry, NASA was also eager to enter into cooperative arrangements with Delhi to "maximize the orientation of Indian scientists towards the US and away from the Soviets in the advanced application of science."[29]

While the INCOSPAR was being constituted the UN Committee on the Peaceful Uses of Outer Space (COPUOS) passed a resolution recommending and sponsoring the creation and use of sounding rocket launching facilities, especially in the equatorial regions in the southern hemisphere. Taking the cue from the United Nations, a possible site in Southern India was discussed by the Indian scientists along with NASA. To help choose the most appropriate location, NASA forwarded volumes of the Wallops Island handbooks, and Frutkin communicated to Bhabha his willingness to host Indian representatives at Wallops for additional discussions and/or to send NASA representatives to India for "possible assistance there in problems relating to site selection and instrumentation."[30] The role played by the Indian pioneers in the selection of this site is often stressed but the extent to which scientists and officials from NASA were also involved has been ignored.[31] Reports indicate the active participation of scientists R. G. Bivin, Jr., Robert Duffy, and Lawrence Cahill of NASA, and of their close relationship with Vikram Sarabhai.[32]

The Thumba Equatorial Rocket Launching Station (TERLS) was established in 1963 at the coastal village of Thumba, in the state of Kerala. Its southern location (8° 33' N, 76° 56'E) close to the magnetic equator (0° 24'S) proved an ideal location for launching sounding rockets to undertake geophysical investigations, particularly those dealing with the interaction of neutral and charged particles in the earth's magnetic field.[33] The advantages of such a site were pointed out by Frutkin. As he noted, the "true potential of sounding rockets as a scientific tool can be realized only if many vertical profiles are obtained—in a wide range of localities and epochs—with correlation of the results. International cooperation is obviously an essential ingredient for sounding rocket work."[34] Cooperative launchings of sounding rockets took place in many countries with shared responsibility from the host countries, mainly ground instrumentation and data analysis.[35] Sarabhai saw the importance of sounding rockets for upper atmospheric studies but also recognized the importance of ground facilities such as those at Thumba. "The study of this region in the equatorial areas is one of the major gaps in the study of our environment today," he wrote, adding that "as far as India is concerned with the facilities that have grown up, we have fantastic opportunities in the years to come to understand many complex phenomena involving the interaction of the ionosphere with the geomagnetic field, problems of the neutral and the ionized atmosphere and the interaction of these two." Consistent with his stress on the significance of basic research for applied and socially relevant problems, Sarabhai went on to emphasize that "these subjects are of importance not only for the understanding of radio propagation, but also from the point of view of meteorology and basic problems of energy and momentum transport into the lower atmosphere where climate is made."[36] These were persuasive claims for an agricultural economy that depended crucially on the weather to feed millions of rural families.

Scholarly research on the origins of the Indian space program often mention the launch from Thumba of a Nike Apache sounding rocket donated by NASA on November 21, 1963, as the starting point of the Indian space program—truly a historic moment for the country. Apart from making scientific measurements in the southern region of India, it was also a visual manifestation of modernity in the tropical skies. When the rocket lit up the twilight sky with an orange trail left by sodium vapor experiments, there was real excitement and jubilation in the subcontinent (see also the section on France in chapter 2). The Legislative assembly of Kerala, a communist-led government,[37] where Thumba is located, was adjourned for a few minutes so that the members could watch the magnificent display left behind in the western sky by the Nike Apache rocket and the sodium vapor trail.[38] This spectacle, displayed thanks to the joint collaboration between NASA and INCOSPAR, was translated into a great achievement for the early Indian scientists and national leaders, who saw space research as a harbinger of modernity in the newly decolonized state and as a symbol of prestige and development.[39]

Site selection was just the first step, of course. Beyond this there were various technological hurdles to establishing a sounding rocket range for launching and retrieving data from the sounding rocket payload. To ease the difficulties the MOU between NASA and INCOSPAR included a provision for the recruitment of a small group of young men affiliated with INCOSPAR to visit NASA for training at the Goddard Space Flight Centre, and at the Wallops Island facility, where they would learn about building and launching sounding rockets. This training was in assembling imported sounding rockets and their scientific payloads, procedures for the safe launch of these rockets, tracking the flight of the rockets, receiving data radioed down during flight, and collecting other scientific information required. Initially, eight Indian representatives appointed by INCOSPAR were trained at NASA field centers for approximately six months in preparation for operations at the Thumba Range. On their return, these men set up the sounding rocket range in Thumba. Subsequently, there was a constant traffic of scientists and engineers, in batches, from India to NASA facilities during the 1960s.

What began as a "bilateral Indo-American launching facility" at TERLS evolved into an international facility, a productive site where different countries, including France and the Soviet Union, could join together for promoting the peaceful uses of outer space in spite of their political differences. Frutkin strongly favored Soviet participation, believing that it "might lift some of the veil of secrecy from Soviet space activities."[40]

Frutkin suggested to Sarabhai that he offer TERLS to international participants and to seek UN sponsorship. A resolution was later introduced by the United States into the Technical Subcommittee of the UN Committee on the Peaceful Uses of Outer Space for UN sponsorship of sounding rocket ranges in "scientifically critical locations," encouraging other countries to use such facilities.[41] COSPAR was also looking for the creation of an equatorial sounding rocket launching facility for two major international programs—the International Indian Ocean Expedition (1962–1967) and the International Quiet Sun Year (1964–1965).[42] Sarabhai decided to make TERLS available and told Frutkin that "you will be glad to learn that India has decided to extend an invitation for the location of a U.N. equatorial launching facility in India, on the lines of the recommendations made at the Geneva meetings of the Scientific and Technical Subcommittee of the U.N Committee on the Peaceful Uses of Outer Space."[43] R. Shroff, deputy secretary, Department of Atomic Energy, government of India, said that "if the United Nations accepts the offer, it is our intention that the launching facility to be set up in collaboration with NASA should be dovetailed into the international facility."[44]

In January 1964 a team of scientists appointed by the UN committee inspected TERLS to determine its compliance with the condition of sponsorship for an international sounding rocket facility, and reported favorably. Sarabhai years later mentioned that "the sponsorship of TERLS by the UN [was] not simply formal; it constituted an umbrella under which over 105 rocket experiments were conducted by various nations like France, Germany, Japan, United Kingdom, USA and USSR, jointly with India, as an example of active co-operation in space research."[45] An International Advisory Panel was formed comprising two representatives each from India, the United States, the USSR, and France to continue operations. TERLS was formally dedicated to the United Nations in 1969 in the presence of various dignitaries including, from NASA, Arnold Frutkin and Leonard Jaffe, director of Space Applications programs, Office of Space Science and Applications. The meeting was presided over by Prime Minister Indira Gandhi.

The sounding rockets provided by NASA during the early 1960s were "low-end" declassified scientific instruments. The case of the transfer of Arcas sounding rockets for the International Indian Ocean Experiment (IIOE) throws light on the sensitiveness of donating advanced sounding rockets. IIOE involved multinational sounding rocket experiments at various points in the Indian Ocean region for the "intensive and coherent investigation of an ocean atmosphere regimen." NASA wanted to organize this joint experiment in cooperation with the National Academy of Science, the US Weather Bureau, and the American Coordinator for Meteorology in the IIOE, along with India and Pakistan. Problems soon emerged. The Atlantic Research Corporation manufactured the Arcas rockets for the Navy and they classified the technology as "confidential." Providing these rockets to Pakistan did not cause any problem because, as was pointed out earlier, Pakistan was a preferred ally of the United States, and a diplomatic framework was in place to enable the transfer with appropriate guarantees. But there was no such framework for dealing with India—and Frutkin felt that it would be "awkward to conduct an Indian Ocean program without the participation of India." He cited the visit of Prime Minister Nehru to the United States in the fall of 1962 and specifically mentioned the joint statement issued by President Kennedy and Prime Minister Nehru, which indicated that space cooperation was among the areas of US/India relationships that were discussed. Frutkin was so determined that this multilateral project should work that he devised alternative arrangements for giving India the Arcas sounding rockets either by "declassification of Arcas or by provision of the classified Arcas under suitable waivers and guarantees."[46]

When TERLS became operational with the launching of foreign sounding rockets Sarabhai actively sought to advance the field by nurturing the development of space technology in India incrementally. Needless to say, without external assistance and training it would have been extremely difficult for India to have built a sounding rocket program at this early stage. In the early 1960s, when rockets had attained the capability of launching satellites, Sarabhai was still developing small sounding rockets. This effort has to be understood within his larger picture of developing a nucleus of capable scientists and technologists around the essentials of rocketry, which would eventually help India if a path was taken to indigenize launch vehicles. Sarabhai noted that "when a nation succeeds in setting up a scientific program with sounding rockets, it develops the nucleus of a new culture where a large group of persons in diverse activities learns to work together for the accomplishment of a single objective."[47] Also, in August 1968, for the first time a concrete effort was made by the United Nations to host an international conference on the Exploration and Peaceful Uses of Outer Space in Vienna. Leading scientists from around the world attended the conference and reported about the activities carried out during the first decade of the space age and the plans for the future. For many developing countries in Latin America and in the Asian region, the space age dawned at Vienna.[48] Founding fathers of many developing countries' space programs saw the immense promise of space science and technologies for socioeconomic development. Sarabhai was the scientific chairman at the conference and in his presentation he talked of there being a "totality about the process of development which involves not only advanced technology and hardware but imaginative planning of supply and consumption centers, of social organization and management, to leapfrog from a state of backwardness and poverty."[49]

The first step in that direction was directed toward the indigenous production of sounding rockets and complementary subsystems—scientific payloads, instrumentation, telemetry, and ground systems. As a result of this conscious attempt, Thumba during the early 1960s witnessed both the transnational traffic of scientific and technological experts and the mushrooming of new firms, facilities, and institutions. A Rocket Propellant Plant (RPP) and Rocket Fabrication Facility (RFP) were established in Thumba. The indigenous production of sounding rockets was gradually scaled up to a satellite launch vehicle that could place a small satellite in low-earth orbit in 1980. Parallel skills were also acquired in satellite technology. A step in the direction of participating in the evolving global satellite communications system was taken through the establishment of the Experimental Satellite Communication Earth Station (ESCES) by INCOSPAR in Ahmedabad with assistance from the United Nations Development Program (UNDP) through the International Telecommunication Union (ITU)—the executive agency of the project. The equipment came from National Electronics Corporation (NEC) of Japan. Through an agreement with NASA this earth station participated in the Application Technology Satellite (ATS-2) Test Plan. ESCES was also foreseen by officials at NASA, the UN, and INCOSPAR as a node for training scientists and engineers from several developing countries in the field of satellite communication and related technologies.[50] When Sarabhai became the head of India's Atomic Energy Commission, after the tragic death of Homi J. Bhabha in an air crash over Mont Blanc in 1966, he was himself thinking of how best to use nuclear power for development needs. By associating itself with the tenets of modernization the nascent space group was able to convince the Indian government of the potential of the space program for socioeconomic benefits and thereby extract financial support for their efforts.

ISRO was formed on August 15, 1969. By this time several other institutional developments had been initiated by Sarabhai and a concrete ten-year plan for future nuclear and space activities was brought out, entitled the *Profile for the Decade*. This 40-odd-page booklet was produced by the Department of Atomic Energy—mainly Sarabhai and his cohorts. The *Profile* stated that

the principal objectives of the space programme in India are to develop indigenous competence for designing and building sophisticated hardware involved in space technology including rockets and satellites for scientific research and practical applications, the use of these systems for providing point-to-point communication and a national television hook-up through a direct broadcast synchronous satellite, and the applications of satellites for meteorology and for remote sensing of earth resources.[51]

The Satellite as a New Plow for Rural Farmers: NASA, Hasselblad Cameras, Coconut Wilt Disease, and the Origins of Remote Sensing in India

Among the developing countries only Brazil and India have advanced remote sensing capabilities. Ideas of using modern remote sensing techniques for observing natural resources began to take shape in the late 1960s. Many scientists were sent by Brazil and India to US institutions, mainly MIT and Stanford, for basic training in the use of remote sensing technology. Beginning in the 1920s, black-and-white aerial photography was used for land survey and river assessments. Multispectral imagery was introduced in the 1970s.[52] The availability of revolutionary Landsat images produced by a series of American earth observation satellites in the 1970s opened new possibilities for the Indian planners to use this technology for the management of natural resources.[53] These images were used extensively for surveys and for tracking natural vegetation.[54] The promise of this new technology led to the institutionalization of remote sensing in India.[55] NASA played an important part in the evolution of the technique by training scientific personnel and providing scientific and technological instruments to promote this new field. This helped impart technological know-how to the Indian scientists enabling them to build the first Indian Remote Sensing (IRS) satellite. The eminent Indian scientist P. D. Bhavsar viewed remote sensing in India to be full of cooperative and collaborative efforts, between scientists and engineers, technologists and bureaucrats, planners and decision-makers, at all levels, within and across national boundaries, between the technically advanced and developing nations, and between developing nations themselves.[56] What follows is a short account of the relationship between NASA and India since the early 1970s in the field of remote sensing.

As stated in the previous chapter, the UN conference on the Peaceful Uses of Outer Space held at Vienna in August 1968 was an important milestone. It was attended by delegates from many countries who presented papers that dealt with applications of aircraft-based remote sensing in agriculture, forestry, soil mapping, watershed inventory and planning, pest and disease detection, mapping of forest fires, range surveys, hydrology and water resources development, and geological applications. The EROS, Earth Resources Observations Satellite program of the US Department of Interior, was discussed for the first time. One of the major objectives described was "to disseminate data collected by the satellite to appropriate scientists in order to facilitate assessment of land and water resources of the U.S. and other nations desirous of this information." The conference also discussed in great detail all facets of international cooperation and opportunities, including economic, legal, and social problems of the exploration and use of outer space.

The decade following this conference saw a great spurt in the international collaborative activities in the field of remote sensing. In its initial phase these activities were almost entirely bilateral. On September 28, 1969, US president Richard Nixon told the UN General Assembly that America would proceed with its earth resources program so as to share the benefits of its work in this field with other nations "as this program proceeds and fulfills its promise." In accordance with UN General Assembly resolution 2600 (XXIV), NASA concentrated on actions to inform the international community about the evolving American program, to offer orientation and training, and to mount aircraft-based programs in preparation for the later use of satellite data.[57]

After this UN meeting Vikram Sarabhai constituted a small group at the space physics division of the Space Science and Technology Centre (SSTC) in Trivandrum to develop remote sensing. This small group was later moved to the Physical Research Laboratory, PRL, in Ahmedabad. It was expanded and later moved to Space Applications Centre, SAC, located in Ahmedabad under the eminent meteorologist and father of remote sensing in India, P. R. Pisharoty.[58]

The first interdepartmental meeting was convened by ISRO in December 1969 for acquainting the policymakers and departmental chiefs about the potentialities of remote sensing for earth resources surveys. About 40 members representing various organizations attended this meeting[59] and several members of parliament and a number of Indian policymakers in the government attended for part of the time. As a result of this, it was decided to conduct a small remote sensing project for the early detection of the blight disease, which affected the coconut plantations. This wilt disease devastated coconut plantations in the Travancore-Cochin area of the Kerala state in Southern India. It affected about one hundred thousand acres of coconut plantations and was estimated to cause an annual loss of about $2 million. Hence any method of early detection was of great economic value to the state of Kerala. It was decided to carry out an aircraft survey for this purpose. It was also decided to conduct this work with minimum expenditure by utilizing the existing facilities and manpower of the ISRO. Coincidentally, ISRO's Thumba Equatorial Rocket Launching Station was also situated in this locality and the detection of coconut wilt disease using an aerial remote sensing technique was taken up as a good opportunity for justifying the usefulness of a space research program to the nation.

ISRO communicated this interest to NASA and their request was forwarded to Edwin Henry Roberts, an expert scientist in agriculture and forestry remote sensing at the University of California, Berkeley.[60] Roberts suggested it was possible to identify diseased trees through aerial remote sensing at an early phase of the disease. Further, at ISRO's request, NASA arranged to send one scientist from Roberts's lab in early 1970, to help in taking the necessary photographs. The program was accommodated in the existing agreement for the conduct of scientific experiments between two space research agencies of India and the U.S.

As a collaborative effort between India and NASA, two 70-mm Hasselblad cameras and films were loaned to India. The helicopter was given to TERLS by the Hydro Meteorological Services (HMS) of the USSR, an agency that collaborated with ISRO on scientific work in rocket meteorology and upper atmosphere studies. It took photographs from a height of about one thousand feet using Kodak infrared films and panchromatic black-and-white films using different color filters. A total of about four hundred infrared false color (these images are produced by coloring the invisible portion of the electromagnetic spectrum) and black-and-white pictures were taken over a period of five days. Most of the photographs showed very fine details and were found to contain very valuable information. The photographic results confirmed that the disease could be detected by the new technique even before visual symptoms appeared.

The success of the aforementioned program led ISRO to plan a continuing future program. As a second step, ISRO took up the project to complete an infrared scanner for aircraft use. The infrared scanner was constructed in France at the laboratories of CNES by an Indian scientist and an engineer in collaboration with a French group. It was used for the thermal mapping of oceans and land areas from an aircraft platform. Many scientists were sent to American institutions along with P. R. Pisharoty to learn the benefits of using remote sensing technology.

To convince the Indian bureaucracy, a test was conducted to show how remote sensing technologies could be used for addressing agricultural problems that were faced by India. In 1973 user agencies participated in a seminar on remote sensing, and specially prepared papers were presented on the role of space technology in various application areas to convince the user departments of their importance. Such efforts not only promoted the applications, but also established a healthy trend where the user agencies defined the sensor needs for the satellite, a key factor in the success of the program. To introduce remote sensing technology for applications in various fields the National Remote Sensing Agency, NRSA, was set up in 1975 under the Department of Science and Technology, which became the nucleus of Indian remote sensing. It was involved in the training and education of scientists.

Six years after NASA had launched Landsat 1 (ERTS-1) in 1972, NRSA negotiated a deal to receive Landsat data directly in India by setting up a receiving station. The governments of the United States and India signed a memorandum of understanding, which covered the services to be offered to India and the terms of payment to the United States. NRSA sent its engineers for training to the United States in order to help set up a Landsat receiving station in Hyderabad, located in the state of Andhra Pradesh, which was commissioned in 1979. The station was expanded in later years to receive data from the French SPOT, the European Remote Sensing Satellite (ERS-1), and the US NOAA meteorological satellites, Canada's Radarsat, and India's own Indian Remote Sensing, IRS, series of satellites. The follow-on second generation IRS satellites, IRS-1C and IRS-1D, with better spectral and spatial resolution, stereo viewing and on-board recording capabilities further added to the country's remote sensing ability.

NASA and an Indian Launcher

The sounding rocket program in India provided an important stimulus to the development of an indigenous capability in rocketry from as early as 1961. In that year G. B. Pant, a scientist based in the Birla Institute of Technology, expressed a desire for assistance in establishing a Department of Rocketry at the university level in India. His request was refused citing the potential strategic military implications.[61] The United States had no security agreement with India under which assurances were given for the protection of sensitive information.[62] However, in 1964, Professor Pant again approached NASA with the "endorsement" of Sarabhai seeking NASA support for the assignment of an American academic expert in solid rocket propulsion theory to spend a year initiating a research program at the Birla Institute. The US Department of State gave a favorable response and NASA arranged with Princeton University to send Maurice Webb to work on the theoretical aspects of rocket propulsion. After the completion of Webb's "tour-of duty" Pant again asked for two experts in the field of propulsion and aerodynamics. By this time Sarabhai was also planning to come over to the United States to recruit fifteen people for a solid rocket development program in India under the auspices of INCOSPAR. India was building French Centaure rockets under license with Sud-Aviation and Hideo Itokawa at Tokyo University (see chapter 9) was providing consulting assistance.[63] Situating Pant's request in this broader context (and aware of even greater Indian ambitions, to be discussed in a moment), Frutkin sent a cautionary confidential memo on August 25, 1965, to J. Wallace Joyce, acting director, International Scientific and Technology Affairs in the Department of State about the risk of supporting such an academic endeavor. As he explained, NASA had "so far carefully avoided contributing to rocket development programs abroad." Several other Asian countries, including Pakistan and Indonesia, were interested in developing rockets and once the agency had helped one it would necessarily become embroiled in helping the others. Frutkin concluded by remarking that while NASA wanted to accommodate the State Department's wishes, it was concerned that "assistance in the Birla program as now understood might compromise NASA's international space responsibilities, involve NASA in a difficult precedent with regard to other countries, and might contribute to nationalistic competition with military implications," most obviously as regards India and Pakistan.[64]

The Chinese nuclear test in October 1964 triggered greater ambitions. Both Homi Bhabha and Vikram Sarabhai discussed the possibility of cooperating with NASA in building a launch vehicle as one response to the loss of prestige to their communist rival. The discussions centered on procuring the technology of the all-solid four-stage Scout rocket. Commonly called the "poor man's rocket," it was capable of launching satellites weighing close to 100 pounds into low-earth orbit. In February 1965, Bhabha asked Frutkin about the cost and time factors for the development of a small satellite booster system. The results were predictable. Frutkin reminded him that whereas the Scout had been approved by the Department of State as available in principle for purchase by other countries in connection with scientific research, the transfer of this technology as such posed a quite different problem. Granted the security aspects, this was "a matter for determination by the Department of State under Munitions Control procedures."[65]

Bhabha's visit to inquire into the possibilities of acquiring Scout rockets triggered a major exchange between Frutkin and Robert F. Packard in the State Department, who was interested in finding ways to assist India regain regional influence without developing nuclear weapons. He sought detailed advice on India's ability to engage in a range of programs, from launching its own satellite outside India with foreign assistance using a foreign launch vehicle to launching an Indian satellite as a solely national enterprise, as France would do in November 1965 with its Diamant/Asterix (launcher/satellite) combination.[66]

Frutkin responded in detail to the queries and did not think that India could do too much in the short term. Regarding the time frame, he pointed out that even if India made fundamental progress in major areas in the development of a booster within five years, US, Japanese, and French experience suggested that India could not complete a *total* booster system in this time. Comparing the Indian case with France and Japan he noted that the Japanese had been working on solid propellant technology for close to ten years with a fairly large industrial base without any concrete results. Similarly, the French had been working for at least six or seven years toward building a satellite launch vehicle without reaching their objective. Frutkin noted that India might also have difficulty with respect to several systems that go with the launcher—telemetry, command, guidance, test, and check out systems. He categorically stated that such an extensive program would "preempt all of the known Indian competence in the necessary areas for a period of years roughly related to the period of time used by France and Japan." As regards cost, this was likely to be $55–65 million—$ 45 million for building a launcher. Add another $11–15 million for launch facilities: Frutkin pointed out that Thumba was small and not a conducive location for satellite launching, so a launch site on the East coast was needed.[67]

Of course cost and schedule could be reduced with foreign assistance. Sarabhai had apparently already done a cost analysis of a "partially independent Indian booster development program for a Scout type vehicle at $ 25 million using French and Japanese technology."[68] He added that an "indigenous" satellite would cost around "2–4 million and would take the Indians three years with foreign assistance." If India sought the help of Japan and France, the country "could probably produce a satellite launch vehicle in 8–10 years." Sarabhai estimated that if US assistance was forthcoming this could be reduced to seven–eight years.[69]

Should the United States help speed up the "Indian National" booster program the time required could be reduced substantially. Frutkin noted that Scout guidance, for example, was not classified and could very likely be made available to India under existing policy (this system is essentially an attitude reference system with limited value for strategic purposes). Nevertheless, substantial numbers of personnel would be required to work in India, with inevitable publicity and high costs.[70] In short, if the United States agreed to cooperate, it would be only "partially an indigenous development" and the whole process would "involve highly visible foreign assistance" so defeating the purpose of boosting India's prestige in the subcontinent using space technology.

There was an alternative: cost and time could be significantly reduced if the Indians were to use a Scout in America. If, as in the case of the Italian San Marco project (see chapter 2), the arrangement were to be a cooperative one between NASA and the Indians, NASA could provide the launch vehicles at a cost of about $3 million to the United States. This latter alternative assumed that the project would be of sufficient scientific or political value to America to justify direct US involvement and expenditure, of course.

Nothing came of these initial approaches. While work at TERLS engaged Indian energies in the latter half of the 1960s, Sarabhai promoted the indigenous production of launch vehicles through the incremental development of sounding rockets. This is evident from his address at the UN conference in Vienna and the institutional developments directed toward the needs of a budding launch vehicle program.

At the UN Conference in Vienna in 1968 Sarabhai spoke about the importance of an indigenous capability, fully aware of the difficulty of getting foreign assistance: "[T]he military overtones of a launcher development program of course complicate the free transmittal of technology involved in these applications."[71] By 1968 he had already done a cost analysis of building a launch vehicle program and the required ground systems, including a launch pad on the eastern sea coast. He factored in the costs of a scientific pool for supporting a fully fledged program.[72] Reports and published sources indicate that at this time India made its first-ever study for developing its own Satellite Launch Vehicle (SLV).

The Chinese launched Long March I (CZ-1) on April 24, 1970, placing the Dong Fang Hong (the East is Red) DFH-1 satellite in low-earth orbit. Though launched a few months after the Japanese launch of the Osumi satellite in February 1970 using the Japanese Lambda rocket, the Chinese launch triggered an outcry in India. The debate in India, soon after launch, centered on whether the country should develop a nuclear deterrent against China—India had refused to sign the Nuclear Nonproliferation Treaty (NPT) brokered by the United States, the USSR, and the United Kingdom in 1968—and the resultant opinion was highly in favor of one. The then defense minister Swaran Singh "reaffirmed" before the Indian parliament that he would "review the possibilities for an accelerated space program."[73] This triggered another effort by Sarabhai to obtain US cooperation in building an Indian launching capability including guidance and control technology.

An April 1970 memo from the American Embassy in Delhi to the State Department, after detailing the situation in India, warned that "US denial would generate serious irritation in Indo-US relations, would turn Indians to other suppliers and would inhibit our capacity to *monitor* Indian space research developments, and our ability to influence developments toward peaceful rather than military applications."[74] Another such dispatch a few months later reiterated these points.[75] However, here the negative arguments far outweighed the pro arguments for any meaningful cooperation. "India's overall economic development could be imprudently retarded by major expenditures in atomic and rocket fields": something else was needed to contain hunger in the rural areas. Helping India would also send the wrong signals to China and Pakistan concerning American policy on international military applications of science and technology. If the United States provided technology to India and not to other interested countries it would have "corrosive effects" on US relations elsewhere. A "premature US commitment" could also "inadvertently nudge Government of India's program into direction Indians might later find fruitless, with possible consequent recriminations against U.S." The US government was also aware of the rhetoric of the Indian political elite that "only a nuclear equipped India can win a rightful place in counsels of major powers." US support would "stimulate advanced rocket development" and enhance the early development of "Indian nuclear weapons system." The United States, as the architect of NPT and an opponent of Indian nuclear weapons development, would not even indirectly wish to facilitate such an Indian decision. In light of these considerations, the embassy recommended a flexible long-range policy of selective cooperation and restraint whereby the United States could provide India unclassified technology and other types of assistance directed toward India's peaceful economic and social development.[76]

The State Department looked into these possibilities from various angles bearing in mind the agreement being reached with Japan over the provision of unclassified Thor-Delta technology (chapter 10). Anthony C. E. Quainton, senior political officer for India in the department, discussed possibilities with U. Alexis Johnson who struck the deal with Tokyo. He favored a joint collaboration with the Indians up through the Scout level in unclassified technology on propulsion systems without financial support and with suitable assurances about peaceful use.[77] In December 1970 Joseph T. Kendrick sent a proposal to Robert A. Clark of Munitions Control (MC) asking him to agree to assist India on similar terms as agreed between the United States and Japan. Clark's reply indicated that he had no policy objection to the substance of the proposal. However, he expressed reservations about sending the proposal to Johnson for approval as it had not been discussed with NASA and the DOD who had been unhappy with the Japanese arrangement. Clark drew attention to the vagueness of the offer to cooperate in the development of a limited space program "up to and including the general level of Scout Rocket Technology." Clark said that he knew "from personal experience that Indian officials are aggressive and persistent individuals who might be more likely to cry foul whenever they believe correctly or incorrectly that their understandings differ from someone else's understandings." Thus, wrote Clark, "the USG position on what Scout technology means should be prepared in advance and not left to chance as has been done with the Japanese and Thor-Delta technology"[78]—a nice example of bureaucratic learning.

Three years later we find that, though critical elements of launch vehicle technology were denied, the declassified State Department papers indicate the approval of some "hardware" related to sounding rockets and satellites, which were "unsophisticated in character." However, the Indian space program was still closely watched for potential ballistic missile activities. As the memo put it, "So far, the Indian program appears peaceful in character—as the Indians claim—but it is developing the technological capability for a missile system should the Government of India opt for this course."[79]

The last available discussion on the subject was in a confidential memo from John Sipes to Joseph Scisco on June 27, 1973, requesting the formulation of a departmental position on whether it would be in the overall interest of the United States to assist India in the development of its space program. The question was prompted because the Office of Munitions Control had received a number of requests from industry for Department of State approval to export space hardware and technology for India's space program. The hardware included components such as gyros and accelerometers, which were essential for the guidance and control of launch vehicles and missiles.[80] Sipes brought up the Japanese case of Thor-Delta technology transfer for comparison and explained how the Japanese had undertaken to use the launch vehicle and satellites developed with US assistance on condition that they would be used for peaceful purposes only and in line with the Intelsat agreements. This was not the case with India. In fact in April the Indian government announced that it was developing missiles for its armed services. Sipes asked rhetorically whether US help to India with satellite and launcher development would further "world peace and the security and the foreign policy of the United States." He concluded that it was not "prudent to

permit the release of space hardware and technology" especially gyros and accelerometers, which were critical for inertial navigation systems.[81]

A Scout license production agreement never made its way into India. The
Indo-Pakistan war of 1971, Sarabhai's sudden demise in December 1971, and
the Peaceful Nuclear Explosion (PNE) by India in 1974 all undermined the
possibilities of cooperation between NASA and India in sensitive technologies.[82] To make matters worse from Washington's point of view, in the light of
the alienation with the United States the-then prime minister Indira Gandhi
sought increased friendship with Moscow, which led eventually to the successful launch of three Indian satellites by the Soviet Union.[83] This is not to say
that NASA had no regrets. A recent interview with Arnold Frutkin captures the
dilemma that NASA and the State Department faced when it came to sharing
launch vehicle technology. The issue, as he stressed, "was slanted by the fear
that India would be using it as a delivery weapon for—a delivery vehicle for a
weapon." Frutkin was discussing the acquisition of a Scout with India at the
time, and was convinced that "in the long run India would have what it wanted
by way of a delivery vehicle or a space vehicle, and either they would have it
with our goodwill and friendship or they would have it over our dead bodies."
His preference was plainly for some kind of collaborative arrangement, and
he personally regretted that the United States had not been more forthcoming, leading to Indian resentment and a decline in relations, all of which, in
Frutkin's view, "could have been sidestepped by working with India to arrive at
just where they are today."[84]

The Indigenous Development of a Launcher

The idea of using a Scout design for India's first SLV persisted ever since Bhabha
and Sarabhai contemplated developing a launch vehicle. Several years of negotiation, and the familiarity Indian scientists and engineers gained with Scout
during their tenure at Wallops Island and other NASA facilities, played a key
role when India opted for a launch vehicle that was at once proven and reliable and within India's reach. Gopal Raj also claims that the Scout model was
chosen because "Indians did not then have sufficient experience for *ab initio*
design of a launch vehicle."[85] In 1968, aeronautical engineer Y. J. Rao along
with electronics engineer Pramod Kale did a detailed study on developing a satellite launch vehicle. The report was in favor of a vehicle configuration based on
four-stage solid propellant rocket, modeled on Scout.[86] Being all solid propellant, a technology easier than complex liquids, this seemed to be a possible route
that Indians could attempt and succeed.

In his report *Profile for the Decade* Sarabhai explicitly spoke of the indigenous
building of a satellite launch capability for "many applications of outer space in
the fields of communications, meteorology and remote sensing." He also gave
the performance specifications of an all-solid four-stage satellite launch vehicle
weighing 20 tons, and capable of launching a satellite weighing 30 kilograms
in a 400-kilometer low-earth orbit. According to the report, the flight testing
of sensitive instruments, electronics, and instrumentation would be done using
sounding rockets. Sarabhai also talks about the follow on program that could
launch 1,200-kilogram satellite into a circular geosynchronous orbit at 36,000

kilometers: This was "the type of capability which is needed to fully exploit, on our own, the vast potential arising from the practical applications of space science and technology."[87]

Since SLV-3 was modeled after Scout, two views have dominated the historiography of its development: indigenous development and technological diffusion. The first viewpoint was expressed by scientists and engineers who orchestrated the SLV-3 program. The second viewpoint comes from Western policy analysts who have denied that there was any indigenous contribution and basically state that SLV-3 was built using the technological "blueprints" freely given by NASA, albeit without any documentary evidence.[88] Granted the dangers of sharing sensitive launcher technology with India it is doubtful whether NASA gave Scout "blue prints" to the Indians. However, the declassified documents at NARA and NASA and the oral histories clearly tip the balance toward what Gopal Raj asserted in his book *Reach for the Stars* on the history of India's launch vehicles, that is, that SLV-3 was built using freely available unclassified reports and that the incremental development of sounding rockets paved the way for developing SLV-3 after a span of 15 years. Though SLV-3 resembled Scout in its morphology, the subsystems and the fuel assembly showed marked difference from Scout architecture. Though the negotiations on the sharing of Scout technology and critical components did not lead to any tangible results, published articles and government reports indicate the importation of several minor subsystems and components from the United States and Europe that were crucial for the development of SLV-3. With these subsystems the engineers and scientists at ISRO incrementally scaled up their sounding rockets to higher configurations. As indicated earlier, an agreement was signed with Sud Aviation of France to produce under license an advanced sounding rocket called Centaure. Working on Centaure helped in building indigenous Rohini sounding rockets, which were advanced further to carry heavier payloads.[89] Many of the subsystems including the heat shield and guidance were tested using an RH-560 prior to incorporating it in the SLV-3 vehicle. During the development of SLV-3, various changes were incorporated and the version eventually launched was entirely different from the originally conceived one.

By 1971 the design phase of the launcher was completed and of six designs Sarabhai chose the third, hence the name SLV-3. It was a vehicle measuring 22 meters in length and weighing 17 tons and it could place a 30-kilogram satellite into near-earth orbit. The Indo-Pakistan war and the untimely death of Sarabhai in December 1971 was a setback to the launch vehicle program. A restructuring of space was initiated by Indira Gandhi and the ISRO was split off from the DAE. A separate Department of Space, directly under the Indian government, was created. Sathish Dhawan, a Caltech graduate and the director of Indian Institute of Science (IISC), situated in Bangalore, became the chairman of ISRO after M. G. K. Menon's brief stint. To lead the SLV-3 project Sathish Dhawan and Brahma Prakash, director of the Vikram Sarabhai Space Center, chose Abdul Kalam. Kalam was one of those who had been handpicked by Sarabhai to get trained at NASA in the earlier 1960s. He had visited the Langley Research Center, the Goddard Space Flight Center, and the NASA facility at Wallops Island, located on Virginia's Eastern Shore. His NASA training facilitated the first sounding rocket launch from TERLS in November 1963.[90]

On July 18, 1980, India placed its 35-kilogram Rohini (RS- Dl) satellite in low-earth orbit, so becoming the sixth nation to accomplish this feat.[91] Experience gained in building SLV-3 was built upon to produce heavier rockets. The Augmented Satellite Launch Vehicle (ASLV) added two strap-on boosters to the existing SLV-3 configuration and could place a 150 kilogram satellite in low-earth orbit. It was followed by the Polar Satellite Launch Vehicle (PSLV), which can launch 1,600-kilogram satellite into 600-kilometer polar orbit (PSLV-C6 mission in May 2005) and about 1 ton into GTO (PSLV-C4 mission in 2002).

Just as the Pokhran-I nuclear test exhibited the visibility of India's nuclear program in 1974, the successful launch of the Rohini satellite made the space program visible. The launch attracted global attention. The US State Department expressed grave concern. The tense situation was only exacerbated when the Defense Department of India, seeing the successful satellite launch, enrolled Abdul Kalam, the project manager of SLV-3, to rejuvenate their ailing missile program. He joined DRDO where he orchestrated the Integrated Guided Missile Development Program in 1983, which led to the organized research and development of guided missiles for different strategic military needs. Chief among those missiles was Agni, an IRBM successfully tested in 1989, which was built using the experience gained on SLV-3 and could carry warheads weighing almost 1,000 kilograms to targets deep inside the People's Republic of China.

The End of the Cold War and Beyond: Chandrayaan-1

With the end of Cold War and the demise of Soviet Union India had to restructure its foreign policy to meet the emerging geopolitical realities. The Indian political elite began to formulate new recipes to begin closer relations with the United States. India's economic liberalization in the early 1990s and the momentum sustained by successive governments created a conducive environment for a closer relationship between India and the United States. The Clinton administration's grand strategy of "engagement and enlargement" was favorably received by the Indian political leadership. However, despite these expanding links, the overall political relationship continued to be undermined by the India-Pakistan dispute over Kashmir and India's nuclear weapon and ballistic missile programs.

Notwithstanding the controls on technology transfer India went alone or worked with others. It managed to keep a steady pace in developing launch vehicles and satellites for India's domestic economic, commercial, and strategic needs. The Polar Satellite Launch Vehicle, PSLV, was followed by the Geosynchronous Satellite Launch Vehicle (GSLV), a technically upgraded version of PSLV. The architecture of the GSLV included a cryogenic stage that replaced the top two stages of the PSLV. Considering the pound per thrust, these were much more superior to ordinary liquid engines that used other propellant combinations.

The 1998 nuclear weapons tests by India attracted worldwide condemnation and onerous sanctions were imposed on India by the United States and many other developed countries. The United States prohibited trade with a long list of Indian entities and curtailed, for a short time, a broad array of cooperative space initiatives. The geopolitical situation that ensued after the terrorist attacks of 9/11 changed the situation again and catalyzed closer cooperation between India and the United States.

The Bush administration lifted the sanctions in September 2001 and a framework was established through the US-India High Technology Cooperation Group (HTCG) for closer technological cooperation between the two countries. Critical civilian technologies that were once out of bounds—space and nuclear—became tools for improved bilateral relations. Kenneth I. Juster, undersecretary of commerce in June 2004, indicated the various steps that were being taken by the US government to foster closer relations with India. He noted that "since the lifting of the U.S. sanctions in September 2001, only a very small percentage of our total trade with India is even subject to controls. The vast majority of dual-use items simply do not require a license for shipment to India." During the fiscal year 2002 (October 2001 through September 2002), 423 license applications for dual-use exports to India, worth around $27 million, were approved by the US government. This was 84 percent of all licensing decisions for India that year. In 2003 the United States approved 90 percent of all dual-use licensing applications for India. These actions were indicative of the new strategic partnership with India.[92]

In March 2005, a US-India Joint Working Group on Civil Space Cooperation was established. The inaugural meeting was held in Bangalore in June 2005. This forum was meant to provide a mechanism for enhanced cooperation in areas including joint satellite activities and launch, space exploration, increased interoperability among existing and future civil space-based positioning and navigation systems, and collaboration on various earth observation projects. At this time a memorandum of understanding was signed for a joint moon mission.[93] Called Chandrayaan-1 it was a continuation of the international efforts to study the lunar surface to understand origins and the evolution of the moon.[94]

The $83 million Chandrayaan-1 had a cluster of eleven instruments, five from the Indian side and six from foreign agencies: three payloads from the European Space Agency (ESA), two from NASA, and one from Bulgarian Academy of Sciences (BAS). The experiments aimed to map and configure the chemical and mineralogical composition of the lunar surface using more enhanced instruments than previously attempted. The spacecraft was launched using India's trusted workhorse, the Polar Satellite Launch Vehicle (PSLV)—C11. Its launch weight was 1,380 kilograms. The two instruments sent by NASA were the Miniature Synthetic Aperture Radar (MiniSAR) prototype developed by the Johns Hopkins University Applied Physics Laboratory and the US Naval Air Warfare Center to look for water/ice in the permanently shaded craters at the lunar poles, and the Moon Mineralogy Mapper (M3). M3 was an imaging spectrometer developed at Brown University and the Jet Propulsion Laboratory, and was designed to assess and map lunar mineral resources at high spatial and spectral resolutions.

November 14, 2008, was an historic day for the Indian space program. A Moon Impact Probe (MPI) with the Indian tricolor, representing the national flag painted on its surface, made contact with the lunar soil. The timing of the MPI was coordinated to coincide with the birthday of Jawaharlal Nehru, the first prime minister of independent India, who gave his passionate support to the growth of science and technology—especially nuclear and space sciences. It was a significant moment for NASA too to see the maturation of a space program that it helped to found with the scientific elite in India in the early 1960s.

Conclusion

Science as an organized national activity gained an important place in Indian national life only after independence. The period from 1962 to 1972 was crucial for developing an institutional and technological base for space research in India. The growth and establishment of a domestic space program, and collaborative relationships with organizations as well as scientists and technologists in foreign lands, was due to the active interest shown by India in the field of space sciences. NASA helped the scientific elite to create bases for sounding rockets and develop institutions along the way to shaping a space program that was geared toward the development needs of the country as defined by Sarabhai. As far as technological collaboration was concerned, US assistance during the early stages of India's rocket program was limited to the donation of sounding rockets and the loaning of launchers; it never shared details of producing the sounding rockets locally. Homi Bhabha's request for more advanced rockets in 1965 for testing and possible technology transfer were rejected. The attempt to acquire Scout technology after India had lost a border war with China in 1962 and the Chinese nuclear test of 1964 was rebuffed: the risk of further destabilizing the region by supporting a rocket/missile program trumped NASA's determination to assist India. Other major prestige projects (such as the SITE—see next chapter) were embarked on to highlight the country's modernizing urge without helping to rearm it, and to realign Delhi with Washington. US denial of advanced launcher technology led India to combine its own resources with help from other countries, mainly France, Germany, and the Soviet Union, to begin a launch vehicle program. By the time of Sarabhai's death in 1971, his *Profile for the Decade* was accepted by the government of India, and his vision was carried further. Within a decade, incremental progress was made toward meteorological, remote sensing, and communication satellites, which were directed toward India's socioeconomic needs. These were later launched on an indigenous Indian rocket that was developed along with a national missile program. By the end of the twentieth century Vikram Sarabhai's famous quote "there is no ambiguity of purpose" had been fulfilled in a full-spectrum national space program.

Chapter 12

Satellite Broadcasting in Rural India: The SITE Project

The Satellite Instructional Television Experiment (SITE) was a major NASA applications satellite program for educational TV in India. The project involved the use of NASA's Application Technology Satellite-6 (ATS-6) to broadcast educational programs directly to television sets placed in different rural clusters. The agreement for SITE was signed between NASA and India's Department of Atomic Energy (DAE) in 1969. The project was executed from August 1975 to July 1976 and received a great deal of media attention in the country. It was touted as a massive experiment in social engineering and was hailed by some enthusiasts as the world's largest sociological experiment.[1] The British science writer Arthur C. Clarke called it the "greatest communications experiment in history."[2]

Praise for the intangible benefits of the SITE project was perhaps best summarized in a report to the United Nations Committee on Peaceful Uses of Outer Space:

> SITE can be considered a pace-setter and fore-runner of satellite television systems particularly of those meant for development. It is an example of technological and psychological emancipation of the developing world. Its most important element was the commitment and dedication of all people and organizations involved to the one overriding goal of rural development in India. From this follows the crucial role of motivation and cooperation for the success of complex and challenging tasks.[3]

The official Indian reaction to SITE was very positive. The immediate visible results of the broadcast, as cited by project evaluators in the rural clusters, was improved school attendance, increased concern for proper nutrition, and an awareness of sanitation and personal hygiene as methods of disease prevention. One of the unanticipated benefits of the program was the electrification of numerous villages, a prerequisite for television reception.[4] For the Indians, the visual demonstration galvanized public opinion in favor of a space program focused on socioeconomic needs. It helped the country gain competence in using satellites for mass communication and was a systems management lesson for managing Indian National Satellite (INSAT) systems.[5] SITE played an important role in the development of mass media in India, and its legacy can

Figure 12.1   Artist's conception of ATS relay.
Source: NASA.

still be seen today when one watches educational programs sponsored by the University Grants Commission (UGC), which are broadcast on national television channels on a regular basis. ISRO's recent launching of EDUSAT, a satellite designed exclusively for educational needs, can be traced back to SITE.[6]

The Origins of the Project

Arthur C. Clarke first conceptualized the idea of a geosynchronous satellite for broadcasting purposes in a trade journal in 1945.[7] By the early 1960s communication satellites such as Echo, Telstar, Relay, and Syncom were developed to transmit communications to different parts of the world.[8] The technological, cultural, and political possibilities offered by these satellites prompted the US military and private corporations, notably AT&T and Hughes Aircraft Corporation, to develop communications satellites to expand America's global outreach. They aimed to create a "single global system" benefiting the entire world but also serving the Cold War interest of the United States.[9]

The idea of a broadcast satellite for India appears in the middle of these developments in the mid-1960s (figure 12.1). The proposal gained momentum soon after the Chinese nuclear test in October 1964. This forced a major revision in US policy toward India, whose policy of nonalignment and hostility to US-ally Pakistan had led Washington until then to keep Delhi at arm's length.

Communist China's nuclear ambitions and its growing popularity among Afro-Asian countries in the 1950s and 1960s exerted constant pressure on the United States to seek alternatives that could minimize the Chinese influence in the Asian region. Citing India as the world's largest democracy, US officials hoped to establish that nation as a showcase for American-backed development in the "third-world" and as an Asian counterweight to the communist model in the People's Republic of China, PRC.[10] In general, there was a pervasive notion that India was a great laboratory that would demonstrate that liberalism and democracy were the way to go, rather than the Chinese model. During 1961, while analysts at the CIA and the other intelligence agencies tried to determine exactly what progress China had made toward an atomic capability, other arms of the administration began to explore the implications of such an eventuality, and what the United States might do to lessen or eliminate its impact. Suggestions from officials in the State Department that the United States should assist India to "beat Communist China to the punch" by helping their nuclear weapons program were immediately vetoed by Secretary of State Dean Rusk who objected that such a step "would start us down a jungle path from which I see no exit."[11] Soon after the Chinese test the United States began to look for alternative programs that it might undertake jointly with India in the fields of science and technology, which could offset the damage done by the Chinese detonation to Indian prestige and self-confidence.

In January 1965 Jerome B. Wiesner, former science advisor to President Kennedy and the dean of science at the Massachusetts Institute of Technology, and Dr. J. Wallace Joyce, International Scientific and Technological Affairs, Department of State, agreed to visit India at the request of US ambassador Chester Bowles. A list of possible proposals was formulated in consultation with the Atomic Energy Commission (AEC) and NASA. They grouped all possibilities under three major headings: nuclear energy, space, and general science.[12] These moves dovetailed with initiatives being taken by Bhabha and Sarabhai in their periodic visits to Washington. Bhabha explained that India needed to make some dramatic peaceful achievement to counteract the "noise" (his term) of communist China's nuclear explosion. He noted that the Chinese were greatly indebted to the USSR for help on their weapon program adding that if India went all out, it could produce a nuclear device in eighteen months; with a US blueprint it could do the job in six months.[13] Bhabha expressed the view that "if India was to maintain its prestige relative to the Chinese in the field of science and technology two things should be done: (1) ways must be found for it to demonstrate to other Asian and African countries India's scientific achievements, (2) a greater awareness of Chinese indebtedness to the Soviet Union for its nuclear achievements must be created."[14]

Bhabha also met with NASA administrator James E. Webb, deputy administrator Hugh Dryden, and with Arnold Frutkin. During the meeting Bhabha swiftly moved away from the idea of a peaceful nuclear explosion (PNE) to discussing the possibility of India developing a satellite orbiting capability. Bhabha stated that if India undertook such a project, it would wish to launch from India and do the largest part of the job itself. Hearing this from Bhabha, NASA presented estimates of cost, technology, and time requirements, all of which suggested that this was not a project well adapted to achieve Indian objectives. NASA also pointed out that by the time India orbited a satellite, several other nations would likely have progressed so far in this field that India's accomplishment would appear relatively insignificant. Webb's line of thought differed with that of Bhabha; he said that a major effort should be made to select projects that would have a meaningful impact on Indian technology and industrial growth, not spectaculars that would drain resources to no useful social effect.

Sarabhai also made a visit to the United States seeking scientific and technological aid in the area of space. As stressed in chapter 11, Sarabhai viewed science and technology predominantly as tools for socioeconomic development. He believed that a poor nation like India could only close the gap with the rich through self-reliance and self-sufficiency: "[W]e do not wish to acquire *black boxes* from abroad but to grow a national capability."[15] He saw high technologies such as nuclear power and space as crucial to leapfrog into modernity. Sarabhai added that there was some pressure within India to build a nuclear bomb, and to deflect this pressure India needed to do something else to demonstrate an advanced scientific capability.[16]

It was in this context that NASA administrator James Webb proposed a satellite broadcasting initiative to U. Alexis Johnson in May 1966. It was not only a technical experiment in direct broadcasting, but could also serve as a pilot project in the social impact of direct broadcasting and, through suitable program content, it would contribute to the attack upon the food and population problems of India. In the memo Webb stated that the United States would build and position a synchronous satellite near India in such a way that broadcasts from it could be received over the major part of the Indian subcontinent. He went on to point out that India, for its part, could use its nascent electronics capability, now focused at the atomic energy center at Trombay, to develop improved television receivers. These could be established in perhaps a thousand rural population centers. Webb waxed lyrical about the multiple advantages the program would have for the country. Indians could learn new technological and management approaches to education and to the uses of informational media to weld together a nation-state. The government could invest in a modern electronics industry that would "materially raise India's technological base and contribute thereby to the development of other, similar industries." Resources would be redirected from nuclear weapons to more socially valuable endeavors. The United States for its part "would learn more about the Indians and their most pressing problems," and improve its global "posture" "through a generous demonstration of its willingness to share the benefit of advanced space technology with underdeveloped nations."[17]

Webb's educational satellite resonated with a scheme that Sarabhai had been playing with for some time. He began to visualize a national satellite program to provide a better way of life to the inhabitants of India's 63,000 villages. He hoped that, thanks to the research and development activities of the space program, television would be available to 80 percent of India's population within ten years. This project was of special significance because by providing entertainment and instruction of high quality, it would be possible to bring about a qualitative improvement in the richness of rural life.[18]

Winning Hearts and Minds

SITE offered the State Department twin benefits: a benign technological tool to offset communist China's influence, and a technology that would help to bring literacy and development to the rural population. This was perfectly in line with what the communication scholars and media experts were promoting in the early 1960s, the idea that television and other media of mass communication would help national development. Stalwarts in communication and development studies such as Daniel Lerner, Wilbur Schramm, and Everett M. Rogers based their theories of development and media efficacy on Walt Rostow's influential *Stages of Economic Growth: A Non-Communist Manifesto*.[19] In the book Rostow stressed that the economic and technological development achieved by the Western nations were the result of increased media use. If the developing countries could follow the path of modernization initiated by the West, they would leapfrog centuries of inaction and underdevelopment and catch up with the modernized West.[20] Rostow who later became the national security adviser to President Lyndon Johnson, was himself interested in putting "television sets in the thatch hutches of the world" to defeat both tradition and communism with the spectacle of consumption.[21] The political value of communication satellites was also emphasized by Arthur C. Clark:

> Living as I do in the Far East, I am constantly reminded of the struggle between the Western World and the USSR for the uncommitted millions of Asia. The printed word plays only a small part in this battle for the minds of the largely illiterate population and even radio is limited in range and impact. But when line of sight TV transmission becomes possible through satellites directly overhead, the propaganda effect may be decisive...the impact upon the peoples of Asia and Africa may be overwhelming. It may well determine whether Russian or English is the main language of the future. The TV satellite is mightier than the ICBM.[22]

India was particularly appropriate for a satellite experiment in the direct broadcasting of TV. First, there was no existing TV distribution network, which could be utilized by conventional means. The population was distributed relatively homogeneously throughout the subcontinent rather than concentrated in a few large cities easily reached by conventional TV, and there was a high level of Indian government support for this kind of experiment. This contrasted with other developing countries, for instance, Brazil. There, a substantial portion of the population was concentrated in coastal cities, all of which already possessed TV networks, while only the scattered inland population lacked TV. So, India stood apart as an ideal laboratory for testing the technology. Wallace Joyce, in the International Scientific and Technological Affairs of the State Department, particularly liked Webb's idea. It had the potential for India to exert "regional leadership" in space-related educational TV for development purposes in the surrounding Asian and other modernizing regions.[23]

For Frutkin, the instructional television project was a constructive step forward in cooperation between one of the world's superpowers and a progressive, neutral, developing nation. "For other developing countries, it should serve on a non cost basis to test the values, the feasibility, and the requirements of a multi-purpose tool which could be critical to accelerating their progress in an increasingly technological world."[24] There is "some measure of generalization, hyperbole, and technological misconception" when it came to direct broadcasting of television, remarked Frutkin. In order to realistically consider the problems and technological hurdles associated with direct broadcasting he sought an "actual experience

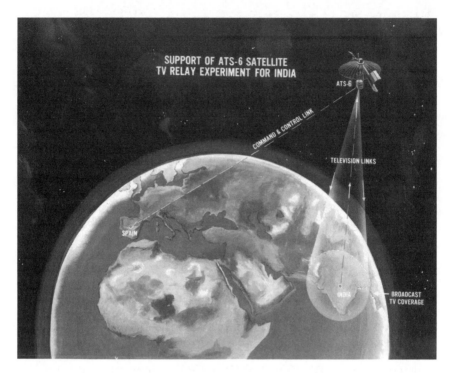

Figure 12.2   Artist's conception of ATS-6 support.
Source: NASA.

with the medium." The experiment represented a "rarely grasped opportunity to use modern technology so as to leapfrog historical development stages."[25]

The Indian space experts too were interested in exploring the potentialities of TV as a means of mass communication in a developing country. In 1967, only Delhi, the capital city of India, had television transmission services. The Indian broadcast planners organized under the Ministry of Information and Public Broadcasting (MIPB) wanted to extend the television services by first focusing on the cities and gradually extending it to rural villages through transmitters. Seeing the cities to be already "information rich" through various other media, Vikram Sarabhai, in contrast to the broadcast agency—which blamed the space agency for unnecessarily encroaching on their domain—wanted the villages to receive the high technology first. In June 1967 Sarabhai sent a team to NASA to study the prospects of using a satellite over a conventional transmission links. After looking at various options, the visitors focused in on a "hybrid system for rebroadcast stations for high population areas, and a satellite for interconnection and transmission to low-population density areas." The interaction between NASA, Indian actors, and the business corporations in America planted the seed for the Indian National Satellite (INSAT), which was developed during the early 1980s.[26]

To test the efficiency of such a massive system for the entire Indian population officials at NASA and the State Department conceptualized a limited one-year SITE project using the ATS-6 satellite (figure 12.2). The SITE project was not without domestic resistance, however. To reach a consensus among different agencies Sarabhai set up an ad hoc National Satellite Telecommunications Committee (NASCOM) in 1968. SITE was finally approved after an extensive debate in the parliament.

An agreement was signed between NASA and ISRO in 1969 wherein NASA agreed to provide this satellite for one year. NASA would provide the space segment while ISRO took charge of the ground segment and programs. NASA helped ISRO by offering training facilities to its engineers at different NASA facilities and by helping in the procurement of critical components when these were urgently required at short notice. Numerous ISRO-NASA meetings held in India and America helped sort out interface problems and in acquainting each other with the progress of the SITE project. In order to plan the for the year-long project, the Indian space agency undertook a small experiment called *Krishi Darshan* (Agricultural TV Program). Around 80 television sets were placed in rural villages around Delhi to test "software development, receiver maintenance, and audience information utilization."[27] To prepare for the future, joint studies were also done by ISRO engineers with NASA and private corporations such as Hughes Aircraft, and General Electric for configuring systems for INSAT. In 1970, ISRO engineers undertook a study at Lincoln Labs at MIT for spacecraft studies of INSAT. Sarabhai planned INSAT as a follow on after the SITE experiment.[28]

The Technological Component: The Hardware

SITE was conceived as a classic communication system consisting of an information source, a channel, a receiver, and the destination. The operation and execution of the experiment was dependent on a network of complex technological systems. Occupying the central node in the whole network was the ATS-6. It functioned as a relay station for receiving and sending signals originating in India; in other words it acted as a channel of communication between the transmitter and the receiver. The retransmitted signals from ATS-6 were received on the ground by a Direct Reception Systems (DRS). The United States provided the satellite while the full responsibility for the ground segment—earth stations, DRS, and television programs (software)—fell on India.

The ATS-6 Satellite

ATS-6 was the most complex and advanced communication satellite in NASA's Applications Technology Satellite series.[29] In 1966 NASA began launching a series of six such satellites manufactured by Fairchild to test and improve satellite communications. They were designed to carry out technological, meteorological, scientific, and communications research. The last of the series, ATS-6 was the largest, most complicated, and powerful of them all. It was a geosynchronous satellite—the orbital period for the 1,402-kilogram satellite around the globe matched the Earth's 24-hour rotation so that the ATS-6 remained over the same spot on the earth. It was designed in such a way that it could be moved along the equator using its onboard thrusters to conduct space-based experiments in any region of the globe. In general it served as a powerful rebroadcasting station in space, capable

Figure 12.3   Testing ATS-6.
Source: NASA.

of transmitting signals directly to many small ground stations over a large area
(figure 12.3). The prime objectives of ATS-6 missions involved demonstrating
a 30-foot deployable antenna in synchronous orbit, providing a three-axis stabilized spacecraft with 0.1 degree pointing capability in all three axes, and providing an oriented platform at synchronous altitude for advanced technological and
scientific experiments. The SITE was made possible when all of these objectives
were achieved.[30] ATS-6 thus represented the kind of satellite system appropriate for
communications within many developing countries, where most of the population
lived dispersed in rural areas, rather than in large population centers.[31] ATS-6 was
launched on May 30, 1974, and it carried approximately 15 scientific experiments
in the field of communication, meteorology, and spacecraft stabilization.

Earth Stations

ISRO was responsible for the installation and maintenance of the earth stations and also the design, installation, and maintenance of the augmented community receivers. The earth stations and the DRS formed the ground segment
of the network. While NASA provided the satellite, the ground segment was
indigenously manufactured by ISRO with little help from foreign countries. The
earth stations helped transmit signals to the satellite, the satellite received these

signals, amplified them, and transmitted them back to earth where they were received by custom-made television sets that were suitably tuned.

Four earth stations located in Ahmedabad, Delhi, Amritsar, and Nagpur were utilized for the SITE project. The central earth station that transmitted the bulk of the programs using the 15-meter parabolic antenna was located in Ahmedabad. The second earth station, a 10-meter parabolic antenna located in Delhi, helped in telecasting national programs—Republic Day, Independence Day, and addresses from the prime minister and president. It also served as a backup facility if the central station in Ahmedabad were to face any technical glitches. The third earth station was located in central India in Nagpur. It housed the "monopulse beacon" instrument. The ATS-6 satellite had the capability of "homing in" to a beacon station located to keep the satellite accurately oriented if its internal pointing systems failed. This was one of the important back up modes to ensure the ongoing functionality of the ATS-6 spacecraft.

Direct Reception Systems

The direct reception systems, DRS, completed the vast network that was put in place for the SITE project. The development of the DRS was started in 1972 at the Electronics Systems Division of the Space Applications Center in Ahmedabad. The system had three main components: the antenna to receive the signals transmitted by the satellite, the front end converter to transform the signals into a form compatible with a normal television receiver, and a television receiver.

The antennas measuring ten feet in diameter, the front end converter, the most complex one in the assembly, and the television sets were first designed in the Space Application Center (SAC) in Ahmedabad. The prototypes were given to a public sector company, the Electronics Corporation of India Limited (ECIL), located in Hyderabad, for mass production. The television monitor itself was basically a commercial model slightly modified for community viewing and rural use. "Seven hundred of the 2400 sets were 'ruggedized' by using higher quality components, as a part of an 'experiment-within-an-experiment' to investigate the tradeoffs between initial cost and maintenance cost."[32] To facilitate transfer of 'know-how' and to expedite production, some ISRO engineers who had developed these units were posted to ECIL.[33]

The direct reception systems were deployed in selected villages and the districts of six states, namely, Andhra Pradesh, Bihar, Karnataka, Madhya Pradesh, Orissa, and Rajasthan. The villages were selected according to the criteria laid down by the Planning Commission of India. The criteria included availability of electricity, public buildings, low population, and so on. To carry out an organized effort of deployment, operation, and maintenance of these television sets, maintenance subcenters and a central cluster headquarters were established in each state. These cluster headquarters acted as nodes for the distribution and maintenance of the community reception system.

Before the SITE mission in India, the satellite was used to perform a variety of health and education television experiments via satellite in the Appalachian area, the Rocky Mountain Region, and Northwest United States including Alaska.[34] In July 1975, while it was being shifted eastward along the equator for the SITE mission, the ATS-6 tracked the docked Apollo and Soyuz spacecraft as they orbited the Earth in the Joint US/USSR manned space mission (see chapter 7). It also relayed live television from these spacecraft to the Earth, thus becoming the first satellite to perform such a feat.[35] After this it was positioned at 35,900 kilometers over east Africa and controlled from the Goddard Space Flight Center through a ground station in Spain. Since the downlink SITE frequency of 860 megahertz could interfere with terrestrial services in Europe, its antenna was pointed eastward toward India and away from Europe, thus avoiding interference with European surface broadcasts.

Technological Component: The Software

The term "software" was generally used for the program content of the satellite broadcast. An enormous amount of programming had to be produced for SITE as the satellite was available to India for approximately 1,400 hours of transmission.[36] All India Radio, later Doordarshan, took the overall responsibility for producing these programs. The educational programs were produced in three base production centers: Cuttack, Hyderabad, and Delhi. The production of such a large number of programs, keeping in view the basic objectives and the specific audience requirements, was a challenging task. Most of the studio facilities available to SITE were small, underequipped, and understaffed. This fact, coupled with the time pressure for production, created a continuing pressure toward easy-to-produce "entertainment" programming, even when audience feedback indicated a preference for the so-called hard core instructional programs.[37] Since the software part demanded a lot of attention from the Indian side Frutkin made every attempt to ensure that it was done properly. "When he visited India in January 1975, seven months before the experiment started, he insisted on a physical examination of television studios and programs."[38]

Doordarshan formed separate committees to assist program production relating to agriculture, health, and family planning. These committees were helped by institutions like the agricultural universities, teachers training colleges, the Indian Council of Agricultural Research, and so on.[39] Other departments and agencies like the Film Division, the National Center for Educational Research and Training (NCERT) under the Department of Education, along with independent producers, contributed to making film material for the software content of the SITE project. SITE broadcasts regularly reached over 2,300 villages. Their size varied from 600 to 3,000 people, with an average of 1,200 inhabitants. Thus, about 2.8 million people had daily access to SITE programming.[40] The programs were available for some four hours a day and were telecast twice, morning and evening.

SITE ended on July 31, 1976. Seeing the success of the project the Indian officials and policymakers requested an extension of the program for one more year, but the request was not granted and ATS-6 was pulled back to the American region.

Frutkin, who orchestrated the SITE project for NASA, said that the one-year experiment proved the possibilities of the use of advanced satellites for mass communication. And he clearly knew that it would bring monetary benefits. "We took the satellite back. What was the consequence? India contracted with Ford Aerospace for a commercial satellite to continue their programs...the

**BROADCAST SATELLITE BRINGS
EDUCATION TO INDIAN VILLAGE**

Figure 12.4   ATS receiver and SITE watchers.
Source: NASA.

point is, this program not only was an educational lift to India and demonstrated what such a satellite could do, but it brought money back into the U.S. commercial contracts for satellites for a number of years."[41] Years earlier in a House Committee report on the implications of satellite communications, he expressed the same view: "I'm quite confident that by virtue of our participation in this experiment, India will look to the U.S. first for the commercial and launching assistance it requires for future programs. And I think this is a very important product of our relationship."[42]

SITE was regarded by many as a landmark experiment in the rapid upgrading of education in a developing country (figure 12.4). It became the most innovative and potentially the most far-reaching effort to apply advanced Western technologies to the traditional problems of the developing world. For the first time NASA and ISRO cooperated very closely in an effort to determine the feasibility of using experimental communication satellite technology to contribute to the solution of some of India's major education and development needs.[43] For NASA the experiment provided a proof that advanced technology could play a major part in solving the problems of less-developed countries. It was seen as an important expression of US policy to make the benefits of its space technology directly available to other peoples and also a valuable test of the technology and social mechanisms of community broadcasting. Seeing Indian states to be linguistically divided, the US State Department hoped that the experiment offered India an important and useful domestic tool in the interests of national cohesion. The experiment also stimulated a domestic television manufacturing enterprise in India with important managerial, economic, and technological implications. It provided information and experience of value for future application of educational programs elsewhere in the world.[44]

Frutkin was emphatic about the value of SITE for other developing countries. "The Indian experiment is, of course, of prime significance for developing countries, those which have not been able to reach large segments of their population, those which have overriding social problems which might be ameliorated through communication and education and particularly those where visual techniques could help to bypass prevalent illiteracy."[45] The SITE experiment played a crucial role for India too. The results of the year-long SITE project were evaluated carefully by the Indian government. The data played a major role in determining whether India should continue to develop her own communication satellite program (INSAT) or fall back on the use of more traditional, terrestrial forms of mass communication in order to transmit educational programs to the populace.[46] Thanks to SITE the first-generation Indian National Satellite (INSAT-1) series, four in total, was built by Ford Aerospace in the United States.[47]

The SITE project represented an important experimental step in the development of a national communications system and of the underlying technological, managerial, and social supporting elements. Following the proposal made by India, Brazil too initiated a proposal for a quite different educational broadcast experiment utilizing the ATS-6 spacecraft. The project was intended to serve as the development prototype of a system that would broadcast television and radio instructional material to the entire country through a government-owned geostationary satellite.[48] Frutkin saw the Indian project and the Brazilian experiment to be a model for other developing countries. In 1976 Indonesia became the first country in the developing world to have its own satellite system, the Palapa satellite system, manufactured by engineers at NASA and at Hughes Aerospace.[49]

SITE showed India that a high technology could be used for socioeconomic development. It became one justification for building a space program in a poor country—the question became "not whether India could afford a space program but can it afford not to have one"?[50] "Modernization" through science and technology was not new to the Indian subcontinent. In more than two centuries of British occupation India witnessed a huge incursion of technologies—railways, telegraph, telephone, radio, plastics, printing presses for "development" and extraction.[51] The geosynchronous satellite in postcolonial India can be seen as an extension of the terrestrial technologies that the British used to civilize/modernize a traditional society. In this case the United States replaced the erstwhile imperial power to bring order, control, and "modernization" to the newly decolonized states through digital images using satellite technologies that were far removed from the territorial sovereignty of nation-states.[52]

Part V

Into the Twenty-First Century

John Krige

Chapter 13

Space Collaboration Today: The ISS

Two major geopolitical changes in the 1990s have had very different impacts on NASA's international relations over the past 20 years. The implosion of the Soviet system and the political will to integrate Russia into the core of what became the International Space Station (ISS) produced an exception to some time-hallowed NASA policies, notably, the notions of clean interfaces and no exchange of funds. By contrast, the "leakage" of sensitive satellite and missile technology to China, and its willingness to work closely with "rogue states" like Iran, gave traction to those who believed that the United States had to be far more prudent in its international posture, above all in sharing technology.[1] This led to a tighter implementation of the ITAR (International Traffic in Arms Regulations) particularly as regards satellites. This added more layers of complexity and bureaucracy to international collaboration with traditional allies, and has stimulated lively debates between diverse stakeholders about the costs and benefits of implementing export controls more rigorously.

This chapter and the next discuss these developments. Since NASA's decision to incorporate Russia into the ISS is treated in chapter 8, this analysis will pay greater attention to the structures of collaboration that were put in place before 1993. Those structures were deeply influenced by the history of NASA's relations with its traditional partners, above all Western Europe. Concerns about European disappointment at the outcome of the post-Apollo negotiations (chapters 4–6) and the ISPM affair (chapter 2) hung over negotiations between NASA and ESA. These past setbacks to the otherwise smooth path of cooperation were consciously drawn on as lessons that were not to be repeated. The classic principles of no exchange of funds and clean interfaces were not, however, put in question. That only happened when Russia was drawn into the project, bringing with it an array of Cold War technologies, record-breaking experience of long-duration human spaceflight, and a disintegrating infrastructure of institutions and industries that were seeking a new role for themselves. The architecture of the ISS was accommodated to incorporate Russian elements into technologies that were critical to mission success. Millions of US dollars, both private and public, flowed to various actors in the Russian space sector in an attempt to revitalize them, and to engage them more tightly with American practices and priorities. The end-result was a space station in which NASA found itself dependent on its partners in ways that were historically unprecedented. A new kind of international cooperation had been imposed, in which NASA's mandate to sustain US leadership had to contend with its loss of autonomy.

The President Commits to the International Space Station

No sooner was the space shuttle declared operational in 1981 than the new NASA administrator, James Beggs, appointed by President Reagan, began to actively promote a space station as the next "logical step" for the agency.[2] He quickly made a major effort to stimulate foreign interest in the new American project, using the shuttle to advertise America's ongoing mastery of space. In June, 1983 Beggs and his deputy administrator, Hans Mark, toured European capitals with the unpowered prototype of the orbiter Enterprise piggy-backing on a specially adapted Boeing 747. It was displayed at the Paris Air Show on June 4 and to a wildly enthusiastic crowd at Stansted Airport near London the next day. It then went to Ottawa where 400,000 people turned out to see it, and the Canadian minster of science and technology announced the creation of a Canadian astronaut program.[3]

The opponents of the space station were not swayed by the excitement.[4] Caspar Weinberger, the secretary of defense was particularly hostile to the project. George Keyworth, the president's science adviser, was skeptical. Two pro-station staff members in the White House, Gil Rye, of the National Security Council Staff, and Craig Fuller, an ardent advocate of the commercial potential of space, decided to take steps to circumvent the opposition. They scheduled a Cabinet meeting on December 1, 1983, at which Beggs could present his case for the station directly to the president in a reasonably hospitable environment.[5] The NASA administrator gave a masterful presentation that skillfully exploited Reagan's concern about the decline of American power and prestige vis-à-vis the Soviet Union. He dramatized the threat by showing the Salyut space station overflying the country, adding that the Soviets were preparing to launch an even bigger facility into space in the near future (Mir). If an American station was begun at once, said Beggs, it could build on the breadth and depth of the country's spaceflight capability to ensure that the United States would "dominate the space environment for twenty years."[6] The president was persuaded that a civilian space station with scientific and commercial potential would be a useful counterweight and complement to his space-based antimissile system called SDI (the strategic defense initiative). A few days later David Stockman in the OMB met with Reagan and Beggs to sanction the appropriation request for the space station.[7]

When Beggs spoke before the president in December he made no mention of international participation. The issue did arise though when the cost came up. Beggs suggested to Reagan's associates that the station would cost $8 billion, a figure that was essentially determined by his judgment as to the maximum figure that the president would accept. He added that international collaboration would provide additional funds. This argument was crucial at the time because the DoD were convinced that the station would drain money away from SDI. Peggy Finarelli, who joined NASA's Office of International Affairs in 1981, worked closely with Rye to push the space station to the top of the president's agenda.[8] She takes up the narrative:

Defense Department objected to the Space Station, period. CIA sided with them because they're part of the national security community. OMB sided with them because OMB hates anything that costs money. State Department sided with them because the Under Secretary in charge of science and technology at State at that time was a fellow who had come from earlier political careers in both DoD and OMB, so even though he was at State, he was siding with the national security community and objecting to the Space Station. So we were alone in our proposal, but adamant that we wanted to do the Station and that we wanted to do it as an international partnership.[9]

The international issue remained a "major battling point" with the other agencies as NASA tried to figure out how to present the president's decision publicly. Rye and Finarelli stuck to their guns, recognizing that "if it was an international project and it was announced as such in the State of the Union, it was going to be far harder to unwrap" than if it was simply a domestic project. They won the day, and it was to Beggs's "surprise and pleasure" that Reagan chose to announce his support for the space station on January 25, 1984.[10] In his annual State of the Union address the president reported that he had directed NASA to develop a permanently manned space station within a decade. Reagan announced that NASA would "invite other countries to participate so that we can strengthen peace, build prosperity, and expand freedom for all who share our goals."[11]

Just before he made his public statement the president alerted the political leaders in Britain, France, Germany, and Italy, as well as in Canada and Japan. He added that Beggs would shortly be coming to meet with senior officials of foreign governments on his behalf to develop the cooperative effort.[12] The space station was thus presidentially sanctioned as an advanced technological platform that would bind together the nations of the free world. This gave it immense social weight both at home and abroad. As one leading British space administrator put it to Finarelli, whereas the decision in principle of whether to build a station was taken in Washington, "we had a very different decision to make. The [decision] that our political authorities had to take was not whether a space station made sense to build. The decision we had to make was: Given that the U.S. has decided to build a space station, and has invited us to join, can we afford not to?"[13]

A further boost to international involvement was provided at the London Economic Summit in June 1984. It was one of the talking points on President Reagan's agenda for private meetings. As the seven heads of state emerged from one of their meetings they were confronted with a model of the space station that included elements that could be built abroad. The communiqué issued after the London summit was positive but prudent, endorsing manned space stations as valuable for industrial and economic development and committing the signatories only to "consider carefully the generous and thoughtful invitation received from the President of the United States to participate."[14]

Beggs made it clear that the United States was looking for significant contributions to the space station project, roughly 10–20 percent of the partners' overall space budgets for the next decade. Technological exchange would be restricted as much as possible. As he put it to the director of the Johnson Space Flight Center in April 1984, the administration was "concerned about careless and

unnecessary revelation of sensitive technology to our free world competitors—
sometimes to the serious detriment of this nation's vital commercial competitive
position."[15] Beggs hoped that participation in the station would draw the sting
from this competition by diverting resources into a major technological project
under American leadership. In fact the NASA administrator admitted as much
in the presence of representatives from foreign industries and space agencies.
The station, he said, lent itself "uniquely to international cooperation," adding
that "if we can attract that cooperation then other nations will be cooperating
with us in the resources that they spend, rather than competing with us."[16]
Beggs's one-sided emphasis on the benefits to the United States of international
participation was probably "particularly galling" (Logsdon) to those present, a
clumsy way to resolve the tension between NASA's joint obligations to lead and
to collaborate.[17]

NASA Prepares for Collaboration

International participation in the space station was not universally welcomed
inside NASA. The benefits were easily defined. International partners would
provide dollars—perhaps as much as 12 percent of the costs of the development
program by ESA and by Canada, and $100 million annually by Japan.[18] They
would also provide added political robustness, and confirm to skeptics that there
was merit to NASA's claim that the time had come to develop the station. There
were drawbacks too, though.[19] Kenneth Pedersen tackled the issue head-on.

Pedersen was keen to get other countries involved in the space station from
the outset. In January 1982 he called a meeting of potential space station partners at the Johnson Space Center. Each participant was invited to undertake
Phase A (conceptual) studies at their own expense to determine what the mission
of such a station should be. NASA's partners were not being asked "to contribute
mere pieces to a U.S. conceived, designed and managed programme but to join
with NASA in developing and *operating* an international space complex fitted
to their collective requirements."[20] This is what had gone wrong in post-Apollo.
As Pedersen explained to the director of NASA's Space Station Task Force, he
objected strongly to encouraging partners to get involved technologically and
financially in Phase A studies like those currently under way, either of separable
components (like a sortie module or a tug, in post-Apollo), or of an integrated
system (like the shuttle itself).[21] This was because he had noticed that, as post-
Apollo had evolved, NASA's priorities had changed. It preferred collaboration
in the *use* of space, not in joint engineering projects. It had concluded that
European industry was five–ten years behind that in the United States. It did
not want to depend on foreign countries for critical parts of the shuttle. It did
not want the tug to use liquid propellants, as Europe was proposing. As a result
in 1972 the US government found itself in the embarrassing position "of having
to walk back from the European perception of the cooperative possibilities" in
the program, creating suspicion and distrust that still persisted in some quarters.[22] The mistake would not be repeated. Foreign partners should focus their
work during Phase A studies on mission requirements rather than hardware contributions. All cooperation should be managed through NASA Headquarters,
and should be exclusively with representatives from foreign governments, who

would keep their national industries informed of developments. Foreign visitors to field centers were to be discouraged for fear that they would become embroiled in intercenter rivalry over mission concepts. There was to be no formal industry teaming.[23]

To build domestic support Pedersen emphasized that NASA should retain close contact with all agencies that had foreign policy responsibilities—and there were many, including the State Department, the National Security Council, the Office of Management and Budget, and the Department of Defense. The DoD was likely to be particularly important, since, thanks to SDI, "the interest and debate over the militarization of space is at an all-time high—much more intense than at the time of post-Apollo planning activities." Pedersen surmised that "the question of how military involvement would infringe on access rights to the station" was likely to be "in the end the single most important factor influencing foreign participation."[24] Opposition to this would probably be least in Canada, who did not object to the DoD's use of the Remote Manipulator System that it had built for the shuttle. By contrast, although Japan was eager to join in the station, feeling that it had missed a key opportunity by not joining in post-Apollo, the science minister of the ruling Liberal Democratic Party had already warned NASA that its participation would be "unavoidably narrowed" if the program had a large military component. The situation in Europe depended on the country concerned. Pedersen felt that this thorny issue was best dealt with by "working to accommodate both civil and military uses within the basic design of the space station, so that one does not make the other impossible."

In August 1982 Pedersen had little new to add to the guidelines for controlling technology transfer that had emerged in the post-Apollo debate. He favored cooperative agreements for discrete hardware pieces with minimal interfaces. He also emphasized that this was an increasingly sensitive issue in the administration. It was essential for NASA to remain in close contact with the export control community. Increasing evidence that the Soviet Union was engaged in a major, centrally coordinated effort to gain access to American high technology by any means possible had led to "closer application of existing guidelines and review of appropriate future steps in staunching the flow of advanced technology."[25] Space industries in Europe were also stronger than they had been in the early 1970s, and Europe had just acquired independent access to space by qualifying its Ariane launcher in December 1981. In short, as McCurdy puts it, as regards cross-border knowledge flows, the guidelines laid down by Pedersen in 1982 "reaffirmed the traditional conservative values that had governed international participation within NASA for more than twenty years."[26] By building the core elements of the station, by excluding collaborators from making contributions to the critical path, and by keeping interfaces as clean as possible, the asymmetry in technical and financial contributions to the project was built into the hardware of the station from the start.

European Reactions to Reagan's Proposal

There was considerable interest in the space station in Europe. Following on Pedersen's invitation, in June 1982 NASA and ESA agreed that the European agency finance Phase A industrial studies on both utilization aspects and potential hardware contributions. Later that year the ESA Council, with some difficulty, drummed up support for studies on "maintaining in Europe an independent launch capability, developing a European in-orbit infrastructure, and pursuing transatlantic cooperation through participation in the future United States space program."[27]

This formulation was meant to be flexible enough to accommodate the diverse needs of the member states, notably the drivers of the European space effort, France and Germany. As Niklas Reinke points out, both were committed to the idea of a space station, although their political motives differed. The federal minister for research and technology, Heinz Riesenhuber, who took office in October 1982 "wanted substantial European participation in the American programme, with Germany in the lead; France was interested in the technical know-how to be gained from a space station but was wary of becoming involved again in such close cooperation with the United States."[28] Germany's prime aim was to build on its Spacelab experience, expanded to include the development of reusable space platforms like the free-flying pallet suitable for commercial and scientific experiments called Eureca (EUropean REtrievable CArrier).[29] It teamed up with Italy to fund industrial studies of pressurized models derived from Spacelab and an unmanned platform that were combined together in a program it called Columbus.[30]

In January 1984, just a week before President Reagan made his State of the Union address announcing that he would support the space station, the German and Italian delegations suggested to their partners in ESA that they might like to participate in the development of Columbus. This was now a generic name for a research module to be attached to the space station plus one or more free-flying platforms for more complex experiments in science and applications, above all microgravity.

Representatives of the member states of ESA, meeting at ministerial level in Rome in January 1985, defined their priorities for the next phase of their joint space effort. The ministers spelt out the principles that should guide their participation in the joint venture. They sought European "responsibility for the design, development, exploitation and evolution of one of several identifiable elements of the space station together with responsibility for their management." They also wanted to have "access to, and use, on a non-discriminatory basis, of all elements of the space station system on terms that are as favorable as those granted to the most-favored users and on a reciprocal basis."[31] The ministers also expressed strong support for Columbus, whose precise content would "depend on the terms and conditions of the partnership agreement concluded with the United States."[32]

The enthusiasm generated by the Phase A studies, and the support of the ministers meeting in Rome in January, quickly led to the signature of a memorandum of understanding (MoU) between ESA and NASA in June 1985. It dealt with the conduct of parallel detailed definition and preliminary design studies (Phase B studies). (Similar agreements were signed with Canada and Japan.) The MoU specifically identified a key milestone in March 1986, about halfway through the planned definition phase, at which NASA and ESA would mutually agree on the composition of the Columbus program that would be carried forward for the remainder of the definition phase. This second Phase B2 was scheduled to run from April 1986 to March 1987. Tough negotiations between the two agencies over the Columbus content delayed the start of Phase B2 by over six months to November 1987.[33] In parallel, starting in 1986, bilateral discussions were begun between the potential European partners and the United States on establishing the legal instruments governing the space station. The European group insisted that these be conducted on two levels. They wanted bilateral MoUs between NASA and the partner agencies for defining how cooperation in the design, the construction, and the operation of the space station and its constituent elements could and should be implemented in practice. The MoUs were subsumed under a single intergovernmental agreement (IGA) defining the policy guidelines and legal principles that would govern collaboration between the United States and the member states of ESA, Canada, and Japan. These various instruments were signed by almost all parties at the end of September 1988. NASA's MoU with its Japanese counterpart was signed in March 1989.[34]

Europe's phase B1 proposals had three main elements. The first was a pressurized module that could either be tethered to the station or detached and used in a human-tended, free-flying mode. The second was a retrievable platform derived from the Eureca concept that would be placed in an orbit near the space station. The third was the polar platform that was intended as a "workhorse" for earth observation missions in polar orbits and whose scientific interest was enhanced by growing concerns about environmental degradation and climate change in the early 1990s.[35]

ESA was particularly interested in the first of these elements. Its dual-configuration, tethered or free-flying, allowed it to be used as a Spacelab-like environment for scientific experiments as well as a small autonomous European space station to acquire capabilities in rendezvous and docking procedures, and in the use of automation and robotics. NASA rejected the scheme—the space station would not be big enough nor would it have enough electrical power for each nation to operate its module both docked and untethered. Europe complied by restricting this component to a permanently attached pressurized module (APM), which was the length of four Spacelab segments and was to be used for materials science, fluid physics, and life-sciences experiments. ESA then successfully demanded that it develop a separate laboratory, the man-tended free-flyer (MTFF), to be operated in a microgravity-optimized orbit.[36] The MTFF fulfilled some of the original mission requirements of the Eureca platform and retained the potential of evolving into a permanent autonomous space station. Thus in the Columbus configuration eventually agreed on in 1987, the MTFF and the polar platform (PPF) "were...the elements that were to carry the banner for Europe's autonomy in space, while the APM, as a fully integrated part of the station, had to be adapted to fit American ideas."[37]

The disagreements between ESA and NASA were not restricted to hardware contributions; they extended to use. It seems that during the negotiations over the final cooperative agreements the United States did not want Europe to perform microgravity research in materials science, even in its own part of the station. Only the United States was to be allowed to use any part of the station for experiments of commercial promise. As McCurdy puts it:

> Because of strong congressional and presidential interest in the commercial potential of space, NASA would eventually insist that it be allowed to build the

materials-processing lab. That would leave the Europeans with the less glamorous task of building the life sciences lab. To conduct materials-processing experiments, the Europeans would have to use a U.S. module. Furthermore, they could not just float in and use it. The experiments would have to be scheduled on the basis of international agreements acceptable to all of the partners and based on their relative contributions to the station.[38]

This situation did not persist. As Peggy Finarelli stressed in an interview with the author, "the utilization plan of any partner, what they wanted to put on the Station, how they used their resources was their call. [...] There was absolutely no carving up like 'You can do this and you can't do that.' We have unilateral rights to do this."[39]

Then there was the question of military use. At the end of 1986 the United States raised the question in general terms of the use of the space station for military research related to SDI. This threatened to derail the whole process. Japan was totally against the idea. ESA's convention specifically committed the agency to peaceful use, and no backsliding would be tolerated by the "neutral" member states—Austria, Sweden, and Switzerland. Indeed this issue caused such consternation that "early in 1987, the view was expressed in German government circles that, although it was perhaps not necessary to think about breaking off the negotiations just yet, the positions had become irreconcilable."[40] Caspar Weinberger attempted to still these fears by submitting a list of possible military experiments to be conducted on the station that he thought should be unobjectionable. It made little difference. When the representatives of the ESA member states, meeting at ministerial level in November 1987, adopted a long-term space plan that committed them to participation in the station, they thought it fit to include a special clause regarding peaceful use in their resolution.[41] In the final agreements the space station was defined as being "civil" and "for peaceful purposes, in accordance with international law" (see also chapter 1). The US chief negotiator placed on record that his country "has the right to use its elements, as well as its allocations of resources derived from the Space Station infrastructure, for national security purposes."[42] This was coupled with a clause in the agreement that allowed any partner (including Japan) to refuse that its attached module be used by a military body.[43]

Peggy Finarelli, who was involved in the negotiation of these agreements on behalf of NASA, provided an insider's perspective in an interview in 2010. She stressed that the "creative ambiguity" over the meaning of the term "peaceful" in the Outer Space Treaty allowed all the adherents to sign on while maintaining their separate perspectives. Put simply, for the United States the term "peaceful" meant "non-aggressive," while for her partners the term meant "non-military" (see chapter 1). The disagreement was so deep that "we cancelled one of the scheduled negotiating sessions because everybody was waiting for government instructions on this. That was closest we came, really, to losing it in the negotiations over that issue." The dispute was resolved when "we finally agreed that each of us would use our own territory on the Station according to our own definition of peaceful purposes." There has been a convergence of attitudes since then, she suspects, "everybody's evolved more to the U.S. perspective" as "space becomes more and more useful for military, nonaggressive purposes."[44]

Another source of friction between the partners arose over the handling of cost increases on the NASA side. As was mentioned earlier, in 1983 Beggs put a figure of $8 billion (in 1984 dollars) on space station development, the amount that the NASA administrator thought the president could accept. In October 1985 NASA officials announced that they had adopted a "dual-keel" design for what would be a multifunctional space station with foreign participation.[45] A year later its cost was estimated to be $14.5 billion (1984 dollars). Then in April 1987, under pressure to reduce costs further, NASA announced a "revised baseline configuration" with a cost estimate of $12.2 billion (1984 dollars). This omitted the cost of operations, an emergency crew return vehicle, and the cost of transporting hardware into space with the shuttle.[46] NASA signed contracts for four "work packages" with aerospace contractors.

President Reagan baptized the new configuration Space Station Freedom, a name that hearkened back to the State of the Union address in January 1984 in which he had said, "We are first, we are the best, and we are so because we are free."[47] As Finarelli remarked, it also made clear that "[t]he Space Station was clearly one of the nation's Cold War high-technology infrastructure projects undertaken at least in part to demonstrate our leadership vis-à-vis the Soviets, and part of that leadership is showing that people will follow your lead in what you choose to do"[48]—as did the Europeans.

The Europeans played a major role in shaping the final agreements on participation in Space Station Freedom. Their financial contributions were substantial: at the time, about twice what was expected from Japan and four times more than Canada. They also brought far more historical baggage to the negotiating table.

What of Canada and Japan? Canada had built the Remote Manipulator System (or Canadarm) for the shuttle. It had established its reputation as a reliable partner that could be trusted to build technological elements that were critical to mission success. Three main reasons determined its decision to join in the station. First, the in-orbit assembly and operations of the station provided it with an opportunity to further valorize its acquired experience in automation and robotics. Second, it was attracted by the polar-orbit earth observation facility, which could provide remote sensing data for many of its needs. Finally, the Canadian authorities were persuaded that the space station would "alter dramatically many of the established ways of operating in space." Joining the American project along with Western Europe and Japan would provide a platform for "new business relationships and cooperative programs with the world's major space nations."[49] For Canada, then, foreign policy concerns were overshadowed by the possibilities for expanding its existing industrial capabilities and markets in high technology, for consolidating space cooperation with partners other than the United States, and for providing remote sensing data that covered its vast geographical space.[50]

Japan's engagement with the space station had a different trajectory.[51] It had long been champing at the bit to develop its own, autonomous space program. Many felt that it had, for too long, been under foreign technological tutelage. Though NASA had helped Japan develop launchers, it had denied it access to cutting-edge technologies and had restricted the payloads that the country could launch with "its" rockets (see chapter 10). It seemed clear that to fully reap the benefits of the conquest of space Japan needed to have its own launcher. Could it afford to do so (at a development cost of $1 billion), and at the same time accept

President Reagan's offer in January 1984 to join in the space station? The famed MITI (Ministry of International Trade and Industry) and a group of major Japanese industries were persuaded that it was imperative not to "miss the boat" on manned space flight, as Japan had done on post-Apollo. However, Japan also wanted an indigenous launcher that would not be subject to US restrictions on use. It eventually adopted a two-track approach. It developed a "made in Japan" H-2 launcher that proved to be neither a commercial nor a technological success.[52] Its contribution to the space station was a Japanese Experiment Module (JEM), also called Kibo (meaning hope).

Formalizing the Collaboration

The legal instruments codifying the design, construction, and use of the space station (bilateral MoUs between agencies and an IGA between the governments) were signed after 15 rounds of negotiations over three years in September 1988. The flexibility available to NASA and the American delegation was constrained by a number of requirements. One of the most contentious of these, as we have seen, was that they had to "explicitly reserve the right to conduct national security activities on the U.S. elements of the Space Station, without the approval or review of other nations." They were also not to "accede to multilateral decision-making on matters of Space Station management, utilization, or operation." Technology transfer was to be controlled by not permitting a "one-way flow of U.S. space technology to participating nations who are also our competitors." And finally, they were to ensure that the concept of "equal partnership" did not "displace either the reality or symbol of U.S. leadership."[53]

The Europeans were reasonably satisfied with the final agreements. Take the question of management. In the midst of the negotiations Pedersen publicly wrote that "perhaps [the] most difficult leadership adjustment for NASA is to learn to share direct management and operational control in projects where it is the largest hardware and financial contributor, especially when manned flight systems are involved." How did the legal instruments respect this? On decision-making procedures, for example, it was agreed that although the United States would be responsible for the overall coordination of the program, the Europeans had jurisdiction and control over their three Columbus elements. The United States and Canada were attributed 49 percent utilization of the APM in return for their contributions to the core elements of the station. Europe also had access to the whole station. And it was allowed to use its space transport system and communications equipment, in addition to having access to those that the United States would provide. This meant that the MTFF and the PFF would be launched by Ariane.[54]

The management practices were shaped by the architecture of the project. At the macro-level this restricted technology transfer to flows across clean interfaces. NASA alone would build the core of the station. This core would be augmented by discrete hardware elements that would be dedicated to scientific and/or manufacturing research of potential commercial interest. Only Canada's robotic arm for assembly was critical to mission success.[55]

What of "genuine partnership"? Peggy Finarelli, who was among those who negotiated these agreements on behalf of NASA, explained that she was emphatically against the "metaphysical" phrase "genuine partnership" being included as such in the legal agreements. Instead she asked for a list of 25 things that constituted "genuine partnership," "then we'll negotiate on each of those twenty-five points, and, god knows we did...and twenty-five more. That's why at the end of the day we were all happy with the agreement, even though it did not include that phrase."[56]

Another traditional area of disagreement concerned the legal ties obliging partners to sustain their commitment to the station once the project was embarked upon. As pointed out in the discussion of ISPM in chapter 2, the Europeans were particularly sensitive to programmatic changes required of NASA by the annual revision of its budget allocation by Congress. They hoped to get around this by raising the space station agreements to the status of an international treaty. Finarelli insisted that this was not in *anyone*'s interest. As she put it:

> What the partners wanted was a mechanism to make the space station agreements 100 percent binding, something that we would never be able to walk away from. Their thought was that a treaty tying in the US Congress would accomplish this goal. But we said: We can't do it. Its impossible in our government. Even if we have a treaty, it's still subject to the availability of appropriated funds [as required by the Antideficiency Act of 1982 that prohibited the incurring of obligations or the making of expenditures in excess of such funds]. So what you're asking for, number one, does not accomplish the end that you would like to accomplish, and number two, you're running the risk of putting a whole new set of players in this thing, many of whom hate the Space Station and don't like NASA much either, meaning there's a very high probability that the treaty would be rejected.[57]

In the event in the final agreement NASA (and all the parties) could still appeal to the lack of availability of funds as a reason for reconfiguring the project, though each signatory did formally undertake "to make its best efforts to obtain approval for funds" to meet its international obligations.[58]

The Crisis of the Early 1990s and the Inclusion of Russia

When President Reagan authorized the space station in 1984 it was to have been completed within a decade for $8 billion. During the next nine fiscal years (FY1985–1993) more than $10 billion had been spent without much to show for it. As of January 1995 only about 25,000 pounds of flight quality hardware had been fabricated, less than 3 percent of what was then projected to be a 925,000-pound facility. This was primarily because "the space station effort for nine years languished in the design phase."[59] The "dual-keel" design of October 1985 was followed by the "revised baseline configuration" of Space Station Freedom, and then a "restructured space station" that was unveiled in March 1991, and scheduled to cost $30 billion.

This redesign did not satisfy Congress. Its threat to terminate the program was strongly opposed by the Bush administration, however. The year before 64 senators had insisted that the Space Station Freedom be sustained as "the cornerstone of our civil space policy and a symbol of our commitment to leadership and cooperation in the peaceful exploration of outer space."[60] British, German, French, and Italian ambassadors to Washington added their voices to the chorus that included President Bush himself and his secretary of state, James Baker. In July 1991 Baker wrote to the chairman of the Committee on Foreign Relations asking the Senate to support Space Station Freedom. As he put it, "The credibility of the United States as a Partner is based on its ability to make durable commitments. We will increasingly need to cooperate with these allies on common endeavors, whether in security, economic, environment, or science and technology areas. A failure by the United States to keep the Space Station Freedom on track," Baker emphasized, "would call into question our reliability."[61]

Space Station Freedom survived Congressional criticism in 1991 partly because its "durability" was indicative of the Bush administration's determination to maintain its leadership of the free world even as the Soviet Union imploded. It also sent a strong signal to Moscow just when the United States was reaching out to engage in closer relationships with its erstwhile rival. On July 31, 1991, President Bush and Premier Gorbachev signed the historic START I treaty in which they agreed to dramatically reduce their stockpiles of nuclear weapons. They also charged a number of joint working groups to negotiate cooperation in various space-related fields (see chapter 8), including an extended stay by an American astronaut on the Soviet Space Station Mir. In 1992 Bush and Russian president Boris Yeltsin extended plans for space cooperation beyond scientific support and an exchange of astronauts to include a rendezvous and docking mission between the Shuttle and Mir.

Mir, which had been launched in 1986, was the "strangest, largest structure ever placed in Earth orbit," "a dragonfly with wings outstretched," "the best and the worst of Soviet technology and science," a "cluttered mess" inside, "with obsolete equipment, floating bags of trash, the residue of dust, and a crust that grew more extensive with the passing years."[62] Mir was also a testing ground for long-duration human spaceflight. Cosmonauts typically spent four–six months, even more than a year on board.

Bill Clinton was inaugurated as the new president in January 1993. He and Vice President Al Gore were determined to continue the process of modernizing and stabilizing Russia, of demilitarizing its high-technology sector, and of remodeling its institutions and industries along American lines. For Clinton and Gore space collaboration was embedded in a broader attempt to encourage Russia and the Newly Independent States (NIS) in their transition to democracy and market economics. It had the programmatic aim of capitalizing on Soviet space technology and know-how. However, it was also seen as an instrument to channel hard currency into a crumbling infrastructure, to retain elite scientists and engineers who might otherwise drift into the arms of rogue states, to encourage government and industry to adhere to the provisions of the Missile Technology Control Regime, and to isolate the opponents of reform by sustaining a high-prestige Soviet activity even as the communist system collapsed.[63] In April Clinton met with Yeltsin in Vancouver and finalized an American aid package of $1.6 billion. He also invited Russia to participate in a renewed space station program. One of the most important by-products of this meeting was

THE INTERNATIONAL SPACE STATION 261

the so-called Gore-Chernomyrdin Commission (Victor Chernomyrdin was the Russian prime minister). It first met in September and then in December 1993 to work out details of bilateral agreements on space, energy, and technology (see chapter 8).

Clinton's efforts did not win universal approval at home. But they played an important role in keeping the project alive in 1993. On entering the White House he told Dan Goldin (who was appointed NASA administrator in 1992 and remained in post throughout his mandate) that he was willing to support a space station. However, he asked the NASA administrator to come up with a leaner design. He was presented with three options. One was a modular concept that would use existing hardware. Another was a derivate of the Space Station Freedom. The third was a station that could be placed in orbit with a single Shuttle-derived launch vehicle. On June 17 President Clinton chose "a medium-sized modular space station" that used a "combination of Freedom hardware and flight-qualified space systems from other sources." Goldin announced that Russian hardware alternatives had been incorporated into the plans where appropriate.[64] He said he needed $12.8 billion for the Space Station: Clinton capped its cost at $10.5 billion over the next five years.[65]

Congress voted on two expensive technological projects inherited from the Reagan years within days of each other in June. Both of them were intended to restore American prestige in the context of Cold War rivalry. One was the Superconducting Super Collider, on which $2 billion had been spent. Work had already begun on digging an oval, 54-mile underground tunnel near Dallas to hold the particle accelerator. The other was Space Station Freedom. Congress voted to kill the SSC; the Senate confirmed the decision a few months later.[66] The Space Station survived by just one vote on a day that Dan Goldin later recalled was his worst ever at NASA.[67] A year later, in summer 1994, the House of Representatives endorsed the station by 123 votes. Saving domestic jobs was one important reason for Congress's support: NASA spread industrial contracts for the space station across 39 states, thereby spawning an estimated 75,000 jobs by 1992.[68] Foreign participation and the diplomatic consequences of being seen as an unreliable partner undoubtedly also carried some weight.

With the Gore-Chernomyrdin commission getting into its stride, NASA drafted a new International Space Station Project. It had three phases and Russia was crucial to all of them. Phase 1, scheduled to last from 1994 to 1997, was a joint Shuttle/Mir program that would enable American astronauts to familiarize themselves with living and working in space for extended periods of time. The station core would be built in Phase 2, that would last the next three years and to which Russia would contribute several critical elements, including guidance, navigation, and control. In Phase 3, lasting from 2000 to 2004, the station would be completed with the addition of research modules from the four countries that were building them. Russia would again provide key elements, like a habitation module (until the United States had built its own), and a crew return lifeboat for emergency evacuation. A comprehensive $400-million contract was signed between NASA and the Russian Space Agency to implement this plan.

These plans evoked criticism both at home and from the foreign partners. One of the major concerns was whether, given the state of the nation, Russia could be relied on to provide items that were critical to mission success. Others complained that the United States was using foreign aid to boost the space infrastructure of Russia and the NIS without being sure that they could deliver and at the expense of American jobs. Indeed NASA paid dearly for making an exception to its policy of clean interfaces and no exchange of funds.

The evolution of the collaborative project with Russia has been described in chapter 8, and will not be repeated here. The difficulties encountered with Zarya (the Functional Cargo Block—FGB) are illustrative. Zarya had to be put in space before anything else. With 16 fuel tanks holding more than 16 tons of propellant, and two solar arrays 35 feet long and 11 feet wide, this pressurized module was to provide orientation control, communications, and electrical propulsion for the station until the Russian-provided crew quarters arrived. It was to be built in Russia under contract and owned by the United States. Schedules slipped. Costs increased. All the partners were infuriated when Moscow, who was supposed to cover all the costs related to Zarya, attempted to drop the module entirely and replace it with a Mir module. In April 1998 an internal NASA report noted

> the anticipated one billion dollar cost savings to the U.S. to be accrued from Russian provision of a Functional Cargo Block . . . and an Assured Crew Return Vehicle capability was a faulty assumption as far back as 1994. The continuous economic situation in Russia has also negated most of the $1.5 billion in schedule savings to be achieved through their involvement.[69]

The Shuttle/Mir program was also a headache. The Russians demanded funds for goods and services that NASA believed had already been paid for, and charged "exorbitant" fees for cosmonaut time on American projects. When Goldin heard that the Russians were getting Mir ready to fly a space tourist, he exploded. "They always seem to have a little extra money around for Mir but not for the International Space Station."[70] In the event the original $400 million that NASA had offered for Shuttle/Mir and Phase 2 space station costs ballooned to double that figure as the Russians added ever more requests for financial support (see table 8.3).

European Reactions

Goldin did not want the existing partners in the space station to drain the momentum from his big-picture vision of a transformed space station that included Russia. He was advised that before moving ahead with Moscow "we needed to consult with our partners. He didn't want to hear it. Those people didn't last long in the agency. His plan had to go forward."[71] As was mentioned a moment ago, he and Clinton adopted a design that included a major Russian contribution in June 1993. Three months later in September 1993 the ESA member states were officially informed of the inclusion of Russia in the space station, now formally referred to as simply the ISS.

European space actors, like their American counterparts, had already moved quickly to build collaborative programs with the Russian Federation.[72] Early in 1993 they signed an agreement with the Russian Space Agency to develop a European Robotic Arm and a Data Management System for the Russian

Service Module. In preparation for their Columbus contribution to the space station they also arranged for European astronauts to live and work on Mir (Euromir 94 and Euromir 95). These missions would prepare their corps for living on the space station, enable them to validate items of Columbus, and provide flight opportunities for the user community before the space station itself was operational.[73]

Lynn Cline was brought in by NASA to negotiate an agreement with Russia once it had accepted an invitation to join the station. The original approach was minimalist, involving as few changes as possible to the previous documents defined with the original partners. Cline explains why that did not work:

> It became clear rather quickly that Russia wanted none of that, that they had very strong opinions about this partnership and what capability they were bringing to the table, and therefore, their desires on what their role should be. So once we crossed that threshold of, "It's not going to be minimal. There are going to be significant changes to this agreement," what happened was, Japan pretty much didn't want to change anything, Canada was rather flexible, and Europe came in with a whole new list of non-negotiable demands of changes that they wanted to have in the agreements as long as we're revising them, or they'd walk away from the partnership. So when I went through these negotiations, I had as hard a time working out the terms and conditions with Europe as I did with Russia.

Still, she insists it went relatively smoothly since in the process

> everybody recognized that Russia was a significant player, that they were bringing substantial capabilities with the launch capability, the cargo resupply capability, power capabilities, the main core of the Station. So there was a recognition that they had a key role. They had a right to certain demands, but also the original partners wanted us to truly bring them into the fold and have us all work multilaterally as a single integrated partnership.[74]

All the existing partners officially endorsed the proposal in May 1994. The programmatic advantages were evident. Russia would contribute its extensive experience of long-duration human spaceflight, and valuable hardware: the heavy-lift Proton launcher and the Soyuz capsule that could be temporarily attached to the ISS during construction. There was a "peace dividend" too. The German chancellor said he was "convinced that this international cooperation will make a major contribution to lasting cooperation world wide and will be a beacon of hope and trust for men and women on every continent."[75] This sentiment was endorsed by the ministers of the member states meeting in Toulouse in October 1995. Here the ministers agreed to fund what was now called a Columbus Orbital Facility (COF), which had been reduced to a third of its original size, with Germany bearing 41 percent of the costs.[76] France agreed to pay 27 percent of the costs of an automated transfer vehicle (ATV). The first of several ATVs called Jules Verne would be launched by Ariane, controlled from Toulouse, and would resupply the ISS with propellant, water, air, and payload experiments every 18 months.[77] Its pressurized cargo bay was based on a "space barge" developed in Italy and flown on the Shuttle, and that carried equipment to and from the station. The ministerial meeting also agreed to fund the design studies of a crew transport vehicle (CTV), a "lifeboat" that could be used to rescue astronauts from the ISS.

In addition to negotiating an additional MoU between NASA and the Russian Space Agency, a new intergovernmental agreement (IGA) was needed to cover the arrival of the new partner. Barter agreements, "equivalent" contributions in kind that required no exchange of funds, were also concluded. ESA would provide the United States with additional hardware for the ISS while its COF would be launched free of charge on the Shuttle, rather than on Ariane. ESA and the Japanese space agency agreed to trade a − 80° laboratory freezer for the ISS for 12 international standard payload racks. ESA persuaded Russia to provide certain services in return for supplying the European robotic arm and the data management system for the Russian segment of the ISS.

The new IGA signed in Washington, DC on January 29, 1998, was based on the first version signed almost a decade before. Thus as before Article 1 of the IGA affirms that "[t]he object of this Agreement is to establish a long-standing international cooperative framework among the Partners, on the basis of genuine partnership, for detailed design, development, operation, and utilization of a permanently inhabited civil international Space Station for peaceful purposes, in accordance with international law."[78] "Genuine partnership" was however parsed to reflect the criticality of the different contributions to overall mission performance. The United States and Russia would produce the elements that served as the "foundation" for the ISS, those provided by the Europeans and Japan would "significantly enhance" its capabilities, while Canada's contribution would be "an essential part" of the system. At the insistence of Russia, the management of the station was placed on a more multilateral basis than in the 1988 agreements. The United States was given the "lead role" for "overall program management and coordination," with the "participation" of the other partners.[79] The other partners were responsible for the management of their own hardware and utilization programs. They would also participate in all important reviews.

The change in the rules on criminal jurisdiction is also interesting.[80] In the 1988 agreements the United States was entitled to exercise jurisdiction regarding accusations of misconduct by non-American personnel anywhere in the ISS—even if they were in or on non-American elements—if that misconduct was deemed to affect the safety of the whole station. In the 1998 agreements each partner state has jurisdiction over the behavior of its nationals in the first instance, though exceptions apply. It was also agreed that both the United States and the Russian Federation could use their elements for national security purposes if they so wished, but that they could not use the European elements without the consent of the European partner.

Throughout its history the space station has combined NASA's determination to sustain its post-Apollo momentum with a multilayered project originally announced in 1969, combined with Congress's willingness to support jobs in the aerospace industry, and with the foreign policy agendas of successive administrations. The fact that it has had strong presidential support at key moments was also crucial, particularly for foreign participants. The partners in this behemoth, once persuaded that the United States was serious about a space station, had very similar domestic aims. Participation in the project would not only release more government funding for space, but it would also provide access to American technology, enhance the national technological base, and stimulate the aerospace and related industries. It also had foreign policy components: notwithstanding their different attitudes to the United States, the 14 European ministers who met in Toulouse in 1995 saw the space station as the "greatest cooperative venture of all time, with significant scientific, technological and political implications."[81]

The form that collaboration took evolved dramatically once Russia came on board. This was partly because NASA had to constantly cut back its ambitions for the station to satisfy a Congress that was increasingly impatient with rising costs and slipping schedules. It was also because Russia seemed to offer one way out of this perpetual crisis by bringing pertinent hardware and experience to the project, which no other nation had to offer, with the added advantage of providing a "peace dividend" for the White House. The architecture of the station allowed for different ways of organizing collaboration with different partners depending on what they brought to the table. By contributing core elements, and by turning its institutional and financial disorder to its advantage, Russia forced NASA to make an exception to its time-hallowed principles of no-exchange of funds and clean technological interfaces. Once the breach was made all could benefit, and all of the partners now contribute elements that are critical to mission success.[82]

The ISS transformed the way that NASA collaborated with its partners. The anxieties over Russian reliability and some resentment about the way in which dollars were spent by Moscow will surely make the agency and Congress extremely reluctant to give others a core role in a mission again without cast-iron guarantees that they can pay for what they do and that they can deliver. Nor is it certain that the much-vaunted foreign policy benefits that Clinton and Gore sought were achieved because Russia was integrated into the space station.[83] Including Russia in the ISS was part and parcel of a wide-ranging initiative to transform the Soviet empire into a democratic, market economy, and might have played little or no role in facilitating the transition. Indeed, the ISS may not be a harbinger of a fundamental revision in NASA's and Congress's approach to international cooperation at all, as Pedersen hoped, but a unique experiment never to be repeated.

Chapter 14

The Impact of the International Traffic in Arms Regulations

The export of space "technology" has always been constrained by the fear that it may compromise American national security or the economic competitiveness of US firms.[1] As we saw in chapter 3, National Security Advisory Memorandum NSAM294 (on ballistic missile/rocket technology) and NSAM338 (on comsats) issued by the Johnson administration in the 1960s were intended to impede undesirable knowledge flows. Fears of technology transfer, and the need to control it, hovered over the debate on European participation in the post-Apollo program, and on the sharing of rocket technology with Japan and India, described elsewhere in this book (chapters 4–6, 10, 12).

Historically NASA has favored a fairly generous policy on technology transfer. The key pillars of the policies put in place by Frutkin in the early 1960s—no exchange of funds, and clean interfaces—shaped the structure of international collaboration and deftly helped NASA kick-start programs all over the world without undermining national security or economic competitiveness. However, as other friendly space-faring nations matured, and as their potential contributions to NASA's program increased, the agency had to navigate between the pressure to deepen scientific and technological collaboration, and the objections of those who wanted more formal restrictions on the sharing of hardware and knowledge. The conflicts emerged with particular intensity in the early 1970s with Western Europe, and with Japan and India in the 1980s. By the 1990s NASA realized that it would have to formalize and streamline its export control system to cope with new international and domestic realities, notably a major scare over the People's Republic of China's appropriation of American weapons and space-related technologies. The more stringent implementation of the International Traffic in Arms Regulations (ITAR) after 2000, and the onerous fines, including imprisonment, imposed for their violation caused some concern to people both in the United States and abroad. Preserving national security across a vast domain of dual-use technologies against the pressure from research and business who favor putting high walls around well-defined sensitive areas involves complex trade-offs and is a topic of ongoing interagency consultation.

Two arms of the executive branch, the Department of Commerce and the State Department, deal with most space-related export controls.[2] The former administers the Export Administration Regulations (EAR), which pertain to "dual-use" commodities, software, and technology, that is, items that have predominantly commercial uses but that can also have military applications and that are to be found on the Commerce Control List (CCL). The Directorate of Defense Trade Controls (DDTC) in the State Department (the Office of Munitions Control in the early 1970s) administers the ITAR. The ITAR are intended to curb the proliferation of sensitive technologies and weapons of mass destruction by preventing the circulation of defense articles and defense services. Defense articles are listed on the US Munitions List (USML).

The USML was described in a brief by Alvin Bass (of NASA's Office of General Counsel) as "a broad enumeration of articles which are considered as having direct or indirect military potential or applicability." When Bass was writing, in 1970, he noted that the list covered almost everything that NASA was concerned with, including

> [s]pacecraft, including manned and unmanned active and passive satellites, spacecraft engines, power supplies, energy sources, launching, arresting and recovery equipment, inertial guidance systems, and all components, parts and accessories of the above-mentioned items. Other categories [Bass went on] include propellants, missile and space vehicle powerplants, launch vehicles, rockets, control devices for any of the above, [various items] designed or modified for spacecraft or space flight, pressure suits, protective garments…space vehicle guidance, control and stabilization systems, and the list continues.[3]

Bass did not enumerate the defense services that could only be supplied if permission was granted. Today defense services are defined as including "the furnishing of assistance (including training) to foreign persons, whether in the United States or abroad in the design, development, engineering, manufacture, production, assembly, testing, repair, maintenance, modification, operation, demilitarization, destruction, processing or use of defense articles."[4]

The term "export" is misleading (as is the phrase "technology transfer") if one wants to understand the scope of the control regime. These terms create the impression that only commodities are regulated. But authorization is also required (in a Technical Assistance Agreement, or TAA) to export technical data (as distinct from "purely theoretical scientific data," which was treated more leniently). The meaning of the term "export" is correspondingly expanded. To quote Finarelli and Alexander, under ITAR, to export was defined in 2008 as "[a]ny oral, written, electronic, or visual disclosure, shipment, transfer, or transmission outside the United States to anyone, including a U.S. citizen, of any commodity, technology (information, technical data, or assistance), or software, or codes."[5] A second clause restricts even the "intent" to make exports of this kind to "a non-U.S.-entity or person wherever located," that is, in the United States or abroad, and a third specifically controls any transfer to a foreign embassy or affiliate. Thus when US entities seek to transfer US technology abroad, they are triggering a process that manages not simply the "export" of commodities or "articles," but that regulates the flow of related data and knowledge, where knowledge is inscribed in many different forms, from the statement and the image to the hardware, and transmission occurs through many different channels, from the spoken word and the visual display, to shipment.

The Arms Export Control Act of 1976, often wrongfully attributed as being the genesis of ITAR, confirmed that the range of space technologies designated by Bass were indeed to be treated as defense articles, and that data exchanged regarding them was a defense service. That granted, it could always be argued by a US entity that specific items did not fall under the ITAR, and should be treated as dual-use technologies to be regulated by the less-restrictive EAR. In the case of the EAR, but not the ITAR, the decision over whether or not to grant an export license takes account of commercial factors, and above all whether or not the client could acquire the item from a foreign source if an American company did not provide it. In practice it is often found that many applications under the EAR do not need an export license, though the item must be evaluated before the determination is made and justifying documentation must be provided.

The reach of the legislation that embodies the control regime is negotiated and renegotiated between arms of the administration that have different and sometimes conflicting mission-objectives. They take account of input and pressure from various stakeholders in space, notably firms interested in expanding their markets, who seek to have their items regulated by the more relaxed EAR on the grounds that they are dual-use commodities, not essentially instruments of war, but also scientists and engineers involved in international projects. Social actors who have to implement the legislation can face stiff penalties—fines, imprisonment, loss of further government contracts—for not respecting its requirements and, in case of ambiguity, spontaneously retreat to a conservative interpretation of the law to protect themselves.

NASA, ITAR, and the Post-Apollo Negotiations

The extensive, blow-by-blow account of the in-house debates over European participation in the post-Apollo program in 1970–1971 (chapters 4–6) demonstrated the shifting perceptions of where the boundary lay between knowledge sharing and knowledge denial. So did the simultaneous debate over upgrading Thor-Delta technology acquired by Japan (chapter 10). Indeed the Nixon administration of the early 1970s is noteworthy for the determination of White House staffers Peter Flanigan and Tom Whitehead, with the support of science adviser Ed David, to rein in what they saw as NASA's profligate attitude to the sharing of knowledge that might undermine national military and/or economic security. Their concerns were reinforced by Bass's brief mentioned earlier, which was forwarded to them amid negotiations over European participation in post-Apollo. The legal counselor argued that although the technologies of interest to NASA's program were largely regulated via the Munitions Controls List, exports of data or articles by government agencies, including NASA, were "specifically exempt from the provisions of the Mutual Security Act and the ITAR." What is more that exemption was extended by the ITAR when the export was in "furtherance of a contract with an agency of the U.S. Government or a contract between an agency of the U.S. Government and foreign persons."[6] In short, according to Bass, in 1970 NASA and its contractors (like the Jet Propulsion

Laboratory in Pasadena) did not need to seek a license or other written authorization from the Department of State to export items on the MCL.

Flanigan and Whitehead were appalled and demanded that Ed David "develop a policy for the transfer of technology developed by NASA."[7] Bass's report confirmed for Flanigan that, as things stood, "NASA had no policy on keeping proprietary technical information developed by it available only to U.S. citizens." NASA's new administrator Jim Fletcher agreed that a new policy was needed to stop NASA "both by its charter and its history" from continuing "to make all its technological developments available nationally and internationally."[8] In the debate that ensued over the next six months Europe found its participation in the post-Apollo program reduced to building a module that fitted in the shuttle's cargo bay, and that restricted transnational knowledge flows to the minimum required for mission success. Japan's access to Thor-Delta technology was also severely restricted.

Revising the Regulatory Regime in the 1990s

In the early 1990s NASA took an important step toward formalizing and streamlining its implementation of the export control regulations affecting space collaboration in all its aspects. Two factors converged to encourage these institutional changes. First, the agency and its contractors were under increasing criticism for being lax in enforcing the statutory regulations controlling exports to foreign partners—for example, they allowed Norway to acquire sounding rockets, which fell squarely under the ITAR, through the less stringent "dual-use" provisions of the EAR that regulate the export of items on the Commerce Control List. Second, new policies were needed to deal with the inclusion of the one-time space rival and communist menace, the ex-Soviet Union, as a significant partner in the International Space Station (see chapter 13). In response to this situation, in 1994/95 NASA replaced its previously fragmented program with a single export control office that handled authorizations required by both ITAR and EAR, to ensure that the different regimes were implemented coherently. Second, an interagency Space Technology Working Group agreed that the civil Space Station should be moved from the USML to the CCL, along with commercial communications satellites. Until that time *all spacecraft* except for comsats were on the USML (but see later). Henceforth (and still today), the ISS could also benefit from the greater clarity, transparency, and flexibility of the EAR over the ITAR.[9] This has undoubtedly contributed to its success as a site for international collaboration.

In 1996 President Clinton ordered that the export controls over commercial comsats be placed on the CCL. This settled an ongoing dispute between the Commerce and State Departments that had simmered for almost a decade. In the late 1980s President Reagan had signed a deal with the People's Republic of China (PRC) authorizing nine launches of American-built comsats on Chinese rockets. The Tiananmen Square sanctions law passed in 1990 (P.L. 101–246) suspended this policy for a few years. However the pressure to secure markets for US manufacturers led to a relaxation in 1992, when the State Department issued a directive transferring some comsats from the USML to the CCL, and so to the jurisdiction of Commerce. This transfer was completed by Clinton's order in 1996.[10] The president was keen to move from a policy of confrontation with the PRC to one of diplomatic and commercial engagement. The sale of supercomputers to China was authorized. Satellite technology for telecommunications was removed from the USML, and from the jurisdiction of the State Department, and placed on the Commerce Department's more lenient CCL. And in summer 1997, at the first US-China summit meeting since the crushed protest in Tiananmen Square in 1989, the president hoped to conclude a nuclear cooperation agreement that would enable American nuclear reactor companies to compete for the Chinese market.

Many in Congress were appalled by this new openness to the PRC. The House's concern was focused on allegations that two American satellite companies, Hughes Space and Communications International, Inc., and Space Systems/Loral, had illegally transferred sensitive missile technology to the PRC. This had occurred during investigations into three unsuccessful launches of their telecommunications satellites for civilian clients on Chinese Long March rockets. The possibility of such leakage led to the passage of the Strom Thurmond National Defense Authorization Act for Fiscal Year 1999.[11] This imposed new restrictions on international exchange before the Justice Department had finished its inquiry against Hughes and Loral.

Fear of irresponsible sharing of missile-related technology also led Congress to establish a bipartisan committee chaired by Representative Chris Cox (R-California) to investigate the matter. The political climate was charged: one observer has remarked that "[a] number of Republican leaders went to the floor of the House and Senate and accused the President of treason for allegedly facilitating this transfer of information."[12] The bipartisan committee's classified report was submitted to the president on January 3, 1999; a declassified version was released on May 25, 1999 (the Cox Report).[13] The account that follows deals first with the specific charges against Hughes and Loral, and then with the more general charges made in the Cox Report.

Hughes, Loral, the PRC, and the Strom Thurmond Act

In December 1992 the Chinese Long March 2E rocket failed to launch the Hughes-built Australian Optus B2 telecommunications satellite due to aerodynamic buffeting of the launcher's fairing.[14] Neither party would at first admit responsibility. Hughes conducted an independent investigation, and divulged information to the PRC suggesting ways in which it should modify the fairing by strengthening its structure. At a subsequent successful launch in August 1994 observers from Hughes noted that the fairing had been modified simply by adding rivets. This proved to be insufficient. The next launch of a Hughes satellite, the Asian Apstar 2 in January 1995, failed for the same reason as had the launch of Optus 2. This time the Chinese members of a joint accident review committee agreed that the cause of the failure was due to weaknesses in their fairing. The marginal improvement achieved by adding rivets was not sufficient to withstand the additional stress caused by the strong upper-altitude winds that buffeted the payload when it was launched in winter. Suitable corrective measures were taken along the lines first proposed by Hughes—corrective measures that, some feared, would be invaluable for improving nose cones that protected nuclear warheads on Chinese ballistic missiles.

A Loral Intelsat 708 satellite was destroyed in the Long March commercial launch failure in February 1996. This time the PRC engineers quickly admitted responsibility. They suspected that the launch failure was probably due to a fault in the inner part of the inertial measurement unit (IMU) of the Long March 3B rocket guidance system, though telemetry data did not fully confirm this. The insurance company that had agreed to cover the imminent launch of an Apstar satellite (typically for about $50 million) demanded that an independent review committee be established. The committee comprised representatives from the PRC, Hughes, Loral, and Daimler Benz, and retired experts that had worked for British Aerospace, General Dynamics, and Intelsat. It placed great weight on the telemetry data, and suggested that the follow-up frame, rather than the inner part of the IMU (the preferred explanation by Chinese engineers), was responsible for the accident. The PRC confirmed that, indeed, a failure in the follow-up electrical servo unit was the cause of the launch failure.

Loral faxed a preliminary report of this finding to the PRC in May 1996. The State Department learned that the firm had disclosed information that some thought would significantly improve the guidance system on Chinese missiles, without first having it reviewed for sensitive content, and without an export license.

The Strom Thurmond Act signed into law in October 1998 took steps to regulate these practices. It devoted 6 pages out of 360 (Title XV.B) to a number of measures designed to control the export of satellite technology to the PRC.[15] One of its most fundamental innovations (in Section 1513) was to remove the president's authority to change the jurisdictional status of satellites and related items even if they had civilian applications. These were, and still are (August 2012), "the only dual-use items that are *required by law* to be controlled as defense articles." Thus whereas normally "the President has the authority to authorize the easing of controls on items and related technologies that transition to predominately civil uses or that become widely available," this did not now apply to satellite-related items. The export of all "satellites and related items" were put back on the US Munitions List and subject to the ITAR, and require Congressional action to remove them.[16] The Strom Thurmond Act also called for new bureaucratic procedures to ensure compliance. It stipulated that any export licenses had to be accompanied by a Technology Transfer Control Plan that had been approved by the secretary of defense and an "encryption technology transfer control plan approved by the Director of the National Security Agency." In response to accusations that security on the launch pad in China (often in the hands of private contractors) had been dismal while the satellite was being installed, the DoD was also called upon to monitor all aspects of the launch of an American satellite in a foreign country, including analyses of launch failure, "to ensure that no unauthorized transfer of technology occurs, including technical assistance and technical data."

In March 2003, Hughes Electronics Corporation and Boeing Satellite Systems, charged with 123 violations of export laws, admitted that they had not obtained the required licenses for their dealings with the PRC. The firms acknowledged "the nature and seriousness of the offenses charged by the Department of State, including the harm such offenses could cause to the security and foreign policy interests of the United States."[17] Their $32 million civil penalty was the largest in an arms export case.

The Cox Report

The declassified version of the Cox Report (or just the Cox Report in what follows) released on May 25, 1999, created a sensation. An editorial in the *Washington Post* caught the mood that day: it quoted Cox as saying that "[n]o other country has succeeded in stealing so much from the United States," with serious and ongoing damage to the country.[18] The Republican House majority leader Dick Armey said, "It's very scary, and basically what it says is the Chinese now have the capability of threatening us with our own nuclear technology."[19] The ensuing sense of urgency led to calls for a transformation in the legal and administrative structure of international cooperation. Tighter controls on hardware and knowledge flows were imperative.

The Cox Report was a three-volume, 872-page glossy publication filled with photographs suitably labeled and a punchy overview that used color and other techniques to highlight key findings and to have them spring to the eye.[20] Though most of the committee's time was devoted to the Hughes and Loral cases, in the latter stages of its hearings it branched out into the "theft" of other sensitive technologies, notably for nuclear warhead design. Volume 1 of the report focused on this domain and on the diversion of High Performance Computers (600 of which had been sold to the PRC) from civilian to nuclear weapons applications. Volume 2 was devoted to the contacts between engineers at Hughes and Loral and their Chinese counterparts, particularly after the launch failures. It also had a section devoted to launch-site security in the PRC or, rather, the lack thereof. The short technological core of the third volume dealt with the efforts made by the PRC to improve their manufacturing processes by acquiring machine tool and jet engine technologies.

The report claimed that the PRC had "stolen design information on the United States' most advanced thermonuclear weapons," including the neutron bomb, from the four major weapons labs (Los Alamos, Lawrence Livermore, Oak Ridge, and Sandia).[21] This would give China design information on such devices "on a par with our own," and would materially assist the country "in building its next generation of mobile ICBMs, which may be tested this year," with "a significant effect on the regional balance of power." China had also "stolen or illegally obtained U.S. missile and space technology that improves the PRC's military and intelligence capabilities." The information passed by US satellite manufacturers to their Chinese clients—without obtaining the requisite licenses, even though they knew they were needed—had "improved the reliability of PRC rockets useful for civilian and military purposes," including ballistic missiles. These security lapses were compounded by poor security at launch pads and by the liberal sharing of technical information with foreign brokers and underwriters of satellite insurance. The knowledge thus acquired would not only strengthen China's military capability, but the PRC was also "one of the leading proliferators of complete ballistic missile systems and missile components in the world," and had helped improve weapons programs in Iran, Pakistan, Saudi Arabia, and North Korea.

Two factors had facilitated these major breaches in the security wall. First, there were recent changes in international and domestic export control regimes that had reduced the ability to control the flow of militarily useful technology.

Supplementing this, there was China's determination to obtain advanced US military technology, which it had actively sought for at least the past two decades. As the report put it, "To acquire U.S. technology the PRC uses a variety of techniques, including espionage, controlled commercial entities, and a network of individuals and organizations that engage in a vast array of contacts with scientists, business people, and academics." In short, the Cox Report emphasized, "The PRC has mounted a widespread effort to obtain U.S. military technologies by any means—legal or illegal."

As regards space policy, the Cox Report urged the executive branch to "aggressively implement the Satellite Export Control Provisions"[22] of the Strom Thurmond Act. It demanded that the State Department be responsible for licensing the export of satellites and any satellite launch failure investigations. The Department of Defense, not satellite firms, was to be responsible for security at foreign launch sites, and had to establish appropriate monitoring procedures to ensure that no information of use to its missile programs was passed to the PRC. The report also insisted that "export controls are applied in full to communications among satellite manufacturers, purchasers, and the insurance industry, including communications after launch failures."[23] Recognizing that the American firms were seeking launch providers abroad because the United States had insufficient domestic launch capability (itself a result of the decision to cut back the production of expendable launchers so as to secure a market for the shuttle), the Select Committee also recommended that steps be taken to stimulate the nation's "commercial space-launch capacity and competition."[24]

The Cox Report proved highly controversial. Joseph Cirincione, the director of the Carnegie Non-Proliferation Project, objected that the report had "taken a real problem and hyper-inflated it for political purposes."[25] The Center for International Security at Stanford University asked four experts (Alastair Iain Johnston, W. K. H. Panofsky, Marco di Capua, and Lewis R. Franklin) to review the report from different angles. Throughout their critique the authors stressed that the analysis was marred by "imprecise writing, sloppy research, and ill-informed speculation," as Johnston put it.[26] In a vigorous riposte Nicholas Rostow, a staff director on the US Senate Select Committee on Intelligence, in turn identified what he called "50 Factual Errors in the Four Essays" that comprised the "Panofsky" critique.[27] And toward the end of 1999 the National Academies' Committee on Balancing Scientific Openness and National Security published its findings on the risks posed by foreign interactions with the national weapons laboratories.[28]

None of the critics denied that the Cox Report had put its finger on a serious issue. But they objected that it had incorrectly elevated security leaks to a privileged position in its analysis of knowledge flows between the United States and China: there were many other ways for PRC scientists and engineers to access the cutting edge of the American research system. They felt that it had incorrectly assessed China's strategic goals, and the urgency with which it sought to update its obsolete nuclear and missile programs.[29] Third, they contended that the combination of these two erroneous convictions had created a climate of crisis in which blanket restrictions on international exchange were being called for. This would be counterproductive and do the United States more harm than good. Tighter controls were thus not the answer to improved security: rather, what was needed was increased funding for R & D, which ensured that the United States always had the technological edge over its rivals. The Committee on Balancing Scientific Openness and National Security took a similar line, "The world is awash in scientific discoveries and technological innovations," it wrote. "If the United States is to remain the world's technological leader, it must remain deeply engaged in international dialogue, despite the possibility of the illicit loss of information."[30]

This is the background against which the demand for a more rigorous application of export control regulations was specifically written into Public Law 106–391, NASA's Authorization Act of 2000 that was passed by both the House and the Senate. This act encouraged international cooperation in space exploration and scientific activities when it served American interests, *and* was "carried out in manner consistent with United States export control laws" (Sect 2.6 (B) (iii)). The point on regulation was picked up again later in a section of the act that twinned international cooperation with American competitiveness. After laying down specific recommendations as regards space cooperation with the PRC, the text went on to stipulate (Sect. 126 (3)) that NASA's inspector general, in consultation with the appropriate agencies of the U.S. government,

> shall conduct an annual audit of the policies and procedures of the National Aeronautics and Space Administration with respect to the export of technologies and the transfer of scientific and technical information, to assess the extent to which [the NASA] is carrying out its activities in compliance with Federal export control laws and with paragraph (2) [relating to the PRC].[31]

NASA had already institutionalized more formalized and systematic procedures for the implementation of export controls in the mid-1990s. P.L. 106–391, however, sent a strong signal that Congress was keeping an eye on the agency to ensure that compliance with ITAR was enforced. The effects have been felt throughout the centers, by NASA contractors such as JPL and by the agency's international partners.

To summarize. The combined effect of the Strom Thurmond Act, the Cox Report, and P.L.106–391 has been to move export controls in the space sector to the very foreground of NASA and the State Department's activities in the international domain. To be sure it was already evident in the mid-1990s that NASA's management of export control needed to be tightened up. The ambiguity over whether or not ITAR or EAR applied to a space-related item was amplified by an internal organization that was fragmented, with different sections dealing more or less independently with the different regulatory systems. A single export control office replaced these in 1994/95. Its importance was fueled by fears that the PRC was gaining ready access to sensitive American defense-related technology. These were generalized by Congress, and embodied in legislation that demanded that NASA and US entities take the restrictions imposed by the ITAR seriously not only in dealings with the PRC but also with traditional allies and partners (subject to some variation for NATO members, for example).

Over the next decade the effects of these restrictions were increasingly felt and resented by NASA's partners, as well by its contractors and US business. For example, John Schumacher, an associate administrator for external relations at NASA who moved into the aerospace industry in the 1990s, stressed the difficulty of deciding if an innovative technology fell under the ITAR or not. Taking the hypothetical example of a nano-ceramic coating for engine blades he noted that neither the firm nor the regulatory authority was sure whether something like this it fell under the ITAR or not. This was particularly frustrating for smaller companies with international clients, like his own, and often led them to "walk back from the edge," and withdraw from regulation-prone innovative research.

NASA contractors such as JPL are also subject to the ITAR, and have established their own in-house ITAR office to ensure compliance, and need to establish Technical Assistance Agreements (TAA) before they can provide defense services to their partners. A TAA is a contract between the parties involved in the technology transfer. It references the ITAR, and defines items such as the roles of the contracting parties, what technology and services are covered, who can access the ITAR-controlled technology, restrictions or exemptions on how the technology can be used, and how long a foreign entity can have access to the technology. Even with these procedures in place there can be friction. Robert Mitchell, a project manager on the Cassini-Huygens mission (see chapter 2), explains:

> [T]he most common thing that comes up now is a problem with an instrument, a European-provided instrument. An example would be the magnetometer, which was provided by and still funded and operated by an organization in the U.K. The magnetometer will from time to time have issues in terms of how it interfaces with the onboard main central processor on the Cassini spacecraft, and there frequently are questions about whether the problem is in our computer, or a problem in the interface, or a problem in the instrument. And of those three things, we understand the first two far better than they do, and we understand the third probably not as well as they do, but we know it pretty well...Now, for us to give them technical assistance in resolving the problem is clearly prohibited except in the presence of a TAA, and even with that we've been cautioned to tread carefully. So when they have a problem— and about once a year they do—we work with them, we get it taken care of, but everybody is very conscious of this issue.[32]

Charles Elachi, the current director of JPL, sees the effect on interpersonal relationships with colleagues abroad as one of the most distressing feature of the ITAR. It undermines the freewheeling climate of mutual trust and respect that is essential to the success of an international project. As he said in an interview in June 2009:

> [T]he bigger impact, in my point of view, was more on the interaction between people, more than actually getting a piece of hardware, because now if we want to talk with the ESA, we have to be careful what we talk about and so on. It's not an issue of do we send a transistor from the U.S. to Europe, even if that's a factor. But it's really the interaction, and, I'm guessing that's where maybe people like us are unhappy, and that's where I'm unhappy also about

this thing, because the strength was in building trust and good relationship and exchange of ideas, and that kind of put a limitation on doing that.[33]

David Southwood, ESA's then director of science and robotic exploration, was blunter in an interview in 2009: "Those of us who want to cooperate with the United States are frustrated by the level of regulation and nonsense we're put through, and indeed the problem we face of trying to explain to people that if we really are cooperating we have to have an understanding of what something does in the partner's piece of equipment."[34]

There can be no doubt that the current (August 2012) ITAR regime is transforming the dynamics of international collaboration with the United States. It is not doing so simply by placing tight constraints on the hardware that can be shared with partners: as this book has made evident, in the domain of satellites and launchers, the components that can be acquired by others has always been subject to close scrutiny. By reaching deep into the daily workings of even non-military cooperation involving scientists and engineers in academia, government laboratories and industry, the regime is making international collaboration more onerous bureaucratically and more risky institutionally, as well as undermining the trust and mutual respect between people that is so essential to the success of any joint project.

It must be stressed that NASA itself is not as tightly bound by ITAR as is a contractor like JPL. Like its contractors it requires a hardware license to export technology, but unlike them it does not require a TAA to supply defense services to foreign partners once an international agreement is in place (that agreement serves as the TAA).[35] The agency has also been engaged for over a decade in discussions with the State Department on ways to improve the implementation of the ITAR, particularly as regards the need for TAAs by its contractors. At the time of writing extensive interagency discussions have also led to major proposals for export control reform. They are guided by the philosophy that the United States must focus limited resources on the threats that matter most, and put in place streamlined procedures, combined with effective safeguards, to control sensitive items appropriately. It is proposed that commercial satellites be put back on the EAR and be regulated by the Department of Commerce. On one issue there is no reform foreseen: export control policies with respect to the PRC and embargoed countries, like Iran.[36]

Chapter 15

Conclusion

What has this book achieved? At one level it has simply filled a yawning gap in the extensive literature dealing with various aspects of NASA's history. Looked at from the perspective of the 1958 Space Act, most of that literature has dealt with one important dimension of NASA's mission: to ensure "the role of the United States as a leader in aeronautical and space science and technology and in the application thereof to the conduct of peaceful activities within and outside the atmosphere (Sec. 2 (c) 5)." The study of this injunction has spawned a huge body of work, much of it celebratory, detailing the many projects that have helped propel NASA and the United States to preeminence in the exploration and exploitation of space. It has also foregrounded human spaceflight, the most expensive, challenging, and spectacular of NASA's activities. The emphasis on these national narratives has, however, tended to eclipse another of NASA's important missions as defined in the Space Act: "Cooperation by the United States with other nations and groups of nations [...] (Sec. 2 (c) 7)." Of course there have been several important studies of international programs undertaken by NASA, and they have been referred to extensively in this volume. None of them can match this book both as regards the range of projects discussed, the variety of countries and regions dealt with, and sweep of time considered. Here then is the first comprehensive, yet necessarily selective study of NASA's international relations covering a span of 50 years.

To shift the intellectual focus from the national to the international means more than just adding a new set of programs to those that were undertaken within a domestic context. The projects described in this book do more than simply expand the huge list of what NASA has done at home. The international dimension is not just a linear extension of the national dimension. It requires that we look at NASA through very different eyes, and that we see the agency not simply as a national actor, but as an actor on the global stage. As Asif Siddiqi stressed recently, "By its physical nature, space exploration has a resonance beyond national borders—at a fundamental level, it is a project that transcends national claims and appeals to the global, perhaps even to the universal." This book rejects "the fetish for nation-centered cold-war geopolitics" as its central organizing framework.[1] It engages with histories that spill over national boundaries.

It inserts NASA in the world.

What does it mean to insert NASA in the world? First, it extends the range of actors engaged in project definition and support to include foreign governments, industries, and space communities. It is no longer enough that diverse American stakeholders throw their weight behind a particular program. Now a multitude of individuals and institutions abroad have to be enrolled in it too, and persuaded that it is also in their interests to invest human, material, and financial resources in a particular (joint) project. NASA does not only have to sell a project at home, it has to sell it abroad too. As many examples in this study have shown, any successful collaborative project requires that each lead partner builds a coalition in its favor, which is strong enough to contain the domestic pressures that would undermine it. An international project can only gain and sustain momentum if it manages to fuse the interests of stakeholders in different countries who sometimes have very different motives for commitment.

The expansion of the range of actors involved in international collaboration also introduces new players in the making of space policy. In the United States, the several arms of the administration that are concerned with national and economic security (the State Department, the Department of Defense, and the Department of Commerce, in particular)—in addition to the White House and Congress—frequently help shape NASA's international space policymaking. These are the lead agencies that mediate between the US government and American firms on the one hand, and foreign entities on the other (as do their equivalents abroad). NASA needs to act coherently with their policies and practices as it goes about its business in the international arena.

This configuration has programmatic implications. As we have stressed repeatedly, as an actor in the world, NASA cannot simply seek to promote American preeminence in space science and technology. It also has to align itself with the needs of American foreign policy, and serve as a vector facilitating the penetration of foreign markets by American industry.

Giving new actors a voice also increases the diversity of sources drawn on. The study makes extensive use of material from NASA's own archives of course. But it also exploits the FRUS (Foreign Relations of the United States), the National Archives at College Park, MD, the Washington National Archives Center in Suitland, Maryland, documents in various presidential libraries, and many other sources that deal with US foreign policy. Foreign archives have also been consulted and members of space communities in both the United States and abroad have been interviewed.

This book also breaks new ground by devoting considerable attention to the material practice of collaboration. It moves beyond the level of programmatic statements and expressions of intent to study how collaboration was administered on a day-to-day basis. This meant putting the interface, the contact zone between NASA and its partners, at the center of many analyses.[2] This is the site in which they come together face-to-face to define a project, to negotiate an agreement, to operate a satellite. This is the site par excellence in which knowledges of many different kinds circulate between scientists, engineers, project managers, administrators, industrialists, and politicians in different countries. In this book the shift from a national to an international framework has involved a shift from a focus on the production and domestic diffusion of knowledge to a study of the international coproduction of knowledge through circulation in the contact zone.[3]

This book adopts an operational concept of knowledge, a concept that was regularly drawn on by key actors *in practice* in defining the framework for international collaboration. It covered everything from propositional to tacit, from data to hardware, from science to technology to management. These regimes also allow for multiple modes of transfer, formal and informal, visual and tactile, by word and by deed, via paper, process, and product. The management of these flows has often been one of central questions for NASA in scientific and technological collaboration (especially as regards advanced technology of any kind). It has always been implicit in international ventures, most obviously in defining from the outset just what projects are candidates for foreign participation and how intimate that participation will be. Clean managerial and technological interfaces were the mantra that guided NASA's relationships with foreign partners. Their privileged position is a symptom of the immense importance that the practical management of knowledge flows across the contact zone has in the projects discussed here.

Several features regarding these knowledge flows need emphasizing. First, they are not unilateral, from the center (the United States) to the periphery (the rest). America is the recognized world leader in space science, technology, and project management. Other countries want to work collaboratively with NASA so as to have access to that knowledge and thereby to strengthen their own space capabilities. The United States, in turn, is ever alert to the benefits that it may draw from working with foreign partners. International collaboration can only work when there is a measure of reciprocity.

That said the contact zone is not a level playing field either: all "centers" are not equal. Time and again NASA has emphasized that the flow of knowledge between it and its partners occurs in an asymmetric field, in which the United States was the leading partner. NASA's challenge has been to define international projects from which both partners benefit without putting national security at risk, or undermining the competitive position of US industry.

This book shows that American scientific and technological leadership does not simply provide NASA with a superior bargaining chip in international collaboration. It is also a resource that can be levered to steer the space programs of other countries along lines that cohere with American interests. It can be used to reconfigure material cultures and have them dovetail with US best practice. It can nudge countries away from developing technological trajectories that threaten US preeminence. In the special case of the Russian Federation, it can even be used to try to completely reconstruct a national space industry. In so doing this particular transformative urge did not simply draw on NASA's scientific and technological strengths: it injected capitalist knowledges and practices of all kinds into the crumbling institutional framework of a collapsing communist empire. In fact, NASA and space collaboration were not simply one player among others in this process of reconstruction. They were in the vanguard of a vast social movement triggered at the highest level of government in the United States and the ex-USSR.

A national history of NASA's activities does not need to look beyond the borders of the United States to construct its narrative. By contrast, this history must of necessity treat the agency as a node in a network of relationships with other countries. The structure of that network, the opportunities it presents, the challenges it poses, the options it excludes change over time and through (geographical) space. The rhythm at which space programs evolve in different countries differs sharply depending on their histories, and on the social weight of those who have a stake in the conquest of space. Surveying the globe from the vantage point of space leadership, NASA had to constantly reposition itself via-à-vis traditional allies (and, in the 1990s, traditional rivals), creatively seeking ways to build appropriate international partnerships that would enhance the goals of American foreign policy even as it strengthened space programs abroad. The global competition that NASA faced, particularly as more free-world countries acquired independent access to space, is at best a backdrop in a national history. Here it is a key concern. The comparative analysis made possible by the treatment of several different countries and regions in this book does not only throw light on the dynamics of this power shift and the different motivations and priorities of the space-faring nations that have brought it about. It also throws into relief the specificity of NASA's historical trajectory, and of the unique local features that have sustained its global ambitions in an age in which we are witnessing not the "decline of America [...] but the rise of everyone else."[4]

There is a temptation to align two aspects of NASA's activity authorized by the 1958 Space Act—the pursuit of leadership in aeronautical and space science and technology, and cooperation with other nations and groups of nations—with national and international histories of the agency. And indeed the international dimension is muted in a study of, say, the ongoing debates within NASA, and between NASA and other arms of the US government, over the configuration of a post-Apollo program that was heralded as a revolution in the exploitation of space and in the means of access to it. Correlatively, that debate only pops up sporadically in a study of NASA's efforts to enroll other countries in the adventure. The distinction between these two approaches, the one drawing exclusively on national sources, the other engaging with what happened in NASA's dealings with foreign entities, is however no more than a convenient way of cutting a complex but seamless cloth to make it more manageable intellectually. The sustained investments in NASA over the last half century, and its determination to remain at the leading edge of space science and technology, are integral to the United States' determination to remain the political, ideological, and military leader on the globe and to play a preponderant role in international affairs. A strong national program is a sine qua non for taking the lead in an international program, and for shaping its trajectory in ways that reinforce, rather than undermine, America's influence in the world. The two missions specified for NASA in the Space Act are not distinct. NASA's challenge has been to reconcile them. This book has described the measures taken to that end, and has revealed the agency's extraordinary capacity to adapt itself to changing geopolitical circumstances, including the dramatic implosion of the rival space power that brought it into being in 1958. It is a history and a tradition that will stand the agency in good stead in the twenty-first century as it carves a niche for itself in a very different national and global context to the one that it has learnt to master so well.

Notes

1 Introduction and Historical Overview: NASA's International Relations in Space

1. James R. Hansen, *First Man. The Life of Neil A. Armstrong* (New York: Simon and Schuster, 2005), 493.
2. Ibid., 393, 503.
3. Ibid., 505.
4. For survey of the historical literature, see Roger D. Launius, "Interpreting the Moon Landings: Project Apollo and the Historians," *History and Technology* 22:3 (September 2006), 225–255. On the gendering of the Apollo program, see Margaret A. Weitekamp, *The Right Stuff, the Wrong Sex: The Lovelace Women in the Space Program* (Baltimore: Johns Hopkins University Press, 2004); "The 'Astronautrix' and the 'Magnificent Male.' Jerrie Cobb's Quest to be the First Woman in America's Manned Space program," in Avital H. Bloch and Lauri Umansky (eds.), *Impossible to Hold. Women and Culture in the 1960s* (New York: New York University Press), 9–28.
5. Hansen, *First Man*, 513–514.
6. Sunny Tsiao, *"Read You Loud and Clear." The Story of NASA's Spaceflight Tracking and Data Network* (Washington, DC: NASA SP-2007–4232), Chapter 5.
7. *Experiment Operations during Apollo EVAs. Experiment: Solar Wind Experiment,* http://ares.jsc.nasa.gov/humanexplore/exploration/exlibrary/docs/apollocat/part1/swc.htm (accessed August 31, 2008).
8. Thomas A. Sullivan, *Catalog of Apollo Experiment Operations* (Washington, DC: NASA Reference Publication 1317, 1994), 113–116. Geiss's team also measured the amounts of rare gases trapped in lunar rocks: P. Eberhart and J. Geiss, et al., "Trapped Solar Wind Noble Gases, Exposure Age and K/Ar Age in Apollo 11 Lunar Fine Material," in A. A. Levinson (ed.), *Proceedings of the Apollo 11 Lunar Science Conference,* Houston Texas, January 5–8, 1970, Vol. 2. Chemical and Isotopic Analysis, 1037–1070.
9. Hansen, *First Man*, 395.
10. This is true of scholarly works like Hansen's *First Man*, Chapter 29, as well of accounts specifically concerned with lunar science, like William David Compton's, *Where No Man Has Gone Before. A History of Apollo Lunar Exploration Missions* (Washington, DC: NASA SP-4214, 1989), autobiographical accounts like Buzz Aldrin's *Return to Earth*, Chapter 8, and semipopular works like Leon Wagener's *One Giant Leap. Neil Armstrong's Stellar American Journey* (New York: Forge Books, 2004), Chapter 14. None of these sources mentions either that the Swiss experiment was deployed before the American flag was unfurled. One has to burrow deep into the official records to extract this data (see *Experiment Operations during Apollo EVAs*). See also Peter Creola interview, Bern, Switzerland, by John Krige, May 25, 2007, NHRC. For Creola's own role in space, see Anon., *Peter Creola. Advocate of Space* (Noordwijk: European Space Agency, ESA SP-1265/E, 2002).

11. Howard E. McCurdy, *Space and the American Imagination* (Washington, DC: Smithsonian Institution Press, 1997).

12. John Logsdon, "Die deutsche Weltraumforschung in transatlantischer Perspective," in Helmuth Trischler and Kai-Uwe Schrogl (eds.), *Ein Jahrhundert im Flug, Luft- und Raumfahrtforschung in Deutschland 1907–2007* (Frankfurt Am Main: Campus Verlag, 2007).

13. Howard E. McCurdy, *The Space Station Decision. Incremental Politics and Technological Choice* (Baltimore: Johns Hopkins University Press, 1990).

14. The Act is available at http://www.hq.nasa.gov/office/pao/History/spaceact.html (accessed January 27, 2005).

15. Gerald Haines, "The National Reconnaissance Office. Its Origins, Creation and Early Years," in Dwayne A. Day, John M. Logsdon, and Brian Latell (eds.), *Eye in the Sky. The Story of the Corona Spy Satellites* (Washington, DC: Smithsonian Institution Press, 1998), 143–156.

16. On the IGY, see Rip Bulkley, *The Sputniks Crisis and Early United States Space Policy* (Bloomington: Indiana University Press, 1991); Fae L. Kosmo, "The Genesis of the International Geophysical Year," *Physics Today*, July 2007: 38–43; and Allan Needell, *Science, Cold War and the American State: Lloyd V. Berkner and the Balance of Professional Ideals* (Chur: Harwood Academic Publishers, 2000).

17. Arnold W. Frutkin, *International Cooperation in Space* (Englewood Cliffs: Prentice Hall, 1965), 31.

18. Eilene Galloway, "Organizing the United States Government for Outer Space, 1957–1958," in Roger Launius, John M. Logsdon, and Robert W. Smith (eds.), *Reconsidering Sputnik. Forty Years Since the Soviet Satellite* (Amsterdam: Harwood Academic Publishers, 2000), 309–325, 322. See also "The Woman Who Helped Create NASA," http://www.nasa.gov/topics/history/galloway_space_act.html (accessed September 20, 2008).

19. Andrew G. Haley, *Space Law and Government* (New York: Appleton-Century-Crofts, 1963), 313–328.

20. John Lewis Gaddis, *The Long Peace. Inquiries into the History of the Cold War* (New York: Oxford University Press, 1987), Chapter 7.

21. NASA Key Personnel Change, June 1, 1979, Record no. 726, Folder 11.2.1, Frutkin, Arnold W., NHRC.

22. Release 85–132, September 20, 1985, Record no. 000137, Folder, Barnes, Richard J. H., NHRC.

23. Special Announcement, February 1, 1979; Release 88–160, November 21, 1988, Record no. 1669, Folder, Pedersen, Kenneth S., NHRC.

24. Release 90–7, August 2, 1990; Release 91–3, January 7, 1991, Record no. 1669, Folder, Pedersen, Kenneth S., NHRC.

25. The organigram is available at http://oiir.hq.nasa.gov/organization.html.

26. Arnold Frutkin, "International Collaboration in Space," *Science* 169:3943 (July 24, 1970), 333–339.

27. Michael O'Brien, *Global Reach. A View of NASA's International Cooperation* (Washington, DC: NASA NP-2008–01–492-HQ, 2008).

28. John Logsdon, "U.S.-European Cooperation in Space Science: A 25-Year Perspective," *Science* 223:4631 (January 6, 1984), 11–16.

29. Roger D. Launius, "NASA and the Attitude of the U.S. Toward International Space Cooperation," in *Les relations franco-américains dans le domaine spatial (1957–1975). Quatrième reconrte de l'IFHE, 8–9 décembre 2005* (Paris: IFHE Publications, 2008), 45–63, at 53.

30. Marcia S. Smith, "America's International Space Activities," *Society* 21:2 (January/February 1984), 18–25.

31. Tsiao, *Read You Loud and Clear.* For a review, see *NASA News and Notes* 25:3 (August 2008), 1–5. The system comprised four tracking programs: mini-track, a

north-south network through the Western hemisphere for scientific satellites; the deep-space network; the manned space flight ground stations; and the Baker-Nunn tracking stations for a Smithsonian astrophysics program. For an account of the last, see Teasel Muir-Harmony, "Tracking Diplomacy: The IGY and American Scientific and Technical Exchange with East Asia," in Roger Launius, Jim Fleming, and David de Vorkin (eds.), *Globalizing Polar Science: Reconsidering the International Polar and Geophysical Years* (New York: Palgrave Macmillan, 2010), 279–306.

32. Frutkin, *International Cooperation in Space*.
33. In the Introduction to *26 Years of NASA International Programs*, signed by the International Affairs Division, then headed by Ken Pedersen, who also wrote the foreword to the booklet.
34. Frutkin, *International Cooperation in Space*, 34.
35. Ibid., 35.
36. Logsdon, "U.S.-European Cooperation," 11–16.
37. Frutkin, *International Cooperation in Space*, 33.
38. For these objections see Logsdon, "U.S.-European Cooperation," 13.
39. For a summary account, see Joan Johnson-Freese, *Changing Patterns of International Collaboration in Space* (Malabar FL: Orbit Books, 1990), Chapter 6; Smith, "America's International Space Activities," 19.
40. See John Krige, "Technology, Foreign Policy and International Cooperation in Space," in Steven J. Dick and Roger D. Launius (eds.), *Critical Issues in the History of Spaceflight* (Washington, DC: NASA-SP-2006–4702, 2006), 239–262.
41. Hermann Pollack, "International Relations in Space. A US View," *Space Policy* 4:1 (February 1988), 24–30, at 24.
42. Kenneth S. Pedersen, "The Changing Face of International Space Cooperation. One View of NASA," *Space Policy* 2 (May 1986), 120–135, at 130. See also Johnson-Freese, *Changing Patterns*, Chapter 9; and Smith, "America's International Space Activities."
43. Kenneth S. Pedersen, "Thoughts on International Space Cooperation and Interests in the Post-Cold War World," *Space Policy* 8:3 (1992), 205–220, at 208.
44. US Congress, Office of Technology Assessment, *U.S.- Russian Cooperation in Space*, OTA-ISS-618 (Washington, DC: US Government Printing Office, April 1995), 81.
45. Pedersen, "Thoughts on International Space Cooperation," at 216.
46. Eligar Sadeh, "Technical, Organizational and Political Dynamics of the International Space Station Program," *Space Policy* 20 (2004), 171–188, at 173; emphasis mine. For the early history of the station, see John M. Logsdon, *Together in Orbit. The Origins of International Participation in Space Station Freedom* (Washington, DC: Space Policy Institute, George Washington University, December 1991); Howard E. McCurdy, *The Space Station Decision. Incremental Politics and Technological Choice* (Baltimore: Johns Hopkins University Press, 2007).
47. *U.S.-Russian Cooperation in Space*, 76.
48. Quoted in Sadeh, "Technical, Organizational," 174.
49. Paul B. Stares, *The Militarization of Space. U.S. Policy 1945–1984* (Ithaca: Cornell University Press, 1985), 42–43.
50. *Final Report to the President on the U.S. Space Program, January 1993*, submitted by NASA administrator Dan Goldin to President H. W. Bush, on January 7, 1993, available at http://history.NASA..gov/33082.pt1.pdf (accessed on December 15, 2008).
51. Demian McLean, "Obama Moves to Counter China with Pentagon-NASA Link," Bloomberg.com, January 2, 2009, available at http://news.yahoo.com/s/bloomberg/20090102/pl_bloomberg/aovrno0oj41g/print (accessed on January 4, 2009).
52. Marcia S. Smith, *U.S. Space Programs: Civilian, Military, Commercial*, CRS Issue Brief for Congress, updated November 7, 2005, available at http://www.unm.edu/~cstp/v_library.html (accessed on January 19, 2013).

2 NASA, Space Science, and Western Europe

1. Arnold W. Frutkin, *International Cooperation in Space* (Englewood Cliffs, NJ: Prentice-Hall, 1965), Chapter 2.

2. On COSPAR, see Newell, *Beyond the Atmosphere. Early Years of Space Science* (Washington, DC: NASA SP-4211, 1980), Chapter 18. For a detailed account of the meeting in March 1959, see H. Massey and H. O. Robins, *History of British Space Science* (Cambridge: Cambridge University Press, 1986), 58–62.

3. For NASA's initiative, see letter Newell to Glennan et al., March 12, 1959. For Porter's offer, see his letter to van de Hulst, March 14, 1959, both available in John M. Logsdon, ed., *Exploring the Unknown. Selected Documents in the History of the U.S. Civil Space Program. Vol. II. External Relationships* (Washington, DC: NASA SP-4407, 1996), Documents I-2 and I-3, 17–19.

4. Frutkin, *International Cooperation in Space*, 38.

5. Reimar Lüst, "Cooperation between Europe and the United States in Space," *ESA Bulletin* 50 (1987), 98–104, at 104.

6. European Science Foundation and National Research Council (hereafter ESF-NSC), *U.S.-European Collaboration in Space Science* (Washington, DC: National Academy Press, 2008).

7. This section makes extensive use of John Krige, "Building a Third Space Power: Western European Reactions to Sputnik at the Dawn of the Space Age," in Roger D. Launius, John M. Logsdon, and Robert M. Smith, *Reconsidering Sputnik. Forty Years Since the Soviet Satellite* (Amsterdam: Harwood Academic Publishers, 2000), 289–307. See also Neil Whyte and Phillip Gummett, "Far beyond the Bounds of Science: The Making of the United Kingdom's First Space Policy," *Minerva* 35 (1997), 139–169.

8. For a detailed study of this program, see Matthew Godwin, *The Skylark Rocket. British Space Science and the European Space Research Organization, 1957–1972* (Paris: Beauchesne, 2007).

9. Jon Agar, *Science & Spectacle. The Work of Jodrell Bank in Postwar British Culture* (Chur: Harwood Academic Publishers, 1998).

10. Quoted in Whyte and Gummett, "Far beyond the Bounds of Science," 152.

11. As is described in more detail in John Krige, *American Hegemony and the Postwar Reconstruction of Science in Europe* (Cambridge: MIT Press, 2006), Chapter 7.

12. John Baylis, *Ambiguity and Deterrence. British Nuclear Strategy, 1945–1964* (Oxford: Clarendon Press, 1995); John Simpson, *The Independent Nuclear State. The United States, Britain and the Military Atom* (New York: St. Martin's Press, 1983).

13. Massey and Robins, *History of British Space Science*, Chapter 5.

14. The formal agreement is attached to letter Glennan to Massey, July 6, 1959, in Logsdon et al., eds., *Exploring the Unknown*, Vol. II, pp. 19–21, Document I-4.

15. Massey and Robins, *History of British Space Science*, 78.

16. http://www.ewh.ieee.org/reg/7/millennium/alouette/alouette_home.html (accessed on March 22, 2010).

17. Jacques Blamont, "International Space Exploration: Cooperative or Competitive?" *Space Policy* 21 (2005), 89–92, at 89.

18. The general argument here is based on Jacques Blamont, "Origines et principes de la politique spatiale de la France," *Rayonnement du CNRS* No. 48 (June 2008), 24–32; "La création d'une agence spatiale: les français a Goddard Space Flight Center en 1962–1963," *Les relations franco-américaines dans le domaine spatial, 1957–1975* (Paris: IFHE Publications, 2008), 157–200; "Les premières experiences d'aéronomie en France," *L'Essor des recherches spatiales en France. 'Des prèmieres experiences scientifiques aux premiers satellites'* (Nordwijk: ESA SP-472, March, 2001), 31–41; Claude Carlier and Marcel Gilli, *Les trente premières années du CNES* (Paris: La Documentation française, 1994); Hervé Moulin, "Les recherches spatiales au temps des comités. Environnement international et contexte national à la fin des années 50

et au début des années 60," *L'Essor des recherches spatiales en France*, 13–17; Jacques Villain, "La cooperation franco-américaine dans le domaine des missiles ballistiques (1944–1960)," *Les relations franco-américaines dans la domaine spatiale*, 114–128. The author is responsible for the translation of these texts and all other French texts cited in this section.

19. France Durand-de Jongh, *De la fusée Véronique au lanceur Ariane. Une historie des hommes, 1945–1979* (Paris: Stock, 1998).

20. Blamont, "Origines et principes de la politique spatiale de la France," 27.

21. Blamont, "Les premières experiences d'aéronomie en France," 35.

22. Marcel Thué and Jean-Pierre Houssin, "Les relations France- États Unis dans le domaine de telecommunication par satellite," in *Les relations franco-américaines dans le domaine spatial*, 99–113. The United Kingdom built a receiver at Goonhilly Downs.

23. Balloon technology was of interest to several constituencies in the 1950s and early 1960s. For a month early in 1956 the Air Force floated over 500 balloons carrying reconnaissance cameras over the Soviet Union (the GENETRIX project)—see Dwayne A. Day, John M. Logsdon, and Brian Latell, eds., *Eye in the Sky. The Story of the Corona Spy Satellites* (Washington: Smithsonian Institution Press, 1998); Echo 1, the first telecommunications "satellite" launched in August 1960 was a reflecting balloon—see the contributions by Donald C. Elder and Craig B. Waff to Andrew J. Butrica, ed., *Beyond the Ionosphere. Fifty Years of Satellite Communication* (Washington, DC: NASA SP-4217, 1997), Chapters 4 and 5.

24. Jean Pierre Causse, "Le programme FR-1," *Les relations franco-américaines dans le domaine spatial*, 214–231, at 231.

25. Claude Goumy, "La cooperation MATRA-TRW dans les satellites," *Les relations franco-américaines dans le domaine spatial*, 246–264.

26. Goumy, *Les relations franco-américaines dans le domaine spatial*, 251.

27. Gilbert Ousley, "French-American Space Relationships," in *Les relations franco-américaines dans le domaine spatial*, 78–94, at 80, 83–84.

28. Jean Pierre Causse, "Le programme FR-1," in *Les relations franco-américaines dans le domaine spatial*, 214–231, at 216.

29. Stephen B. Johnson, *The Secret of Apollo. Systems Management in American and European Space Programs* (Baltimore: John Hopkins University Press, 2002).

30. For the quotes from Bonnet, see Roger Bonnet, interview with John Krige, February 10, 2005, 3, 5, available at http://wwwarc.eui.eu/ESAint/pdf/INT790.pdf (accessed on March 15, 2010).

31. See also John Krige, "NASA as an Instrument of U.S. Foreign Policy," in Steven J. Dick and Roger D. Launius, eds., *Societal Impact of Spaceflight* (Washington, DC: NASA SP-2007–4801, 2007), 207–218.

32. The account that follows is based almost exclusively on Michelangelo de Maria and Lucia Orlando, *Italy in Space. In Search of a Strategy, 1957–1975* (Paris: Beauchesne, 2008), Chapter 3.

33. Ibid., 84.

34. "Italian Satellite Payload Tested," *Flight International*, May 2, 1963, 652.

35. For the interest of the US military in the ionosphere, see David H. DeVorkin, *Science with a Vengeance. How the Military Created the U.S. Space Sciences After World War II* (New York: Springer-Verlag, 1992).

36. Arnold W. Frutkin, interview with John Krige, August 19, 2007, NHRC.

37. On the V2, see Michael J. Neufeld, *Von Braun. Dreamer of Space, Engineer of War* (New York: Knopf, 2007).

38. For these developments, see John Krige, "The Peaceful Atom as Political Weapon. Euratom and American Foreign Policy in the Late 1950s," *Historical Studies in the Natural Sciences* 38:1 (2008), 5–44 and references therein.

39. Peter Fischer, *The Origins of the Federal Republic of Germany's Space Policy 1959–1965—European and National Dimensions* (Noordwijk: ESA HSR-12, 1994), 4.

40. Niklas Reinke, *The History of German Space Policy: Ideas, Influences and Interdependence 1923–2002* (Paris: Beauchesne, 2007), translated from Niklas Reinke, *Geschichte der deutschen Raumfahrtpolitik. Konzepte, Einflußfaktoren und Interdependenzen 1923–2002* (Munich: Oldenbourg Verlag, 2004), 57–58.

41. Helmut Trischler, *The "Triple Helix" of Space. German Space Activities in a European Perspective* (Noordwijk: ESA HSR-28, 2002), 7.

42. Ibid.

43. Wolfgang Finke, "Germany and ESA," *The History of The European Space Agency. Proceedings of an International Symposium, London, 11–13 November, 1998* (Noordwijk: ESA SP-436, 1999).

44. Reinke, *The History of German Space Policy*, 58.

45. The following account is from Reinke, *The History of German Space Policy*, 117–129.

46. Ibid., 120.

47. See ibid., table XI.

48. Trischler, *The "Triple Helix" of Space*, 13.

49. Quoted by Reinke, *The History of German Space Policy*, 108.

50. Letter Von Braun to Frutkin, September 4, 1963, Record No. 14618, Series International Cooperation and Foreign Countries, West Germany, 1956–1990, NHRC.

51. For what follows, see Reinke, *The History of German Space Policy*, 106–113; Trischler, *The "Triple Helix" of Space*; and John Logsdon, "Die deutsche Weltraumforschung in transatlantischer Perspektive," in Helmuth Trischler and Kai-Uwe Schrogl, eds., *Ein Jahrhundert im Flug, Luft- und Raumfahrtforschung in Deutschland 1907–2007* (Frankfurt Am Main: Campus Verlag, 2007). For a review, see Michael J. Neufeld, H-German@H-net.msu.edu, May 20, 2009. The original version in English, kindly given to me by Hans Oberlechner at ESA, is John Logsdon, "Astronautical Research in the Transatlantic Perspective," 2.

52. Quoted by Reinke, *The History of German Space Policy*, 111.

53. Quoted in ibid., 112.

54. Fischer, *The Origins of the Federal Republic of Germany's Space Policy*, 51.

55. This is a revised version of Krige, "NASA as an Instrument of U.S. Foreign Policy," 207–218.

56. "The White House. Exchange of Toasts" between Johnson and Erhard, December 20, 1965, LBJL, National Security Files, Country File Europe & USSR, Germany/Erhard Visit [12–65], folder 12/19–21/65.

57. Logsdon, "Astronautical Research in the Transatlantic Perspective," 3.

58. For the details of the trip see Homer E. Newell, *Beyond the Atmosphere. Early Years of Space Science* (Washington, DC: NASASP-4211, 1980), 315–316.

59. Newell, *Beyond the Atmosphere*, 315, 316.

60. Newell, *Beyond the Atmosphere*, 316.

61. This is how Logdson has characterized his response, in his "Astronautical Research in the Transatlantic Perspective," 3.

62. Quoted by Reinke, *The History of German Space Policy*, 122.

63. Francis M. Bator, *Memorandum for the President. Subject: Erhard Visit, September 26–27, 1966*, September 25, 1966, LBJL, National Security Files, Country File Europe & USSR, Germany 9/66, Erhard Visit, folder Papers, Cables, Memos [9/66].

64. Thomas Alan Schwartz, *Lyndon Johnson and Europe. In the Shadow of Vietnam* (Cambridge: Harvard University Press, 2003), 116.

65. For all quotations in this paragraph, see *Memorandum of Conversation. Part III of III. Subject: Offset*, December, 20 1965, LBJL, National Security Files, Country Files Europe and USSR, Germany, Box 192, Folder Germany. Erhard Visit [12/65], 12/19–21/65.

66. George McGhee, *Memorandum for the President. Your Meeting with Chancellor Erhard, September 26–27*, LBJL, National Security Files, Country File Europe & USSR, Germany 9/66, Erhard Visit, folder Papers, Cables, Memos [9/66].

67. "Text of Cable from Ambassador McGhee (Bonn 3361), Subject: The Offset and American Troop Level in Germany," September 20, 1965, LBJL, National Security Files, Country File Europe & USSR, Germany 9/66, Erhard Visit, folder Papers, Cables, Memos [9/66].

68. "The White House. Remarks by the President at Vehicle Assembly Building, Cape Kennedy, Florida," September 20, 1966, LBJL, National Security Files, Country File Europe & USSR, Germany 9/66, Erhard Visit, folder Papers, Cables, Memos [9/66].

69. James E. Webb to Dean Rusk, October 14, 1966, NHRC, Record no.14465, International Cooperation and Foreign Countries, Foreign Countries, West Germany, folder Germany (West), 1956–1990.

70. Webb to Rusk, October 14, 1966.

71. Reinhard Loosch, Interview with the author in Bonn, June 29, 2007, available in NHRC; emphasis in the original.

72. The MoU also covered an upper atmosphere satellite called Aeros.

73. A third flight model is exhibited in the Deutsches Museum in Munich. See *Helios Solar Probe*, available at http://www.deutsches-museum.de/en/collections/transport/astronautics/satellites/helios-probe/ (accessed on March 11, 2010).

74. Arthur W. Frutkin, "International Cooperation in Space," *Science* 169:3943 (July 24, 1970), 333–339.

75. Ibid.

76. Reinke, *The History of German Space Policy*, 137.

77. John Krige and Arturo Russo, *A History of the European Space Agency, 1958–1987. Vol. I. The Story of ESRO and ELDO* (Noordwijk: ESA SP-1235, 2000), Chapter 8.

78. Ibid., Chapter 11.

79. Roger M. Bonnet and Vittorio Manno, *International Cooperation in Space. The Example of the European Space Agency* (Cambridge: Harvard University Press, 1994), 106.

80. Robert W. Smith, "The Biggest Kind of Big Science: Astronomers and the Space Telescope," in Peter Galison and Bruce Hevly, eds., *Big Science. The Growth of Large-Scale Research* (Stanford: Stanford University Press, 1992), 184–211, at 187. See also Robert W. Smith, *The Space Telescope. A Study of NASA, Science, Technology and Politics* (Cambridge: Cambridge University Press, 1989).

81. Smith, "The Biggest Kind of Big Science," 187.

82. Smith, *The Space Telescope*, 246. For an acerbic account of this visit, see Bonnet and Manno, *International Cooperation in Space*, 76.

83. Bonnet and Manno, *International Cooperation in Space*, 76.

84. Ibid.

85. Ibid., 98. What follows is based largely on their account. See also Joan Johnson-Freese, "Cancelling the US Solar-Polar Spacecraft. Implications for International Cooperation in Space," *Space Policy* 3 (February 1987), 24–37; Joan Johnson-Freese, *Changing Patterns of International Cooperation in Space* (Malabar: Orbit Books, 1990), Chapter 7.

86. Bonnet and Manno, *International Cooperation in Space*, 99.

87. Ibid., 101.

88. Ibid., 106.

89. Lynn Cline, interview with John Krige, March 30, 2009, NHRC.

90. Reimar Lüst, "The [ESA Member State] Ministerial Conference and Beyond," *ESA Bulletin* 53 (February 1988), 9–13, on 9.

91. ESF-NRC, *U.S.-European Collaboration in Space Science*, 64, for the quote.

92. The information in this paragraph is from ESF-NRC, *U.S.-European Collaboration in Space Science*, 60–64. The quote is at 60.

93. ESF-NRC, *U.S.-European Collaboration in Space Science*, 60–64; Arturo Russo, "Parachuting Onto Another World. The European Space Agency's Mission to Titan," unpublished MSS, 2009.

94. Russo, "Parachuting Onto Another World."

95. Roger Bonnet, "The New Mandatory Scientific Programme for ESA," *ESA Bulletin* 43 (1985), 8–13; "European Space Science—in Retrospect and in Prospect," *ESA Bulletin* 81 (1995), 6–16; Bonnet and Manno, *International Cooperation in Space,* 151–153.

96. Bonnet, "European Space Science," 6.

97. For the following ESF-NRC, *U.S.-European Cooperation in Space,* 24; Bonnet and Manno, *International Cooperation in Space,* 80–81.

98. Bonnet and Manno, *International Cooperation in Space,* 81.

99. The Inter Agency Consultative Group (IACG) was another institutional measure taken to improve coordination. It was first set up to coordinate the activities of the four agencies engaged in the observation of Halley's comet when it approached the earth in 1986 (ESA, ISAS-Japan, IKI-USSR, and NASA). It continued afterward with the aim of optimizing the use of space science resources around the world—see ibid., 88–95.

100. ESF-NRC, *U.S.-European Collaboration in Space Science,* 64.

101. The letter is cited by Russo, "Parachuting Onto Another World," 25.

102. Howard E. McCurdy, *Faster, Better, Cheaper. Low-Cost Innovation in the U.S. Space Program* (Baltimore: Johns Hopkins University Press, 2001), 45–50, for this material.

103. Ibid., 120, 121.

104. Ibid., 19, 2.

105. Roger Bonnet, interview with John Krige, February 10, 2005, 10, available at http://wwwarc.eui.eu/ESAint/pdf/INT790.pdf (accessed on March 15, 2010).

106. Cline, interview with John Krige, March 30, 2009, NHRC.

107. Cited in Russo, "Parachuting Onto Another World."

108. ESF-NRC, *U.S.-European Collaboration in Space Science,* 63.

109. "Titan Team Claims Just Deserts as Probe Hits Moon of *Crème Brûlée,*" *Nature* 433 (January 20, 2005), 181.

3 Technology Transfer with Western Europe: NASA-ELDO Relations in the 1960s

1. Memo, Frutkin to Homer J. Stewart, October 13, 1960, "Material for Ten Year Program Document," Record no. 14459, International Cooperation and Foreign Countries, International Cooperation, Folder Miscellaneous Correspondence from Code I—International Relations 1958—1967, NHRC.

2. John Krige and Arturo Russo, *A History of the European Space Agency, 1958–1987. Vol. I. The Story of ESRO and ELDO—1958–1973* (Noordwijk: ESA, SP-1235, 2000), Chapters 3, 4.

3. Active satellites are to be distinguished from balloons or other "passive" surfaces that are used as "reflectors" in space to bounce off telecommunications signals. Echo I was a project of this kind: see Andrew J. Butrica, ed., *Beyond the Ionosphere: Fifty Years of Satellite Communication* (Washington, DC: NASA SP-4217, 1997).

4. Memo from Frutkin to Eugene M. Emme, "Comments on COMSAT History, 'NASA's Role in the Development of Communications Satellite Technology,'" December 20, 1965, NHRC Record no. 15726. See also Marcel Thué and Jean-Pierre Hussin, "Les Reltions Franco—Etats-Unis dans les Domaines de Telecommunications par Satellite," *Les relations franco-américaines dans la domaine spatiale (1957–1975),* Quatrième Rencontre de l'IFHE, Paris, 99–113.

5. David J. Whalen, "Billion Dollar Technology: A Short Historical Overview of the Origins of Communications Satellite Technology, 1945–1965," in Butrica (ed.), *Beyond the Ionosphere,* 95–127. See also David J. Whalen, *The Origins of Satellite Communications, 1945–1965* (Washington, DC: Smithsonian Institution Press, 2002).

6. John Krige, "Building Space Capability through European Regional Collaboration," in Steven J. Dick (ed.), *Remembering the Space Age. Proceedings of the 50th Anniversary*

Conference (Washington, DC: NASA-2008–4703, 2008), 37–52. See also Walter A. McDougall, "Space-Age Europe: Gaullism, Euro-Gaullism, and the American Dilemma," *Technology and Culture* (1985), 179–203.

7. "Draft U.S. Position on Cooperation with Europe in the Development and Production of Space Launch Vehicles," attached to internal correspondence between Frutkin and Milton W. Rosen, October 15 and October 30, 1962. Frutkin told Rosen that the "Draft" position developed by the State Department was in fact now firm US policy, Record no. 14548, International Cooperation and Foreign Countries, Europe, Folder US-Europe 1965–1972, NHRC. The arguments here resonate with those used to justify sharing civilian nuclear energy with Euratom: see John Krige, "The Peaceful Atom as Political Weapon: Euratom and American Foreign Policy in the Late 1950s," *Historical Studies in the Natural Sciences* 38:1 (2008), 9–48.

8. Jean-Jacques Servan-Schreiber, *Le défi americain* (Paris, 1967), published in English as *The American Challenge* with a foreword by Arthur Schlesinger Jr. (New York: Atheneum,1968).

9. See, e.g., the lead article by Philip H. Abelson, "European Discontent with the 'Technology Gap,'" *Science* 155:3764 (February 17, 1967). For a summary of Servan-Schreiber's argument, see "West Europeans Attribute Continuing Technological Lag Behind U.S. to Inferior Management," *New York Times*, December 13, 1967, 6. See also Henry R. Nau, "A Political Interpretation of the Technology Gap Dispute," *Orbis* XV:2 (Summer 1971), 507–524.

10. Arnold W. Frutkin, "The United States Space Program and its International Significance," *The Annals of the American Academy of Political and Social Science* 366 (July 1966), 89–98, at 90. See also *Space Business Daily* 25:35 (Monday, April 18, 1966), 285–286.

11. Ibid., 91, 92.

12. Central Intelligence Agency, Office of Current Intelligence, *Special Report. Western European Space Programs*, May 1964, Record no. 15707, International Cooperation and Foreign Countries, European Launcher Development Organization, Folder ELDO, NHRC, 13.

13. NSAM 294 is available online at http://www.lbjlib.utexas.edu/johnson/archives.hom/NSAMs/nsam294.asp

14. The first Soviet Communications satellite was launched shortly after *Early Bird* (Intelsat I). In 1968 the Soviet Union and seven other socialist countries submitted a draft agreement to the United Nations establishing Intersputnik; nine states signed the revised agreement in November 1971. Cold War rivalry thus called forth two independent international Comsat organizations almost simultaneously. For more on the Soviet program, see Jonathan F. Galloway, "Originating Communications Satellite Systems: The Interactions of Technological Change, Domestic Politics and Foreign Policy," in Butrica (ed.), *Beyond the Ionosphere*, 171–192.

15. See Whalen, *The Origins of Satellite Communications* and David J.Whalen, "Billion Dollar Technology: A Short Historical Overview of the Origins of Communications Satellite Technology, 1945–1965," in Butrica (ed.), *Beyond the Ionosphere*, 95–127. In this paper Whalen points out that shares in Comsat were oversubscribed by carriers eager to get into the business: AT&T was allowed to buy stock worth $57.9 million (29 percent of all shares), while ITT was given 11 percent, GTE 4 percent, and RCA 3 percent, leaving 5 perecnt for all other private operators (more than 200 were actually interested in having shares in Comsat).

16. John M. Logsdon, *John F. Kennedy and the Race to the Moon* (New York: Palgrave Macmillan, 2010), 113–114.

17. This is described in Whalen, *The Origins of Satellite Communications*. Whalen sees this as government "intervention" in business. Lee, on the other hand, sees it as an embodiment of the symbiotic relationship between corporations and the US government described in Galbraith's *New Industrial State*. William E. Lee, *The Shaping of an American Empire. Negotiating the Interim Intelsat Agreements*, paper presented at the

sixtieth Annual Meeting of the Association for Education in Journalism, Madison, Wisconsin, August 1977, available at http://eric.ed.gov/ERICWebPortal/, ERIC #ED151783 (accessed on November 11, 2009).

18. Lee, *The Shaping of an American Empire*, 30; Steven A. Levy, "Intelsat: Technology, Politics and the Transformation of a Regime," *International Organization* 29:3 (1975), 655–680. See also Galloway, *The Politics and Technology of Satellite Communication* and Richard R. Colino, *The Intelsat Definitive Arrangements. Ushering in a New Era in Satellite Communications* (Geneva: European Broadcasting Union, 1973).

19. This principle was derived from the policies in force for the use of cables. Its projections of future use thus favored those countries that were already connected by international cables, and made no allowance for those who might want to leapfrog this technological phase. Nor should one think that the national telecommunications authorities believed that undersea cables would be rendered obsolete overnight by satellites. The British Post Office had extensive plans for a global cable network in the 1970s, believing the two technologies would complement each other: see Nigel Wright, "The Formulation of British and European Policy Toward an International Satellite Telecommunications System: the Role of the British Foreign Office," in Butrica (ed.), *Beyond the Ionosphere*, 157–169.

20. Lee, *The Shaping of an American Empire*, 34.

21. The initial signatories were Austria, Belgium, Canada, Denmark, The Federal Republic of Germany, France, Ireland, Italy, Japan, the Netherlands, Switzerland, the United Kingdom, the United States, and the Vatican City. The Soviet Union was invited to participate. It refused to participate in what it called an "American inspired" organization. Ibid., 39.

22. In April 1964 the Interim Agreement was signed by the United States and 12 countries; Comsat's share was 61 percent, Western Europe's 30.5 percent. These decreased as other countries joined. The figure given here for Western Europe is an estimate.

23. Charles Johnson to Walt Rostow, July 13, 1967, National Security Files/NSAM, NSAM Box 7, Folder NSAM 338, LBJL.

24. Ibid.

25. NSAM 338 is available online at http://www.lbjlib.utexas.edu/johnson/archives. hom/NSAMs/nsam338.asp.

26. McGeorge Bundy, memorandum for the president, "Helping Others to Use Communications Satellites," September 13, 1965, National Security Files/NSAM, Box 7, Folder NSAM 338, emphasis in the original, LBJL.

27. Ibid., emphasis in the original.

28. "Policy Concerning U.S. Assistance in the Development of Foreign Communications Satellite Capabilities," unsigned, August 23, 1965. Bundy remarked that it was the result of O'Connell's negotiations with NASA, and the Departments of Commerce, Defense, and State.

29. The launch of ELDO is described in detail in Krige and Russo, *A History of the European Space Agency, Vol. 1*, Chapter 3. See also Michelangelo De Maria and John Krige, "Early European Attempts in Launcher Technology," in John Krige (ed.), *Choosing Big Technologies* (Chur: Harwood Academic Publishers, 1993), 109–137; Also McDougall, "Space-Age Europe."

30. For an account of the British program in Australia, see Peter Morton, *Fire across the Desert. Woomera and the Anglo-Australian Joint Project* (Canberra: Australian Government Publishing Service for the Department of Defence, 1989).

31. Alfred M. Nelson, undated memo, "Trip to European Launcher Development Organization (ELDO) and Member Countries, France, England, and Germany (December 4–17, 1962)," Record no. 15707, International Cooperation and Foreign Countries, European Launcher Development Organization, folder ELDO, NHRC.

32. More precisely, as Frutkin put it, official US policy was that "[e]xchanges on a purely national basis will be discouraged or rerouted through ELDO," memorandum for Mr.

Milton W. Rosen, October 30, 1962, Record no. 14548, International Cooperation and Foreign Countries, Europe, folder US-Europe 1965–1972, NHRC.

33. "U.S. to Give Europe Space Rocket Details," *Washington Post*, January 3, 1963. On Euratom, see Krige, "The Peaceful Atom."

34. Stephen B. Johnson, *The Secret of Apollo: Systems Management in American and European Space Programs* (Baltimore: Johns Hopkins University Press, 2002), Chapter 6, describes the failure of ELDO's management system in detail.

35. CIA *Special Report. Western European Space Programs*, 6.

36. Memo Frutkin to Robert N. Margrave, director, Office of Munitions Control, State Department, "ELDO interest in high energy upper stages," June 6, 1965, Record no. 14465, International Cooperation and Foreign Countries, International Cooperation, folder International policy manual material from Code I, NHRC.

37. Letter Webb to O'Connell, May 24, 1965, Record no. 14459, International Cooperation and Foreign Countries, International Cooperation, folder Miscellaneous Correspondence from Code I—International Relations 1958–1967, NHRC.

38. Ibid.

39. Memo Frutkin to Margrave, June 6, 1965.

40. This section draws heavily on the story told in John Krige, "Technology, Foreign Policy and International Collaboration in Space," in Stephen J. Dick and Roger D. Launius, *Critical Issues in the History of Spaceflight* (Washington, DC: NASA SP-2006–4702, 2006), 239–260. For details on the crisis in Europe, see Krige and Russo, *A History of the European Space Agency, Vol. I*, Chapter 4.3.2.

41. Britain's partners did not accept these arguments of course. They insisted that ELDO's aim was not to produce a competitor to American launchers in the immediate future. It was rather intended to build up the industrial infrastructure and experience in Europe, which would enable its member states to develop their own launcher in the long term, and so cooperate with the United States from a position of relative autonomy and technological and industrial strength.

42. For general accounts of US relations with France during the Cold War, and with de Gaulle in particular, see, e.g., Frank Costigliola, *France and the United States. The Cold Alliance Since World War II* (New York: Twayne Publishers, 1992); Christian Nuenlist, Anna Lochner, and Garret Martin, eds., *Globalizing de Gaulle. International Perspectives on French Foreign Policies, 1958–1969* (Lanham: Lexington Books, 2010); Robert O. Paxton and Nicholas Wahl, eds., *De Gaulle and the United States. A Centennial Reappraisal* (Oxford: Berg, 1994); John Newhouse *De Gaulle and the Anglo-Saxons* (New York: Viking Press, 1970).

43. For the term "subordination," Ted Van Dyk to the vice president, July 7, 1965, National Security File, Country File Europe and USSR, Box 192, Folder Germany Erhard Visit [12/65], 12/19–21/65, LBJL.

44. Department of State to Amembassy Bonn 1209, outgoing telegram, November 18,1965, signed [George] Ball, National Security File, Country File Europe and USSR, Box 192, Folder Germany Erhard Visit [12/65], 12/19–21/65, LBJL.

45. SC No. 00666/65B, "US Investments in Europe," CIA Special Report, April 16, 1965, National Security File, Country File, Europe, Box 163, Folder Memos [2 of 2], Vol. II, 7/64–7/66, LBJL. Servan-Schreiber's *The American Challenge*, is, of course, the *locus classicus* of this argument.

46. NSAM 357, "The Technological Gap," November 25, 1966, available online at http://www.lbjlib.utexas.edu/johnson/archives.hom/NSAMs/nsam357.gif. Hornig's official title was the special assistant to the president for Science and Technology.

47. "Preliminary Report on the Technological Gap Between U.S. and Europe," attached to David Hornig's letter to the president, January 31, 1967, National Security File, Box 46, Folder Technological Gap [1 of 2], LBJL.

48. Webb to McNamara, April 28, 1966, Record no. 14459, International Cooperation, International Cooperation and Foreign Countries, folder Miscellaneous Correspondence from CODE I—International Relations 1958–1967, NHRC.

49. Diamant-A used a mixture of N2O4 /UDMH (storable liquid fuels) in its first stage and solid fuel in the second and third stages.

50. "US Cooperation with ELDO," position paper, July 21, 1966, Charles Johnson File, National Security File, Box 14, Folder Cooperation in Space—Working Group on Expanded International Cooperation in Space. ELDO #1 [2 of 2], LBJL.

51. Department of State to Amembassy Bonn 1209, outgoing telegram, November 18, 1965, signed [George] Ball, National Security File, Country File Europe and USSR, Box 192, Folder Germany Erhard Visit [12/65], 12/19–21/65, LBJL.

52. NSAM 354, "U.S. Cooperation with the European Launcher Development Organization," July 29, 1974, available online at http://www.lbjlib.utexas.edu/johnson/archives.hom/NSAMs/nsam354.gif.

53. This paragraph is derived from "Policy Concerning US Cooperation with the European Launcher Development Organization (ELDO)," attached to U. Alexis Johnson's "Memorandum," June 10, 1966, Record no. 15707, International Cooperation and Foreign Countries, European Launcher Development Organization, Folder ELDO, NHRC.

54. In summer 1965, ELDO had asked for help from NASA on "designing, testing and launching liquid hydrogen/liquid oxygen upper stages" (Frutkin to Margrave, June 6, 1965.)

55. Richard Barnes to Scott George, chairman, NSAM 294 Review Group, Department of State, April 15, 1966, Record no. 14459, International Cooperation and Foreign Countries, International Cooperation, Folder Miscellaneous Correspondence from CODE I—International Relations 1958–1967, NHRC.

56. NSAM 312, "National Policy on Release of Inertial Guidance Technology to Germany," July 10, 1964, available online at http://www.lbjlib.utexas.edu/johnson/archives.hom/NSAMs/nsam312.gif.

57. Barns to Scott George, April 15, 1966.

58. Webb to McNamara, April 28, 1966, and reply, Bob [McNamara] to Jim [Webb], May 14, 1966, Record no. 14459, International Cooperation and Foreign Countries, International Cooperation, Folder Miscellaneous Correspondence from CODE I—International Relations 1958–1967, NHRC.

59. Frutkin to Hilburn, "Memorandum for Mr. Hilburn—AAD. Policies Relevant to '69 Revision of Intelsat Agreement," April 11, 1966, Record no. 14459, International Cooperation and Foreign Countries, International Cooperation, Folder Miscellaneous Correspondence from CODE I—International Relations 1958–1967, NHRC.

60. "RJHB" to "AWF," "The 'Webb Commission,'" May 5, 1967, Record no. 14459, International Cooperation and Foreign Countries, International Cooperation, Folder Miscellaneous Correspondence from CODE I—International Relations 1958–1967, NHRC.

61. Charles Johnson to Walt Rostow, July 13, 1967.

62. "RJHB" to "AWF," "The 'Webb Commission.'"

63. "Communications Satellite Technology," undated and unsigned memorandum, but obviously written around April 1966, Charles Johnson Files, National Security Files, Box 14, Folder Cooperation in Space—Working Group on Expanded International Cooperation in Space, ELDO #1 [2 of 2], LBJL.

64. Frutkin to Hilburn, "Memorandum for Mr. Hilburn."

65. Frutkin to Webb, "Memorandum for Mr. Webb," August 11, 1966, Record no. 14618, International Cooperation and Foreign Countries, Foreign Countries, Folder Germany (West), 1956–1990, LBJL.

66. "Exchange of Toasts between President Lyndon B. Johnson and Chancellor Ludwig Erhard of the Federal Republic of Germany (In the State Dining Room)," December 20, 1965, National Security Files, Country File Europe and the USSR, Box 192, Folder Germany Erhard Visit [12/65], 12/19–21/65, LBJL.

67. Frutkin, "Memorandum for Mr. Webb."

68. "RJHB" to "AWF," "The 'Webb Commission,'" May 5, 1967, Record no. 14459, International Cooperation and Foreign Countries, International Cooperation, Folder Miscellaneous Correspondence from CODE I—International Relations 1958–1967, NHRC.
69. Krige and Russo, *A History of the European Space Agency, Vol. I*, Chapter 4.3.2.
70. J. Bouvet and A. H. Schendel, "Le Projet Symphonie," presented to the UN Conference on peaceful exploration and use of outer space, A/CONF.34/I.12, June 24, 1968, NHRC, Record no. 6408, Spaceflight—Satellites and Probes, Symphonie Satellites, Folder Symphonie (France/West German satellite), 1969–1983.

4 European Participation in the Post-Apollo Program, 1969–1970: The Paine Years

1. The post-Apollo program was dealt with briefly by John Logsdon, *Together into Orbit. The Origins of International Participation in the Space Station* (Honolulu: University Press of the Pacific, 2005); and by Howard E. McCurdy, *The Space Station Decision. Incremental Politics and Technological Choice* (Baltimore: The Johns Hopkins University Press, 1990). There is some crucial primary source material reproduced in John M. Logsdon, ed., *Exploring the Unknown. Selected Documents in the History of the U.S. Civil Space Program. Vol. II. External Relationships* (Washington, DC: NASA SP-4407, 1996). Excellent work using European archives has been done by Lorenza Sebesta, e.g., "U.S.-European Relations and the Decision to Build Ariane, the European launch Vehicle," in Andrew J. Butrica (ed.), *Beyond the Ionosphere. Fifty Years of Satellite Communication* (Washington, DC: NASA SP-4217, 1997), 137–156; Lorenza Sebesta, "The Availability of American Launchers and Europe's Decision 'To Go it Alone,'" in John Krige and Arturo Russo, *A History of the European Space Agency, Vol. II, ESA, 1973–1987* (Noordwijk: ESA-SP1235, 2000), Chapter 10; "The Politics of Technological Cooperation in Space: US-European Negotiations on the Post-Apollo Programme," *History and Technology* 11 (1994), 317–341.
2. See Joan Hoff, "The Presidency, Congress, and the Deceleration of the U.S. Space Program in the 1970s," in Roger D. Launius and Howard E. McCurdy, *Spaceflight and the Myth of Presidential Leadership* (Champagne, IL: University of Illinois Press, 1997), 92–132.
3. Prior to fiscal year 1977, the Federal fiscal year began on July 1 and ended on June 30. It is designated by the year in which it ends: so fiscal year 1972 began in July 1971 and ended in June 1972. From FY1977 onward it has begun on October 1 and ended on September 30. To effect this change, in calendar year 1976 the July–September period was a separate accounting period (known as the transition quarter, TQ), to bridge the period to the start of FY1977.
4. Hoff, in Launius and McCurdy, *Spaceflight and the Myth*, 98. See also John Logsdon, *From Apollo to Shuttle: Policy Making in the Post-Apollo Era*, Document HHR-46, Spring 1983, NHRC.
5. *The White House Press Conference of the Vice President, Dr. Robert C. Seamans, [...], Dr. Thomas O. Paine, [...] Dr Lee DuBridge, [...] and Lieutenant Colonel Bill Anders [...]*. September 17, 1969, Record no. 12584, Federal Agencies-Presidents, Nixon Administration, Folder Nixon Space Study and Post-Apollo Planning—News Clippings 1969–1970, NHRC.
6. For the material that follows, see Letter Agnew to Nixon, September 15, 1969; *Summary on Space Task Group Report*, Office of the Vice President, September 17, 1969, and accompanying memorandum for Ron Ziegler from Herb Thompson, September 16, 1969; "Statement of Dr Thomas O. Paine [...] before the Committee on Science and Astronautics, House of Representatives," October 1, 1969, all in Record no. 12581, Federal Agencies-Presidents, Nixon administration, Folder Space Task Group (STG), 1969, NHRC. There is a draft of NASA's report to the STG, *America's Next Decade*

in Space, in Record no. 12583, Federal Agencies-Presidents, Nixon administration, Folder Space Task Group (STG) Publications, 1969–1970. See also Marti Mueller, "Task Force Presents Space Options," *Science* 165:3900 (September 26, 1969), 1335; and William J. Normyle, "Mars Options Key to 1971 Budget," *Aviation Week & Space Technology,* September 22, 1969, 22–24.

7. Letter Paine to Agnew, September 8, 1969, Record no. 12581, Federal Agencies-Presidents, Nixon administration, Folder Space Task Group (STG), 1969, NHRC.

8. "Statement of Dr Thomas O. Paine," October 1, 1969.

9. Paine's address in Australia, February 26, 1970, Accession Number 255–74–734, Box 14, Folder ID WNRC.

10. The quotation is from the *White House Press Conference.* Seamans was not particularly interested in the space station: in August he wrote to the vice president to say that he did not think "we should commit ourselves to the development of a [large manned] space station at this time," no matter how "logical" it may appear: Letter Seamans to Agnew, August 4, 1969, Record no. 12581, Federal Agencies-Presidents, Nixon administration, Folder Space Task Group (STG), 1969, NHRC.

11. John Logsdon, "The Decision to Develop the Space Shuttle," *Space Policy* 2 (May 1986), 101–119; "The Space Shuttle Program: A Policy Failure?" *Science* 232:4654 (May 30, 1986), 1099–1105.

12. Minutes of the Fourth NASA/USAF STS Committee Meeting, October 2, 1970, and *Summary of Activities for 1970,* NASA and the Department of the Air Force, forwarded to Frutkin by Dale D. Myers on June 8, 1971. Accession Number 255–74–734, Record Group NASA 255, NASA Division I, Escalation Files, Box 14, Folder II.E, WNRC.

13. http://www.nixonlibrary.gov/virtuallibrary/documents/nationalsecuritystudymemoranda.php.

14. Ibid.

15. Letter Paine to Nixon, August 11, 1969, Record no. 12578, Federal Agencies-Presidents, Nixon administration, Folder Nixon-General (1968–1971), NHRC.

16. Draft Report, *America's Next Decade in Space.*

17. Memo for the STG from the Department of State, *International Implications of the Space Program for the Next Decade,* Record no. 12580, Federal Agencies-Presidents, Nixon Administration, Folder Space Task Group (STG), 1969, NHRC.

18. Letter Paine to Nixon, November 7, 1989, Record Group NASA 255, Box 14, Folder II.B, WNRC.

19. For the address and an outline plan, see text "Paine-ESC, Paris, 10/14/69," Record Group NASA 255, Box 17, Folder IX, WNRC. The address itself, "NASA's Future Plans and Programs. Address by Dr. T.O. Paine, NASA Administrator, before the E.S.C. Committee of Senior Officials on 14 October 1969," was published in *ESRO ELDO Bulletin,* January 1970, 12–16. See also Paine's address in Australia on February 26, 1970, both in Record Group NASA 255, Box 14, Folder I.D, WNRC.

20. "NASA's Future Plans and Programs," Paris, October 1969.

21. *ESC Report on the Ninth Meeting of the Committee of Senior Officials, 15 December, 1969,* Neuilly, December 24, 1969, and, for the quote, *Report to the Committee of Senior Officials,* CSE/HF(70)13, 16 April 1970, Record Group NASA 255, Box 17, Folder VI.D.7, WNRC.

22. *ESC Report on the Ninth Meeting of the Committee of Senior Officials, 15 December, 1969,* Neuilly, December 24, 1969, Record Group NASA 255, Box 17, Folder VI.D.7, WNRC.

23. Arnold W. Frutkin, "Memorandum to the File," November 12, 1969, Record Group NASA 255, Box 17, Folder VI.D.2 WNRC.

24. Arnold W. Frutkin, "Memorandum to Escalation File (Organization Section), Reference: Discussion with Causse, Feb. 12," March 5, 1970, Record Group NASA 255, Box 17, Folder VII.A, WNRC.

25. George M. Low, "Memorandum for the Record. Space Shuttle Discussions with Secretary Seamans," January 28, 1970, Record Group NASA 255, Box 17, Folder VII.A, WNRC.

26. *Agreement between the National Aeronautics and Space Administration and the Department of the Air Force Concerning the Space Transportation System,* signed by Thomas O. Paine and Robert C. Seamans, February 17, 1970, Record Group NASA 255, Box 17, Folder I.F, WNRC.

27. Letter Paine to Nixon, August 22, 1969.

28. Frutkin, "Memorandum for Dr. Mueller" and Arnold W. Frutkin, "Notes on European Trip—October 1969. Items for Discussion," October 29, 1969, Record Group NASA 255, Box 17, Folder IX, WNRC.

29. ELDO, *Call for Tender for a Preliminary Study on a Special Space Tug,* May 6, 1970, Record Group NASA 255, Box 17, Folder VI.D.1, WNRC. The primary mission of the "special tug" referred to here was an "unmanned, economical, either expendable or reusable transportation system for transferring payloads from low orbits (as delivered by the Space Shuttle) to geostationary orbit."

30. For the following paragraphs, see Arnold W. Frutkin, "NASA Briefing on Space Station, ESRO, Paris, June 3–4, 1970, Closing Remarks," and "Organizing International Participation in the Post-Apollo Program," June 4, 1970, Record Group NASA 255, Box 17, Folder I.E, WNRC.

31. Letter Paine to Nixon, March 26, 1970, Record no 14462, International Cooperation with Foreign Countries, International Cooperation, Folder, International Space Documents From Dwayne Day, 1959, 1975, NHRC.

32. Memorandum, John M. Bowie to Captain Minkkinen, *McDonell Douglas Request to Release Space Shuttle Proposal (MC Case 7974),* June 1, 1970, Record Group NASA 255, Box 14, Folder I.G, WNRC.

33. Cited in Alvin S. Bass (Office of General Counsel, NASA), *Dissemination of Scientific and Technical Information Abroad,* June 4, 1970, Record Group NASA 255, Box 14, Folder I.G, WNRC.

34. As cited in ibid.

35. Ibid. A draft letter from the State Department granting the license to McDonnell Douglas stated that, in so agreeing, the state was not implying that it would "act favorably on any follow-on requests for technology or assistance related to the Space Shuttle or any of its components, systems or sub-assemblies," nor was the license to be interpreted "as representing a policy decision by the Department on the question of foreign participation in the Space Shuttle Program," Draft letter to Shaver, McDonnell Douglas, around May 20, 1970, and memo from Bowie to Minkkinen, June 1, 1970, Record Group NASA 255, Box 14, Folder I.G, WNRC.

36. Memorandum, "Some Delays and Uncertainties in Assessing Munitions Control Cases," attached to Frutkin's "Purpose of Discussion with Dr. Kissinger," July 7, 1979, Record Group NASA 255, Box 14, Folder II.B, WNRC. The telecom satellite was Symphonie.

37. Frutkin, "Purpose of Discussion with Dr. Kissinger," July 7, 1970.

38. Frutkin, "Draft Paper Expressing President's Wish for Positive Solution to Technical Exchange Problems," July 8, 1970, Record Group NASA 255, Box 14, Folder II.B, WNRC.

39. A. Frutkin, "Notes for Discussion with Dr. Kissinger," June 10, 1970, Record Group NASA 255, Box 14, Folder I.F, WNRC.

40. Frutkin "Draft" memo, June 26, 1970, Record Group NASA 255, Box 14, Folder I.F, WNRC. This argument was frequently used to justify a generous policy on scientific and technological sharing. See John Krige, *American Hegemony and the Postwar Reconstruction of Science in Europe* (Cambridge: MIT Press, 2006), Chapter 1.

41. Frutkin, "The Case for Establishing a Procedure," July 8, 1970.

42. Memo, "Technology Transfer in the post-Apollo Program," unsigned, undated, but clearly around July 1970, Record Group NASA 255, Box 14, Folder I.G, WNRC.

43. They comprise part IV of the study "Technology Transfer in the post-Apollo Program," of about July 1970.

44. Ibid.

45. Available online at http://nixon.archives.gov/virtuallibrary/documents/nationalse-curitymemoranda.php.

46. Memo, Jacob E. Smart to Charles Mathews, "Space Station Utilization Symposium, Ames Research Center, September 1970," Record Group NASA 255, Box16, Folder VI.D, WNRC.

47. The minutes of its meetings on August 5, 11, and 18 are available. NSDM-72 had to take into account the findings of two parallel studies that would result in NSSM 71, dealing with Advanced Technology and National Security and NSSM 72, dealing with International Cooperation in Space, both discussed earlier: see letter Packard to Paine, June 3, 1970. All material in Record Group NASA 255, Box 14, Folder I.G, WNRC.

48. "Memorandum for Record, Ad Hoc Interagency Group on NSDM-72," August 19, 1970, Record Group NASA 255, Box 14, Folder I.G, WNRC.

49. Arnold W. Frutkin, Memo for Dr. T. O. Paine, "First Report of the NSDM 72 Ad Hoc Interagency Group," July 31, 1970, and Attachment, Record no. 12573, Federal Agencies-Presidents, Nixon administration, Folder International Cooperation (1970), NHRC.

50. "Memorandum for Record, Ad Hoc Interagency Group on NSDM-72" August 19, 1970; Frutkin, Draft Airgram on "Release of Technical Information for pre-Agreement Phase, Post-Apollo Program," July 22, 1970, Record Group NASA 255, Box 14, Folder I.G, WNRC.51. This section is based on Richard R. Colino, *The Intelsat Definitive Arrangements: Ushering in a New Era in Satellite Telecommunications. EBU Legal and Administrative Series, Monograph No. 9* (Geneva: European Broadcasting Union, 1973); Jonathan F. Galloway, *The Politics and Technology of Satellite Communications* (Lexington, MA: Lexington Books, 1972); Pascal Griset, "Fondation et Empire: L'Hégémonie Américaine dans les Communications Internationales 1919–1980," *Reseaux. Communication, Technologie, Société* No 49 (September–October 1991), 73–89; William E. Lee, *The Shaping of an American Empire: Negotiating the Interim Intelsat Agreements*, paper presented at the Annual Meeting of the Association for Education in Journalism (60th, Madison, Wisconsin, 1977), Report ED151783, available at www.eric.ed.gov/ERICWebPortal/, accessed on May 15, 2009; David J. Whalen, *The Origins of Satellite Communications, 1945–1965* (Washington, DC: Smithsonian Institution Press, 2002).

51. Memo Attendance, with list of representatives from State and NASA, March 27 (?), 1970. Record Group NASA 255, Box 14, Folder I.G, WNRC.

52. Informal minutes of a meeting attended by Alexis Johnson and seven other officials in the State Department, with Paine and Frutkin from NASA on March (?), 1970 (month obscured), Record Group NASA 255, Box 14, Folder I.G, WNRC.

53. Draft Memo, "Progress, Prospects and Policy Requirements for Major International Participation in Space Programs of the 1970's," Memo Packard to Frutkin and Loy, May 29, 1970, Record Group NASA 255, Box 14, Folder I.G, WNRC.

54. Frutkin to Hilburn, "Memorandum for Mr. Hilburn—AAD. Policies Relevant to '69 Revision of Intelsat Agreement," April 11, 1966, Record no. 14459, International Cooperation with Foreign Countries, International Cooperation, folder Miscellaneous Correspondence from Code I—International Relations 1958–1967, NHRC.

55. Galloway, *The Politics and Technology of Satellite Communications*, 143–144. It is difficult to know what to make of these figures, quoted by Galloway as indicative of a state subsidy to Comsat. Whalen remarks that NASA was barred by Congress in 1963 from research in communications satellites. The Agency got around this restriction by funding a generation of "applications" satellites (the ATS program), without however any technological advantage to the commercial domain, in Whalen's view (Whalen, *The Origins of Satellite Communications*, 162–163). See also Whalen in Butrica (ed.), *Beyond the Ionosphere*. On the distribution of Comsat's funds between in-house research and external contracts, see Stephen A. Levy, "Intelsat: Technology, Politics and the Transformation of a Regime," *International Organization* 29:3

(Summer 1975), 55–680. Much of the money was spent in-house, until it was agreed in 1972 that the ratio should be 50 : 50. This was another thorn in the side of the United States' partners in Comsat. The DOD developed its own, independent comsat system.

56. Tables of the US, USSR, and Intelsat communication satellites are supplied in Galloway, *The Politics and Technology of Satellite Communications*, Appendix A.

57. Quoted in Galloway, *The Politics and Technology of Satellite Communications*, 156.

58. Marcellus S. Snow, *International Commercial Satellite Communications. Economic and Political Issues of the First Decade of Intelsat* (New York: Praeger, 1976), 123–124. BAC (British Aircraft Corporation) was the major subcontractor on Intelsat IV, News Item, *BAC Build World's Largest Commercial Satellite*, January 1971, Record Group NASA 255, Box 16, Folder VI.C, WNRC.

59. Quoted in Galloway, *The Politics and Technology of Satellite Communications*, 156.

60. Joshua Barker, "Engineers and Political Dreams. Indonesia in the Satellite Age," *Current Anthropology* 46:5 (December 2005), 703–727, at 713.

61. R. Sueur, "The Symphonie Project and its Application," Fourth Eurospace US-European Conference, September 22–25, 1970, Venice Italy, Record Group NASA 255, Box 15, Folder V.B, WNRC. Herman Bondi also gave an account of Symphonie at this meeting. It seemed to suggest that the regional program was a greater threat to Intelsat than Sueur implied: we will deal with Bondi's Symphonie in the next chapter.

62. These are described in detail in Colino, *The Intelsat Definitive Arrangements*. See also Galloway, *The Politics and Technology of Satellite Communications*, chapter 8.

63. Colino, *The Intelsat Definitive Arrangements*, 63–66. Colino, a senior Comsat official, did not think that this provision would have much weight in practice.

64. Letter Joseph Charyk, president of Comsat, to U. Alexis Johnson, December 29, 1970, Record Group NASA 255, Box 16, Folder V.C, WNRC.

65. Colino, *The Intelsat Definitive Arrangements*, 93.

66. Decisions on "substantive" matters in the board required the support of at least four governors representing two-thirds of the investment shares, or the support of all but three governors, regardless of total investments share. The maximum vote that any representative could have, regardless of investment and use, was 40 percent. See ibid., 43–45 and 90–95.

67. Ibid., 94.

68. Ibid., 96.

69. Memos "Attendance" of March (?) 27, 1970, "Summary" of May 20, 1970, and Draft Memo, "Progress, Prospects and Policy Requirements."

70. Draft Memo, "Progress, Prospects and Policy Requirements." This entire section of text was underlined in the original.

71. Memo, "Statement on Shuttle Access," attached to Memo Packard to Frutkin, July 13, 1970. Record Group NASA 255, Box 14, Folder II.D, WNRC. On DOD withdrawal, see Frutkin memo "Purpose of Discussion with Dr Kissinger," July 8, 1970, Record Group NASA 255, Box 14, Folder II.B, WNRC.

72. Memo Packard to Frutkin and Low, May 29, 1970.

73. Diary note, Arnold Frutkin, "Coordination with State on Post-Apollo Participation," Record Group NASA 255, Box 14, Folder I.D, WNRC.

74. Memorandum, Frank R. Hammill, Jr. to George P. Miller, "Trip Report—International Cooperation in Space," August 5, 1970, Record Group NASA 255, Box 14, Folder II.E.1, WNRC.

75. See also the summary attached to memorandum Frutkin to Low, September 11, 1970, Record Group NASA 255, Box 16, Folder V.B, WNRC.

76. Europa III was a two-stage rocket using a storable liquid propellant in its first stage (UDMH and N2O4), in contrast to the kerosene-liquid oxygen mixture in Europa II's Blue Streak first stage. The second stage used a liquid hydrogen-liquid oxygen mixture. Europa III could place 1,800 kilograms into geostationary transfer orbit.

See "Eldo's New Rocket," *Flight International,* May 21, 1970, 871; "Pour l'Europa III," *Air et Cosmos,* No. 339, 1970, 20.

77. (Draft) State Department Report, *Discussion with the European Space Conference of Eventual European Participation in the Post-APOLLO Program,* September 11, 1970, Record Group NASA 255, Box 16, Folder V.E, WNRC.

78. Memo from the acting administrator, "Policy and Progress on International Cooperation in the Post-Apollo Program," October 7, 1970, Record Group NASA 255, Box 14, Folder I.D, WNRC.

79. (Draft) State Department Report, *Discussion,* September 11, 1970.

80. *Meeting between an ESC Delegation and a US Government Delegation, (Washington—16–17 September 1970, Summary Record of Statements,* Record Group NASA 255, Box 16, Folder V.B, WNRC.

81. Annex II, III, and IV to ibid.

82. Letter Johnson to Lefèvre, October 2, 1970, Record Group NASA 255, Box 16, Folder V.B, WNRC.

83. Ibid.

84. (Draft) State Department Report, *Discussion with the European Space Conference of Eventual European Participation,* September 11, 1970.

85. (Transcript of taped) Memo Frutkin to Morris "to be used in conjunction with State Department response to the Lefevre Mission," September 21, 1970, Record Group NASA 255, Box 16, Folder V.B, WNRC.

86. (Draft) State Department Report, *Discussion,* September 11, 1970.

87. Johnson to Lefèvre, October 2, 1970, for all quotes in this paragraph.

88. First Session of the *Meeting between an ESC Delegation and a US Government Delegation, 16–17 September 1970, Summary Record of Statements,* Record Group NASA 255, Box 16, Folder V.B, WNRC.

89. First Session of the *Meeting [...] September 1970, Summary Record of Statements;* Johnson to Lefèvre, October 2, 1970.

90. First Session of the *Meeting [...] September 1970, Summary Record of Statements.*

91. (Draft) State Department Report, *Discussion,* September 11, 1970.

92. Report, *Discussion with the European Space Conference,* September 11, 1970, Record Group NASA 255, Box 16, Folder V.B, WNRC.

93. Johnson to Lefèvre, October 2, 1970.

94. His address is available verbatim as an annex to *European Space Conference. (Draft) Minutes of the Meeting Held on the Morning of 4th November 1970,* CSE/CM(November 70)PV/1, November 4, 1970, Record Group NASA 255, Box 15, Folder V.B, WNRC.

95. In footnote 9 in letter Johnson to Lefèvre, October 2, 1970. It was also in Question 11 posed to the US delegation at the meeting on September 16.

96. (Draft) State Department Report, *Discussion with the European Space Conference,* September 11, 1970.

97. Johnson reported that the STG had suggested that the shuttle and the space station be developed in parallel. NASA was now considering delaying the start of the station, and of building it up from modules small enough to be carried by the shuttle. Development costs remained about the same: $6 billion for the shuttle, and $4.6 billion for the station, but they would be smoothed so that both did not peak at the same time. NASA had also drastically simplified the tug concept, eliminating the option of a manned tug that could take people to the moon, and restricting it to a craft able to transfer satellites from low-earth to geostationary orbit. Its cost would be reduced to about one-third of the originally envisaged $3.1 billion. Johnson used the cost estimates provided in the STG report a year earlier. For doubts on what cost data to use, see note from Donald R. Morris to Low, September 29, 1970, Record Group NASA 255, Box 16, Folder V.B, WNRC.

98. Annex to *European Space Conference. (Draft) Minutes of the Meeting Held on the Morning of 4th November 1970*, CSE/CM(November 1970), November 4, 1970, Record Group NASA 255, Box 15, Folder V.B, WNRC.

99. To emphasize the point the British also remarked that Japan had obtained launch assurances from the United States, and the right to manufacture a US launcher under license. "British therefore failed to see why Europeans should have to participate in post-Apollo program for same benefits Japan has obtained for nothing," telegram, Amembassy Brussels to the State Department, "European Space Conference," November 6, 1970, Record Group NASA 255, Box 16, Folder VI.B.1, WNRC.

100. Part II of *(Draft) Minutes of the Meeting Held on the Morning of 4th November 1970*.

101. Telegram, Amembassy Brussels to the State Department, November 6, 1970. Donald Fink, "British Reject Post-Apollo Participation," *Aviation Week and Space Technology*, November, 9 1970; "Britain Rejects Post-Apollo Role," *New York Times*, November 5, 1970.

102. Excerpt from *Proceedings of the Fourth Eurospace U.S.-European Conference, Venice, September 22–25, 1970*, 60, Record Group NASA 255, Box 16, Folder V.C, WNRC.

103. *Conclusions and Recommendations of the IVth Eurospace U.S.-European Conference, Venice, 22nd–25th September, 1970*, Record Group NASA 255, Box 15, Folder V.B, WNRC.

104. Memorandum, Frutkin for Low, "Balance of Factors in post-Apollo cooperation," December 1, 1970, Record No. 14469, International Cooperation and Foreign Countries, International Cooperation, Folder General, 1960, NHRC.

105. The most important other major program was Skylab, a space station derived from a modified third stage of a Saturn V moon rocket. Skylab was a product of NASA's Apollo Applications program that was called on to find long-term uses for Apollo program hardware. It was placed in orbit in May 1973, and was inhabited three times (for 28, 59, and 84 days) over the next nine months. See T. A. Heppenheimer, *The Space Shuttle Decision, NASA's Search for a Reusable Space Vehicle* (Washington, DC: NASA SP-4221, 1991).

106. Note, Frutkin to Behr, November 9, 1970, Record Group NASA 255, Box 14, Folder II.B, WNRC.

107. Letter Frutkin to Pollack, November 16, 1970, Record Group NASA 255, Box14, Folder II.D, WNRC.

108. Memorandum, Herman Pollack, "Prospects for European Participation in the Post-Apollo Space Program," November 30, 1970, attached to internal State Department Reference Slip from Robert F. Packard, November 30, 1970. Johnson's memo was sent to Kissinger on December 1, 1970, Record Group NASA 255, Box 17, Folder IX, WNRC.

109. Memorandum, Johnson to Kissinger, "Post-Apollo Space Program," attached to internal State Department Reference Slip from Robert F. Packard, November 30, 1970.

110. Letter Low to Kissinger, October 30, 1970, Record Group NASA 255, Box 14, Folder II.B, WNRC. Low wrote again to Kissinger on November 13, 1970, to tell him of the United Kingdom's position at the ESC meeting, which, he said, "confirmed" NASA's argument that a "clear expression of our own national intention to carry forward the space shuttle" and Skylab was needed, Record Group NASA 255, Box 14, Folder II.B, WNRC.

111. Frutkin, "Memorandum to the File. NASA Budget Considerations in International Area," December 4, 1970, Record Group NASA 255, Box 15, Folder III.B, WNRC.

112. Letter Kissinger to Low, January 4, 1971, Record Group NASA 255, Box 17, Folder IX, WNRC.

113. For the budget data, see *NASA Historical Data Book: Volume IV*, NASA SP-4012, at http://history.nasa.gov/SP-4012/vol4/t4.4.htm (accessed on June 9, 2009). The

budget requests for FY71 and FY72 were $3,376.9 million and $3,312.7 million, respectively.

114. Concluding remarks by Paine at a NASA long-range planning conference held at Wallops Island on June 14, 1970, as quoted by Howard E. McCurdy, *Space and the American Imagination* (Washington, DC: Smithsonian Institute Press, 1997), 50.

115. For one such letter, Paine to the president, June 23, 1970, Record Group NASA 255, Box 14, Folder II.B, WNRC.

116. Hoff, in Launius and McCurdy, *Spaceflight and the Myth*, 105.

117. Bob Diegelman, *Congressional Opinion Survey (January–November 1970)*, Record Group NASA 255, Box 15, Folder IV.A.3, WNRC.

118. See Long, this volume, Chapter 8, for more on US-USSR cooperation. An agreement for the rendezvous and docking of a Soviet and an American spacecraft was signed in Moscow on October 28, 1970, *Background Press Briefing. U.S. and USSR Cooperation in Space*, October 29, 1970, Record no. 12701, Federal Agencies-Presidents, Folder Nixon-Speeches (1974), NHRC.

119. Letters Paine to Kissinger and to U. Alexis Johnson, September 15, 1970, Record Group NASA 255, Box 14, Folders II.B and I.D, respectively, WNRC.

120. Hoff, in Launius and McCurdy, *Spaceflight and the Myth*, 100.

121. Letters di Carrobio to Paine, August 19, 1970, Lefevre to Paine, September 7, 1970, and Paine to Johnson, September 15, 1970, Record Group NASA 255, Box 14, Folder I.D, WNRC.

122. Letter di Carrobio to Paine, August 19, 1970.

5 European Participation in the Post-Apollo Program, 1971: The United States Begins to Have Second Thoughts—And So Do the Europeans

1. Extract from the Report of the President to Congress on Foreign Policy for the 1970s, February 25, 1971, Accession Number 255–74–734, Record Group NASA 255, NASA Division I Escalation Files Box 17, folder IX, WNRC.

2. Nixon's talk echoed some of the themes defined by Low in his reflections on "International Cooperation in Space," sent to Kissinger late in January. Letter Low to Kissinger, January 29, 1971, with attachment, and reply, February 4, 1971, Record Group NASA 255, Box 14, folder II.B, WNRC.

3. Memorandum Whitehead to Flanigan, February 6, 1971, attached to memo Flanigan for Erlichman, in John Logsdon, *Exploring the Unknown. Selected Documents in the U.S. Civil Space Program. Vol. II. External Relations* (Washington, DC: NASASP-4407, 1996), Doc. I-19.

4. Letter Charyk to Johnson, December 29, 1970, Record Group NASA 255, Box 16, Folder V.C, WNRC.

5. As pointed out in the previous chapter, both the Board of Governors and the Assembly of Parties were concerned. Thus, "[t]he Board of Governors will render its advice to the Assembly of Parties in accordance with the voting procedures for substantive matters [...]. i.e. an affirmative vote of at least four Governors representing two-thirds of the voting participation on the Board or an affirmative vote of all Governors, less three, regardless of total voting participation. The Assembly will decide upon its findings and make its recommendations in accordance with its own voting procedures, i.e. an affirmative vote of two-thirds of those present and voting with each representative possessing one vote." Richard R. Colino, *The Intelsat Definitive Arrangements: Ushering in a New Era in Satellite Telecommunications. EBU Legal and Administrative Series, Monograph No. 9* (Geneva: European Broadcasting Union, 1973), 93–94.

6. Frutkin, "Memorandum to File. Telecon between Dr Low and Undersecretary U. Alexis Johnson, January 13," January 15, 1971, Record Group NASA 255, Box 17, folder IX, WNRC. For the notes of the same occasion taken by NS, see "Notes on

Telephone Conversation. Participants: Dr. Low and Secretary U. Alexis Johnson," 13 January, 1971, Record Group NASA 255, Box 17, folder IX, WNRC.

7. Letter Johnson to Charyk, January 21, 1971, Record Group NASA 255, Box V, Folder C, WNRC.

8. "Memorandum of Telephone Conversation. Participants: Dr. Low and Undersecretary of State Alexis Johnson," January 23, 1971, by "NS (Read and Corrected by Dr. Low)," Record Group NASA 255, Box 16, Folder V.C, WNRC.

9. For this paragraph, see Frutkin, "Memorandum to File. Telecon [...] January 13," January 15, 1971.

10. French *Aide-Mémoire on the Operational System of European Communications Satellites*, January 15, 1971, Box 16, folder V.C, WNRC; *Analysis of Bondi Proposal for a European Telecommunications Satellite Program*, undated, unsigned, but around January–February, 1971, Record Group NASA 255, Box 16, Folder V.C, WNRC; Frutkin, "Diary Note, Bondi Paper on Eurosat (Venice, September 1970)," Record Group NASA 255, Box 15, Folder III.B, WNRC.

11. *Analysis of Bondi Proposal.*

12. Frutkin, "Memorandum to the File—January 25, 1971. Lefevre Meeting Preparations—Johnson/Charyk Discussions," January 28, 1971, Record Group NASA 255, Box 14, Folder II.I.2, WNRC.

13. Letter Lefèvre to Johnson, January 21, 1971, Record Group NASA 255, Box 16, Folder V.C. Frutkin's draft reactions are dated February 1, 1971, Record Group NASA 255, Box 16, Folder V.C, WNRC.

14. ESC, *Address by Mr. Lefevre*, February 10, 1971, Record No. 14549, International Cooperation and Foreign Countries, International Cooperation, Europe, Folder US-Europe 1973. NHRC.

15. State Department brief for the meeting, *Second Discussion with Representatives of the European Space Conference...*, February 8, 1971, Record Group NASA 255, Box 16, Folder V.C, WNRC.

16. State Department brief for the meeting, *Second Discussion*, February 8, 1971.

17. Document dealing with several ESC questions, written by Frutkin, February 1, 1971, Record Group NASA 255, Box 16, Folder V.C, WNRC.

18. National Security Council Memorandum, February 10, 1971, "Post-Apollo Space Cooperation," available at http://www.state.gov/r/pa/ho/frus/nixon/e1/46119.htm.

19. "Summary Record of Statements," undated, unsigned, Record Group NASA 255, Box 16, Folder V.B, WNRC. See also ESC document, *Address by Mr. Lefèvre*, February 10, 1971.

20. French "Verbal Note," February 10, 1971, Record Group NASA 255, Box 16, Folder V.C, WNRC.

21. *Notes from US/ESC Discussions, Summary of Discussions, Friday Afternoon (2/12/71)*, Record Group NASA 255, Box 16, Folder V.C, WNRC.

22. Telegram, Amembassy Brussels to State Department, March 25, 1971, Record Group NASA 255, Box 16, Folder VI.B.1, WNRC.

23. See contributions by Bondi, Demerliac, and Causse at AAS Goddard Memorial Symposium, Washington, DC, March 10–11, 1971, Record Group NASA 255, Box 17, Folder IX, WNRC.

24. Notes on the Goddard Symposium prepared by Jay Holmes, March 15, 1971, Record Group NASA 255, Box 17, Folder IX, WNRC.

25. William Cohen, *Results of Conference Among A. Frutkin, D. Myers, C. Donlan and Staffs, February 1, 1971*, document dated March 11, 1971, Record Group NASA 255, Box 17, Folder VI.D.7, WNRC.

26. *Report of the Mission to Washington*, CSE/Comite ad hoc (71)8, March 4, 1971, Record Group NASA 255, Box 16, Folder V.C, WNRC.

27. *Notes from US/ESC Discussions, February 11, 1971.*

28. Telegram Drafted by Packard Approved by Johnson from the State Department to various Embassies, February 24, 1971, Record Group NASA 255, Box 16, Folder V.C, WNRC.

29. Telegram, Amembassy Brussels to State Department, "Post-Apollo: Lefevre Letter to Under Secretary," March 6, 1971 (letter dated March 3, 1971), Record Group NASA 255, Box 16, Folder VI.B.1, WNRC.

30. Frutkin, "Basis for Proposal on US Launch Vehicle Availability Abroad," attached to his memo to Pollack, February 22, 1971; Frutkin, Memo to File, "Matters at Stake in Post-Apollo Launch Availability, Phone Call to Colonel Behr, February 25," February 26, 1971, Record Group NASA 255, Box 17, Folder VI.D.7, WNRC.

31. Memo Behr to Kissinger, "Post-Apollo Space Cooperation," March 4, 1971.

32. Memo Pollack to Frutkin, "Proposed Modifications in our Position of Post-Apollo Negotiations," March 6, 1971, Record Group NASA 255, Box 17, Folder IX, WNRC.

33. Memo Behr to Kissinger, "Post-Apollo Space Cooperation," March 4, 1971.

34. Memo David to Kissinger, March 17, 1971, http://www.state.gov./r/pa/ho/frus/nixon/el/46123.htm, Briefing Memorandum from Pollack to Johnson, June 5, 1971, http://www.state.gov./r/pa/ho/frus/nixon/el/46134.htm.

35. Memorandum for the Record, "Meeting between Dr. Kissinger and Dr. David, March 24, 1971," dated March 26, 1971, http://www.state.gov./r/pa/ho/frus/nixon/el/46126.htm. For Behr's talking points for that meeting, see his memo to Kissinger, "Post-Apollo Space Cooperation," April 23, 1971, Record no. 12578, Federal Agencies-Presidents, Nixon administration, Folder Nixon—General (1968–1971), NHRC.

36. Memo Behr to Kissinger, "Post-Apollo Space Cooperation," March 4, 1971.

37. Memo by Clare Farley, "Meeting with Prof Leussink—April 21, 1971," dated April 29, 1971, Record Group NASA 255, Box 14, Folder II.H, WNRC.

38. Memo Behr to Kissinger, "Post-Apollo Cooperation. Leussink Views on Space Cooperation," April 23, 1971, Record no. 12578, Federal Agencies-Presidents, Nixon Administration, Folder Nixon—General (1968–1971), WNRC.

39. Summary of Discussions, Post Apollo Space Cooperation, White House Situation Room, April 23, 1971, http://www.state.gov./r/pa/ho/frus/nixon/el/46128.htm.

40. For the memo to Rogers that Behr had prepared for Kissinger's signature early in April, see http://www.state.gov./r/pa/ho/frus/nixon/el/46127.htm.

41. "Alternatives to Post-Apollo Participation," May 21, 1971, attached to Farley's "Note for Mr Morris," June 1, 1971, Record Group NASA 255, Box 14, Folder II.B, WNRC.

42. Memorandum Pollack to Johnson, "NASA Presentation on Post-Apollo," June 4, 1971, http://www.state.gov./r/pa/ho/frus/nixon/el/46133.htm.

43. "Technology Transfer in the Post-Apollo Program," undated, unsigned, 7pp, and July 15, 1971, 4pp, unsigned, Record Group NASA 255, Box 14, Folder II.B, WNRC. See also Doc I-20 in Logsdon, *Exploring the Unknown. Vol. II.*

44. Frutkin, "Diary Note. OST Meeting on Post-Apollo Technology Flow," May 5, 1971. Attendees included David, Drew, and Neureiter for OST, and Fletcher, Low, and Frutkin for NASA. Present too were two representatives from each of the three aerospace companies. See also Frutkin's report of his meeting with Russell Drew in the Office of Science and Technology on May 4, 1971, Record Group NASA 255, Box 14, Folder II.G, WNRC.

45. Memo attached to letter Lawrence M. Mead (Grumman Aerospace) to Freitag, April 2, 1972, Record Group NASA 255, Box 16, Folder VI.C, WNRC.

46. Letter NAR to Handel Davies, BAC, undated but numbered #320–70, Record Group NASA 255, Box 16, Folder VI.C, WNRC.

47. The presentation of May 7 is attached to memo from SH (Sam Hubbard) dated May 19, 1971, and further information is provided in Memo from Sam Hubbard, *McDonnell Douglas Aircraft Company's Activities in International Cooperation,* June 18, 1971. See also the firm's report, *Space Shuttle Program. International Implementation Development,* revised June 29, 1971, Record Group NASA 255,

Box 17, Folder VI.D.3, WNRC. See also Memo from Hanger, Manager Marketing, Space Shuttle Program to multiple recipients, "SNIAS Space Shuttle Participation," May 18, 1971, and "ERNO Participation in Space Shuttle," May 18, 1971, Record Group NASA 255, Box 17, Folder VI.D.V, WNRC. Memo Freitag, NASA OMS, "SNIAS Visit to NASA," May 14, 1971, Record Group NASA 255, Box 16, Folder VI.D.1, WNRC.

48. *MBB Group Space Tug System Study. Pre-Phase A/Ext, Final Presentation to ELDO, NASA,* and *European Space Tug. Final Presentation July 71, Pre Phase A study, part 2, Prepared for the European Launcher Development Organisation by Hawker Siddeley Dynamics Limited Leading a group of European companies,* Record Group NASA 255, Box 16 Folder VI.D.1, WNRC.

49. Memo Sam Hubbard to Messrs Frutkin, Morris, and Barnes, *Phase B RAM Study,* April 15, 1971, Record Group NASA 255, Box 17, Folder VI.D.IV, WNRC.

50. At this stage NASA was limiting its tug design to a single stage concept, i.e., a "retrievable, reusable machine" that did "not include the manned synchronous or lunar operations but would hopefully be of a character that could 'grow' into broader usage of this nature." Memo for the Record by Freitag, *NASA-ELDO Technical Discussions of Tug Phase A Studies, May 10, 1971,* May 14, 1971, Record Group NASA 255, Box 14, Folder II.I.2, WNRC.

51. For this approach, see, in particular, *Technology Transfer in the Post-Apollo program,* July 15, 1971, Record Group NASA 255, Box 14, Folder II.H, WNRC.

52. *Technology Transfer in the Post-Apollo Program.* The cover page is viewgraph NASA HQ MF71–6399, 7–27–71. The other viewgraphs mostly have the same date but of course have different four-digit numbers, Record Group NASA 255, Box 14, Folder II.H, WNRC.

53. Memo Pollack to Johnson, *Meeting with Dr Kissinger, June 8, 5.00 pm on Post-Apollo Cooperation,* June 5, 1971, http://www.state.gov/r/pa/ho/frus/nixon/el/46134.htm.

54. Memo David to Kissinger and Flanigan, *Post-Apollo Space Cooperation with the Europeans,* July 23, 1971, http://www.state.gov/r/pa/ho/frus/nixon/el/46135.htm.

55. NASA Memo, *Summary Statement on Post-Apollo Alternatives,* July 15, 1971, Suitland, Record Group NASA 255, Box 14, Folder II.B, WNRC.

56. Memo Pollack to Johnson, *Meeting with Dr Kissinger, June 8, 5.00 pm on Post-Apollo Cooperation,* June 5, 1971, http://www.state.gov/r/pa/ho/frus/nixon/el/46134.htm.

57. Memo Pollack to Johnson, *House Discussions on Post-Apollo Cooperation with Europe,* May 5, 1971, http://www.state.gov/r/pa/ho/frus/nixon/el/46129.htm.

58. Memo Kissinger to Rogers, *Post-Apollo Space Cooperation with the Europeans and Launch Assurances,* August 18, 1971, http://www.state.gov/r/pa/ho/frus/nixon/el/46389.htm. See also Doc I-21 in Logsdon, *Exploring the Unknown. Vol. II.*

59. T. A. Heppenheimer, *The Space Shuttle Decision. NASA's Search for a Reusable Space Vehicle* (Washington, DC: NASA SP-4221, 1999), 364–366; Roger D. Launius, *NASA: A History of the U.S. Civil Space Program* (Krieger: Malabar, 1994), Reading No.19, reproduces the memorandum.

60. Letter Johnson to Lefèvre, September 1, 1971, http://www.state.gov/r/pa/ho/frus/nixon/el/46395.htm. The letter is "copied" on a release from the State Department dated November 1, 1971, along with a short explanatory text under the title "Launch Assistance to Europeans," Record no. 14549, International Cooperation and Foreign Countries, Europe, Folder US-Europe 1973–, NASA Historical Reference Collection, Washington, DC. For preliminary drafts and supporting material, see "Draft, August 6, 1971," "Draft, August 13, 1971," and Draft "Statement of US Views on Participation in the Post-Apollo Program," August 6, 1971, Record Group NASA 255, Box 16, Folder V.C, WNRC. See also Doc I-22, Logsdon, *Exploring the Unknown. Vol. II.*

61. Memorandum of conversation between Johnson, David, Fletcher, Pollack, and others on *Post-Apollo Space Cooperation and US Assurance of Launch Assistance,* October 6, 1971,

http://www.state.gov/r/pa/ho/frus/nixon/el/46396.htm. David did not accept this, but was overruled: the procedure to be followed in order to broaden the discussion beyond the STS would be decided after the first meeting of experts in late October.

62. "Telegram to Paris. Post-Apollo Technical Discussions. Paris 16010," undated and unsigned, Record Group NASA 255, Box 14, Folder I.D, WNRC.

63. Memo, "Questions and Answers at the Mathews Presentation, October 22, 1971, Paris, France," November 2, 1971, Record Group NASA 255, Box 14, Folder I.D, WNRC.

64. Charles J. Donlan, "Memorandum for the Record" on "Memorandum for M/Mr. Myers," both dealing with his European Visit to XXII International Astronautical Congress and Organizations and Firms Interested in Post-Apollo Participation, both dated November 1, 1971, Record Group NASA 255, Box 14, Folder II.G, WNRC.

65. *Report of the Meeting of the Joint Group of Experts on U.S./European Cooperation in Space Programs in the Post-Apollo Period, Held in Washington, D.C., 30 November–3 December, 1971*, Record Group NASA 255, Box 17, Folder IX, WNRC.

66. This spacecraft, favored by Low, resembled a shuttle but would not be able to propel itself into orbit. Its payload bay would be 12 x 40 feet, its payload weight about 30,000 pounds, and it would be lofted into space on a two-stage Titan IIIL class rocket. For the evolving shuttle design in this period, see Ray A. Williamson, "Developing the Space Shuttle," in John M. Logsdon (ed.), *Exploring the Unknown. Selected Documents in the History of the U.S. Civil Space Program. Vol. IV. Accessing Space* (Washington, DC: NASA SP-4407, 1999), Chapter 2. See also John M. Logsdon, "The Space Shuttle Program: A policy Failure?" *Science* 232:4754 (May 30, 1986), 1099–1105; "The Decision to Develop the Space Shuttle," *Space Policy* 2 (May 1986), 103–119; Heppenheimer, *The Space Shuttle Decision*, 368–369, 373, 385.

67. *Report of the Meeting of the Joint Group of Experts*, 7.

68. Ibid., 8.

69. Ibid., 11.

70. For the aforementioned details, see European Space Conference *Report by the Joint ESRO/ELDO Working Group on the Post-Apollo Programme*, WG/COOP/US/23, October 1971, and WG/COOP/US/23, Annex 1, October 1971, specifically dedicated to the Space Transportation System, in Record Group NASA 255, Box 16, Folder VI.A.2, WNRC.

71. *Report of the Meeting of the Joint Group of Experts*, 12.

72. Robert F. Freitag, "Memorandum of a Telecom," January 6, 1972, Record Group NASA 255, Box 14, Folder I.C.4, WNRC.

73. Report WG/COOP/US/23, 16.

74. Douglas R. Lord, *Spacelab. An International Success Story* (Washington, DC: NASA SP-487, 1987), 5.

75. This terminology is derived from *An Introduction to Shuttle Sortie Missions*, released by NASA's Office of Manned Space Flight in April 1972, Record Group NASA 255, Box 17, Folder VI.E, WNRC. There was considerable confusion over just what these terms meant. see letter Fletcher to Anders, April 11, 1972, Record Group NASA 255, Box 14, Folder II.B, WNRC.

76. See John Krige, "The Decision Taken in the Early 1970s to Develop an Expendable European Heavy Satellite Launcher," in John Krige, Arturo Russo, and Lorenza Sebesta, *A History of European Space Agency. 1958–1987. Vol. 2. The Story of ESA* (Noordwijk: ESA SP 1235, 2000), Chapter 9.

77. Arnold Frutkin, Memorandum to the File, "ELDO Assistance Clearance with Walsh," December 10, 1971, Record Group NASA 255, Box 17, Folder X, WNRC.

78. The following account is derived from Lorenza Sebesta, "The Aeronautical Satellite System: An Example of International Bargaining," in Krige, Russo, and Sebesta, *A History of the European Space Agency*, 357–386.

79. Ibid., 370.

80. Ibid.

81. Ibid., 374–376.

82. Frutkin, "Diary Note," October 14, 1971, Suitland, Record Group NASA 255, Box 14, Folder II.G. In Telegram Frutkin to Bernier, September 16, 1971, Frutkin had detailed for NASA's European Representative in Paris the progress made in defining alternative approaches to the design and development of the space shuttle, Record Group NASA 255, Box 17, Folder IV.D.8, WNRC.

83. As quoted by Sebesta, "The Aeronautical Satellite," 375.

84. Memorandum from Rogers to Nixon, January 19, 1972, *Post-Apollo Space Cooperation with the Europeans and Launch Assurances*, available at http://www.state.gov/r/pa/ho/frus/nixon/e1/46418.htm.

6 European Participation in the Post-Apollo Program, 1972: Disentangling the Alliance—The Victory of Clean Technological Interfaces

1. Weekly Compilation of Presidential Documents, January 10, 1972, Record no. 12595, Federal Agencies—Presidents, Nixon administration, Folder Nixon-Space Shuttle Statement, January 5, 1972, NHRC.

2. Office of the White House Press Secretary, Press Conference with Fletcher and Low, San Clemente Inn, January 5, 1972, Record no. 12594, Federal Agencies-Presidents, Nixon Administration, Folder Nixon-Space (1972), 1972–1986, NHRC.

3. Transcript of a Conversation between Johnson and Fletcher on January 7, 1972, available at http://www.state.gov/r/pa/ho/frus/nixon/e1/46416.htm.

4. Memorandum for the Record, George M. Low, "Meeting with the President on January 5, 1972," January 12, 1972, Record no. 12575, Federal Agencies-Presidents, Nixon administration, Folder Nixon Correspondence (NASA), 1972–1974. Nixon laid such store by the foreign policy advantages of international collaboration that he asked John Ehrlichman specifically to mention this to Henry Kissinger, along with the possibility of a docking mission with the Soviets in space.

5. Arnold Frutkin, "Diary Note. Post-Apollo Coordination (Walsh meeting #1, Jan. 19)," January 20, 1972, with attachments "Possible Modalities for European Cooperation in post-Apollo," and "Possible Spectra of Alternatives for European participation [...]," January 24, 1972, Record Group NASA 255, NASA Division I, Escalation Files Box 14, Folder II.B, WNRC, Accession Number 255–74–734; Arnold Frutkin, "Diary Note. Post-Apollo Coordination (Walsh meetings #2, Jan 21)," dated January 24, 1972; Arnold Frutkin, "Diary Note. Third John Walsh Meeting on Post-Apollo—Aerospace Contractor Comments on Foreign Subcontracts," dated January 28, 1972; Arnold Frutkin, "Memorandum to Post-Apollo Coordination File, Walsh meeting #4, January 27," dated February 2, 1972, Record Group NASA 255, Box 14, Folder II.H, WNRC.

6. Memo for Herman Pollack, "NSSM 72 Subcommittee," from John B. Walsh, February 18, 1972, Record Group NASA 255, Box 14.II.B, WNRC. Handwritten comments in the margin indicate some points of disagreement by an unidentified official.

7. "Report of the Meeting of the Joint Group of Experts on U.S./European Cooperation in Space Programs in the Post-Apollo Period Held at Neuilly, 8 February–10 February 1972," Box 17.IX; There are also "Personal Notes of R.E. Bernier," NASA's European Representative, on the meeting, and a summary document prepared for the meeting by the head of the US delegation, Philip. E. Culbertson, "General U.S. Position as we go into the February 8–10 meeting on European participation in the 'Post-Apollo Program,'" dated February 4, 1972, all in Record Group NASA 255, Box 14, Folder I.C.IV, WNRC.

8. Unsigned memo, "Possible Spectra of Alternatives," January 24, 1972.

9. Memo, Arnold W. Frutkin, "Approaches to Foreign Participation in the Shuttle Program," January 18, 1972, Record Group NASA 255, Box 14, Folder II.H, WNRC.

10. Frutkin, "Post-Apollo Action Alternatives," February 29, 1972.

11. *Report of the Meeting of the Joint Group of Experts*, February 8–10, 1972.

12. Unsigned Memo, "Possible Spectra of Alternatives," January 24, 1972.

13. Memo, Frutkin, "Post-Apollo Action," February 24, 1972.

14. Memo, Frutkin, "Post-Apollo Action Alternatives," February 29, 1972.

15. Information memorandum from Pollack to Rogers, *Post-Apollo Cooperation in Jeopardy*, March 17, 1972, http://www.state.gov/r/pa/ho/frus/nixon/el/46425. htm. See also Doc I-25, John M. Logsdon, Dwayne A. Day, and Roger D. Launius, eds., *Exploring the Unknown. Select Documents in the History of the U.S. Civilian Space Program. Vol. II. External Relations* (Washington, DC: NASA, 1996).

16. Memo Low to Fletcher, "Position Paper on Post Apollo International Cooperation," March 27, 1972, Record no. 14462, International Cooperation and Foreign Countries, International Cooperation, Folder International Space Documents from Dwayne Day, 1959–1975, NHRC. See also Doc I-24, Logsdon, *Exploring the Unknown, Vol. II*.

17. Transcript of a Conversation between Johnson and Fletcher, January 7, 1972, http://www.state.gov/r/pa/ho/frus/nixon/el/46416.htm.

18. Transcript of a Conversation between Johnson and Fletcher, January 7, 1972.

19. Memorandum William P. Rogers, secretary of state, to the president, April 29, 1972, http://www.state.gov/r/pa/ho/frus/nixon/el/46427.htm. See also Doc I-26, Logsdon, *Exploring the Unknown, Vol. II*.

20. Memo "NASA's Comments on Secretary Rogers' Memorandum of April 29, 1972," under cover of letter Fletcher to Kissinger, May 5, 1972. Record Group NASA 255 Box 14, Folder II.B, WNRC. See also Doc I-27, Logsdon, *Exploring the Unknown, Vol. II*.

21. Memorandum Edward E. David to Henry Kissinger and Peter Flanigan, *Post-Apollo Relationships with the Europeans*, May 18, 1972, http://www.state.gov/r/pa/ho/frus/nixon/el/46428.htm.

22. "Concluding Remarks by Mr. Herman Pollack Meeting with ESC Delegation on Post-Apollo Cooperation. June 16, 1972," verbatim section, No.1, Record Group NASA 255, Box 14, Folder I, WNRC. At the start of the meeting it was left open that Europe might participate in shuttle development: see *Report of the ESC Delegation on discussions held with the U.S. Delegation on European participation in the Post-Apollo program, Washington, 14–16 June 1972*, CSE/CS(72)15, June 22, 1972.

23. Interview, Jean-Pierre Causse with John Krige, Paris, May 18, 2007, NHRC, Washington, DC.

24. "Concluding Remarks by Mr. Herman Pollack," June 16, 1972.

25. *Report of the Meeting of Joint Group of Experts on U.S./European Cooperation in the post-Apollo Period held at Neulliy, 8 February to 10 February, 1972*, Record Group NASA 255, Box 17, Folder IX, WNRC.

26. This is according to the notes "informally taken" during the meeting by Robert E. Bernier, NASA's European representative in Paris, headed "Personal Notes of R.E. Bernier," under cover of a memo dated February 17, 1972, Record Group NASA 255, Box 14, Folder I.C.4, WNRC.

27. Memo Frutkin, "Post-Apollo Action Alternatives," February 29, 1972.

28. European Space Conference, *Report on European Participation in the Post-Apollo Programme*, WG/COOP/US(72)2, March 1972, Record Group NASA 255, Box 17, Folder IX, WNRC.

29. U. John Sakss, Memorandum for File, "Causse/Dinkespiler Visit April 19 to Discuss Post-Apollo Questions," May 2, 1972, Record Group NASA 255, Box 14, Folder I.1, WNRC.

30. Memo Culbertson, "General U.S. position as we go into the February 8–10 meeting on European participation in the 'Post-Apollo Program,'" February 4, 1972, Record Group NASA 255, Box 14, Folder I.C.4, WNRC.

31. Sakss, Memorandum for File, May 2, 1972.

32. The briefing charts spelling out the Air Force's arguments for and against using a tug developed in Europe was attached to Memo, "Air Force Use of European Tug," from Sam Hubbard to Barnes, Morris and Frutkin, May 17, 1971, Record Group NASA 255, Box 16, Folder VI.D.1, WNRC.

33. Memo Hubbard to Barnes et al., May 17, 1971.
34. Douglas R. Lord, *Spacelab. An International Success Story* (Washington, DC: NASA SP-487, 1987), tells the history of its development.
35. Telegram State Department to Amembassies in Europe, June 16, 1972, Record Group NASA 255, Box 14, Folder I.1, WNRC.
36. French goals were widely known: this is as reported by a German industrialist who visited Martin Marietta in Denver in April: Robert F. Freitag, Memorandum for the Record, May 4, 1972.
37. See Lord, *Spacelab*, Appendix A.
38. Ibid. describes the working relationships at the technical level for the construction of the laboratory; the European view is in *Proceedings of the Workshop on the History of Spacelab, ESTEC, 22–23 April, 1997* (Noordwijk: ESA SP-411, 1997).
39. In *Proceedings of the Workshop on the History of Spacelab*.
40. Ibid.
41. P. R. Sahm, M. H. Keller, and B. Schieve, eds., *Research in Space. The German Spacelab Missions* (Köln: Wissenschaftliche Projektführung D-2, 1993). Niklas Reinke, *The History of German Space Policy. Ideas, Influences and Interdependence, 1923–2002* (Paris: Beauchesne, 2007), 165–167, lists all of the Spacelab missions.
42. Reinke, *The History of German Space Policy*, 160.
43. Wolfgang Finke, "Germany and ESA," *The History of the European Space Agency. Proceedings of a Symposium, London, November, 1998* (Noordwijk: ESA SP-436, 1999), 37–50, at 43.
44. Ibid., 45.
45. Reinke, *The History of German Space Policy*, 167.
46. Reimar Lüst, cited by Roger M. Bonnet and Victtorio Manno, *International Cooperation in Space. The Example of the European Space Agency* (Cambridge: Harvard University Press, 1994), 79.

7 Sustaining Soviet-American Collaboration, 1957–1989

1. Melvyn Leffler, *For the Soul of Mankind: The United States, The Soviet Union, and The Cold War.* (New York: Hill and Wang, 2007); Walter McDougall, . . . *the Heavens and the Earth* (New York: Basic Books, 1985).
2. Edward Clinton Ezell and Linda Neuman Ezell, *The Partnership: A History of the Apollo-Soyuz Test Project* (Washington, DC: NASA Scientific and Technical Information Office, 1978); Dodd Harvey and Linda Ciccoritti, *US-Soviet Cooperation in Space* (University of Miami: Center for Advanced International Studies, 1974); Yuri Karash, *The Superpower Odyssey: A Russian Perspective on Space Cooperation* (Reston, VA: American Institute of Aeronautics and Astronautics, 1999); Matthew VonBencke, *The Politics of Space: A History of US-Soviet/Russian Competition and Cooperation in Space* (Boulder, CO, 1997).
3. John Logsdon, *John F. Kennedy and the Race to the Moon* (New York: Palgrave Macmillan, 2010).
4. Howard McCurdy, *Space and the American Imagination* (Washington, DC, 1997).
5. John Logsdon, ed., *Exploring the Unknown: Selected Documents in the History of the US Civilian Space Program Volume II: External Relationships* (Washington, DC: NASA History Office, 1996), 148 and 151.
6. Arnold Frutkin, *International Cooperation in Space* (Englewood Cliffs, NJ: Prentice-Hall), 100–101; emphasis in the original.
7. Ezell and Ezell, *The Partnership*, 56–58.
8. Repeated offers and rebuttals are documented in depth in a number of texts: Ibid.; Harvey and Ciccoritti, *US-Soviet Cooperation*; VonBencke *The Politics of Space*; Karash, *The Superpower Odyssey*.
9. Mose L. Harvey, "An Assent of US-USSR Cooperation in Space," in Michael Cutler, ed., *International Cooperation in Space Operations and Exploration* (Tarzana, CA:

American Astronautical Society, 1971), 157. Logsdon's history of space policy, *John F. Kennedy and the Race to the Moon*, analyzes the history of Kennedy's interest in space exploration and, in particular, provides evidence that his much-debated offers for a joint expedition to the moon were offered with a sincere desire for collaboration and not simply as a political ploy.

10. McDougall,...*the Heavens and the Earth*, 349.
11. Harvey and Ciccoritti, *US-Soviet Cooperation*, 211; emphasis added.
12. Arnold Frutkin, "The United States Space Program and Its International Significance," *Annals of the American Academy of Political and Social Science* 366 (July 1966), 89–98.
13. Ibid., 93.
14. Ibid.
15. Boris Chertok, *Rockets and People Volume III: Hot Days of the Cold War* (Washington, DC: NASA History Series, 2009), 277. See also Asif Siddiqi, *The Soviet Space Race with Apollo* (Gainesville: University of Florida Press, 2000).
16. US solar physicist Herbert Friedman in the preface to Iosif Shklovsky, *Five Billion Vodka Bottles to the Moon: Tales of a Soviet Scientist* (New York: W. W. Norton & Company, 1991), 7.
17. Ibid.
18. Ibid., 17.
19. Ezell and Ezell, *The Partnership*, 125.
20. Robert Divine, "Lyndon B. Johnson and the Politics of Space," in Robert Divine, ed., *Johnson Years VII: Vietnam, the Environment, and Science* (Lawrence, KS: University Press of Kansas, 1987), 239.
21. Yuri Karash, *The Superpower Odyssey*, 89–90.
22. Ezell and Ezell, *The Partnership*, 126–127.
23. Frutkin's five points for collaboration are listed and analyzed in detail in chapter 1.
24. Skylab was a converted Apollo-Saturn-IVB stage that served as a space station. It could accommodate a crew of three for short periods.
25. For details, see Henry Lambright's "James Webb and the Uses of Administrative Power," in Jameson Doig and Erwin Hargrove (eds.), *Leadership and Innovation: A Biographical Perspective on Entrepreneurs in Government* (Baltimore: Johns Hopkins University Press, 1987).
26. Roger Launius, *NASA: A History of the US Civil Space Program* (Malabar, FL: Krieger Publishing Company, 1994), 189.
27. Ibid, 197.
28. Ibid.
29. "Public Knowledge and Attitudes Regarding Space Programs, September 1974," 25, Box 155, Folder 3, James Fletcher Papers MS 202, University of Utah J. Willard Marriott Library Manuscripts Division (hereafter Fletcher Papers).
30. Emphasis added. "Space Goals after the Lunar Landing," October 1966. Obtained originally from LBJ library by Dwayne Day. Record Number 14462 LEK 10/9/1, NHRC.
31. Ibid., 16–17
32. Ibid.; emphasis added.
33. Library of Congress Science Policy Research Division, *World Wide Space Activities: National Programs Other Than the US and Soviet Union; International Participation in the US Post-Apollo Program; International Cooperation in Space Science, Applications and Exploration; Organization; and Identification of Major Policy Issues*. Report prepared for the Subcommittee on Space Science and Applications of the Committee on Science and Technology, US House of Representatives, ninety-fifth Congress, first session by the Science Policy Research Division Congressional Research Service, LOC, September 1977 (Washington: US Government Printing Office, 1977), 429.
34. Erik Conway interview with Morris Tepper, April 26, 2003, Record no. 18945 LEK I/J/6, NHRC.

35. See also Erik Conway, *Atmospheric Science at NASA: A History* (Baltimore: Johns Hopkins University Press, 2008).

36. Most notable among these is the program Earthwatch, intended to "preserve and enhance the human environment," providing data on pollution, drought, earth-quakes, climate change, insect infestation, and major changes in the world's oceans. Ibid., 434.

37. Dr. Douglas Brooks to Morris Tepper regarding NASA's marine and atmospheric programs, May 17, 1972, Record no. 9758 LEK 4/9/2, NHRC.

38. "NASA Policy for Meteorological Programs," Record no. 3458 LEK 6/1/4, NHRC; emphasis added.

39. Bruce Murray and Merton Davies, "Détente in Space," *Science* 192, June 11, 1976, 1067–1074.

40. *SPACE Daily*, June 5, 1965.

41. "Soviets Prepare for New Metsat Series," *SPACE Daily*, June 10, 1968, 193.

42. "Soviet Meteor System to be Expanded and Improved," *SPACE Daily*, March 27, 1968, 150.

43. Ibid.

44. "Soviets Cite Weather Satellite Maritime Savings," *SPACE Daily*, February 26, 1975, 318.

45. Murray and Davies, "Détente."

46. "Meteorological and Remote Sensing Satellites," Record no. 15286 LEK 10/12/8 File "Meteor Satellites," NHRC.

47. Conway interview with Tepper, April 26, 2003.

48. Ibid.

49. E/Breene M. Kerr, "Memorandum for Mr. Webb—A: Dr. Newell's recommended National Program on Weather Modification" January 18, 1967, Record no. 9758 LEK 4/9/2 NHRC.

50. This may have been the case for the late 1960s, but by the turn of the decade, public knowledge-opinion surveys indicated that a considerable number of Americans iden-tified atmospheric studies, meteorology, and pollution monitoring at least in part with NASA. Administrator James Fletcher possessed at least two studies indicating such (additionally, the 1971 and 1974 studies make occasional comparisons to 1972 and 1973 studies, indicating at least a total of four similar surveys). Several questions were structured in such a manner that they asked respondents to prioritize among fields within and without NASA's policy objectives. For instance, "Education" and "Lowering Inflation" might be listed as priorities alongside "space exploration," "space technology," and "helping air/water pollution," and respondents would be asked to rank all such fields. Other questions explicitly asked for ranking *within* the space program. In 1974, when people were asked what initiatives they preferred among Earth Resource satellites, the Nimbus-G weather satellite, the first orbiting solar observatory, the Shuttle, Pioneer and Mariner, Viking, None of the Above, and No Opinion, 43 percent chose earth resources and 2 percent indicated the Nimbus weather satellites. That same year, when asked to select NASA's single most important mission, 31 percent selected "Studying the earth's resources and environment from space," 21 percent selected "Working on airplanes and air travel," 18 percent "Don't know," 12 percent "Establishing a permanent station in space for the manufacture of vaccines and other activities that can only be done in space," 7 percent "Searching for intelligent life," 5 percent "Exploring the planets," and 3 percent each to "Exploring the moon," "Establishing a permanent colony in space," and "Other." Ibid., 35.

51. The 1972 Agreement included what soon became standard fare for Soviet-American cooperation: sharing meteorological observations, exchanging data on life sciences, environmental observation, and plans to share data on planetary exploration.

52. George M. Low and James C. Fletcher to President Richard Millhouse Nixon, described in cover letter as "book on the objectives of our assignments as Administrator and

Deputy Administrator," January 31, 1973. Nixon had requested that the two prepare the book in November of 1972. Box 16, Folder 3, Fletcher Papers.

53. Ezell and Ezell, *The Partnership*, ix.

54. Ibid.

55. Olin Teague to James Fletcher, May 1, 1973. Letter hand delivered to Fletcher by Jack Swigert, executive director of the Committee on Science and Astronautics, US House, Box 22, Folder 4, Fletcher Papers.

56. Olin Teague to James Fletcher, October 15, 1973, Folder 4, Box 22, Fletcher Papers.

57. James Fletcher to Olin Teague, May 17, 1973, Folder 4, Box 22, Fletcher Papers.

58. Teague to Fletcher, October 15, Folder 4, Box 22, Fletcher Papers.

59. Swigert was removed from his position and replaced by Vance D. Brand.

60. James Fletcher to Olin Teague, November 19, 1973, Fletcher Papers and John Swigert to Captain Chester (Chet) Lee, November 14, 1973, Folder 4, Box 22, Fletcher Papers.

61. Ezell and Ezell, *The Partnership*, 222.

62. Ibid., 223.

63. Notes for Meeting Congressman Teague and Dr. Fletcher, January 25, 1975 Folder 4, Box 22, Fletcher Papers.

64. Ezell and Ezell, *The Partnership*, 223.

65. NASA Briefing for New Senate Authorization Committee Members, Briefing Plan, January 31, 1973, Folder 1, Box 15, Fletcher Papers.

66. Ezell and Ezell, *The Partnership*, 307. Soviet administration kept to its old ways when dealing with the Salyut 2 space station and Kosmos-557 satellite failures: the Soviet public never heard of either incident, while the American press expended a great deal of ink in speculation. Siddiqi, *Soviet Space Race*, 814.

67. Siddiqi, *Soviet Space Race*, 794.

68. Emphasis in the original; NASA Briefing for New Senate Authorization Committee Members, Briefing Plan January 31, 1973, Box 15, Section C-3 Folder 1, MS 202 Fletcher Papers.

69. Ibid., D-1, 2. In spite of my efforts to explain the context of ASTP in the history of human spaceflight, it is important to recall the importance of balance to Fletcher's philosophy. Explains Roger Launius: "The tangible response [to Fletcher] was the transformation of NASA into a much more diverse and practically oriented agency during Fletcher's first term with an emphasis on applications satellites to assist in making the planet a better place on which to live" (*NASA. A History*, 236).

70. Ibid., D-3 NASA Briefing for New Senate Authorization Committee Members, Briefing Plan January 31, 1973, Box 15, Section C-3 Folder 1, MS 202 Fletcher Papers.

71. James Fletcher to Roy Ash, July 13, 1973, Box 16, Folder 23, Fletcher Papers.

72. Ibid.

73. For examples of Fletcher using ASTP to link Apollo and Shuttle infrastructures, see James Fletcher, "The Space Flight After Apollo," speech delivered at the National Security Industrial Association, and "Salute to Apollo," Box 12, Folders 1 and 2, Fletcher Papers.

74. James Fletcher, "The Space Flight After Apollo" speech for Hawthorne Engineers' Week, Western Electric, February 22, 1973, Box 12, Folder 5, Fletcher Papers; emphasis in the original.

75. "Space Flight After Apollo," 6, Box 12, Folder 1, Fletcher Papers; emphasis in the original.

76. VonBencke, *The Politics of Space*, 84–86.

77. McDougall,... *the Heavens and the Earth*, 431–433.

78. Information available at: http://www.astronautix.com/craft/bion.htm; http://lis.arc.nasa.gov/lis/Programs/Cosmos/overview/Cosmos_Biosat.html.

79. NASA's Biosatellite III was the last such spacecraft launched, following the death of its single primate inhabitant. The primate's declining health led NASA officials to

truncate the mission and bring him back after only 9 days. The mission had initially budgeted time and resources for a 15–30 day orbit. Photo 69-H-1025 for release July 14, 1969. Record no. 18571 LEK 10/4/1 NHRC.

80. Cosmos 1514, the first to carry primates, required instrument training for the Soviets: http://lis.arc.nasa.gov/lis/Programs/Cosmos/Cosmos_1514/Cosmos_1514.html (accessed January 7, 2009).

81. Kristen Edwards, "The US-Soviet/Russian Cosmos Biosatellite Program," *Quest* 7:3(Fall 1999), 20–35; emphasis added. Preceding paragraph taken from 23, 24, and 33.

82. Rodney Ballard and Karen Walker, "Flying US Science on the USSR Cosmos Biosatellites," ASGSB Bulletin 6, October 1992.

83. Kenneth Souza, Guy Etheridge, and Paul Callahan, eds., *Life into Space: Space Life Science Experiments Ames Research Center Kennedy Space Center 1991–1998*, NASA/SP-2000-534, 200.

84. Ibid.

85. US Congress, Office of Technology Assessment, *US-Russian Cooperation in Space*, OTA-ISS-618 (Washington, DC: US Government Printing Office, April, 1995).

86. Odd Arne Westad, ed., *The Fall of Détente: Soviet-American Relations during the Carter Years* (Oslo: Scandinavian University Press, 1997), 328.

87. VonBencke, *The Politics of Space*, 88.

88. "Sino-US Cooperation in Science and Technology: a Political Overview", File: NASA 8/30/77–11/27/79, Box 6 Aid to Egypt 3/24/79 through US-China Science and Tech 5/77–8/79, Jimmy Carter Presidential Library, Atlanta, GA (Hereafter Carter Library).

89. "Space Summary: PRC Domestic Communications Satellite," File: NASA 8/30/77–11/27/79, Box 6 Aid to Egypt 3/24/79 through US-China Science and Tech 5/77–8/79, Carter Library.

90. Ibid. While ground stations such as this were available for $5–10 million, and export to the PRC would probably have been acceptable by Munitions Control, the Committee did speculate that the Coordinating Committee for Export Controls and Department of Commerce would require review and licensing.

91. "Sino-US Scientific and Technological Cooperation: a Political Overview."

92. Ibid.

93. In the preceding summer, Reagan had delivered a speech, calling for a "renewed US effort to revive or strengthen economic, cultural, consular as well as scientific contact...environmental protection, fishing, housing, health, agriculture, and in discussions of maritime problems and joint oceanographic research." US Congress, Office of Technology Assessment, *US-Soviet Cooperation in Space* (Washington, DC: US Government Printing Office, 1985), 71.

94. Ibid., 41.

95. OTA, *US-Soviet*, 44.

96. Ibid., 47.

97. Edwards, "US-Soviet/Russian *Cosmos*," 29.

98. Souza et al., *Life into Space*, 196.

99. NASA News December 14, 1992, "NASA Scientists Participate in Russian Space Mission." December 14, 1992, Record Number 15967, NHRC.

100. "Cosmos Program," Record Number 15967, NHRC.

101. http://lis.arc.nasa.gov/lis3/Hardware_Appendix/Bion11_Overview.html (accessed January 7, 2009).

102. Due to public relations complications regarding the death of a primate on Bion 11, Bion 12 never flew. Of additional interest, Edwards explains that due to the two-decade run of Bion satellites, "[h]undreds of US scientists and engineers were able to hone their skills while participating in the Cosmos missions." She goes on to suggest that "they have since applied these skills to other US space programs such as the Space Shuttle and International Space Station." Edwards, "US-Soviet/Russian *Cosmos*," 32.

103. US Congress, *US-Soviet*, 55.
104. "Washington News Initiative Special Report #2," 2.
105. Edwards, "US-Soviet/Russian *Cosmos*," 56.
106. US Congress, *US-Soviet*, 1995, 53.
107. "Washington News Initiative," 2.
108. Edwards, "US-Soviet/Russian *Cosmos*," 55.

8 Russian-American Cooperation in Space: Privatization, Remuneration, and Collective Security

1. "Washington News Initiative," 2.
2. Spot was the French equivalent to (and commercial competition for) the US Landsat program. On the symposium, see John McLucas, "The Opportunity in Soviet Space: 'Yes' to Increased Cooperation Between the US and USSR," *Washington Technology*, September 12, 1991, in appendix to "Washington News Initiative."
3. Edward Crawley and Jim Rymarcsuk, "US-Soviet Cooperation in Space: Benefits, Obstacles, and Opportunities," *Space Policy* (February 1992), 36.
4. Ibid.
5. J. Johnson-Freese, "Alice in Licenseland: US Satellite Export Controls Since 1990," *Space Policy* 16 (2000), 197. Freese indicates that the act itself only dates to 1976.
6. Later several new nations, including Spain, Canada, Australia, Denmark, Germany, Greece, Italy, Norway, Portugal, Japan, and Turkey, joined.
7. Crawley and Rymarcsuk, "US-Soviet Cooperation," 34–35.
8. James Asker and Breck Henderson, "Purchase of Russian Space Hardware Signals Shift in US Trade Policy," *Aviation Week and Space Technology* (hereafter *AWST*) 136 (April 6, 1992), 25.
9. In exchange for slightly lower 0.009 pound thrust, electric propulsion promised to halve spacecraft mass and save an approximate $60 million. Their design life of 3,500 hours, too, promised savings for the US Department of Defense. The plutonium and reactor amounted to $14 million (ibid.).
10. See chapter 2 for details. Assembled from information available in: Rodney Ballard and Karen Walker, "Flying US Science on the USSR Cosmos Biosatellites," *ASGSB Bulletin* 6 (October 1992), 121–128; Kenneth Souza, Guy Etheridge, and Paul Callahan, eds., *Life into Space: Space Life Science Experiments Ames Research Center Kennedy Space Center 1991–1998* NASA/SP-2000–534, available at http://articles. adsabs.harvard.edu//full/2000NASSP.534.....S/0000002.000.html; http://lis. arc.nasa.gov/lis/Programs/Cosmos/overview/Cosmos_Biosat.html; and http:// www.astronautix.com/details/cos21763.htm (accessed December 31, 2009).
11. One final set of examples regarding the esteem Americans held for Russian equipment include the series of equipment slated for use by the Department of Defense's space program. Central to these were the $8 million Topaz nuclear reactor and four Hall thrusters priced at $300,000. Leonard David, "The Rush to Buy Russian," *Aerospace America* (June 1992), 40.
12. Since the inception of Space Station Freedom, NASA and its partners had considered a number of options for such a life boat. This need was more pressing in the early years when SSF was supported by only one vehicle—the US Shuttle—and later, provided added incentive for moving the space station's orbit from the original 28.5 degrees to its current orbit of 51.6—an inclination at which Russian launchers could also reach the vessel. The Soyuz ACRV was at the time considered a temporary measure to be replaced later.
13. Thor Hogan, *Mars Wars: the Rise and Fall of the Space Exploration Initiative,* NASA SP-2007–4410 (Washington, DC: Government Printing Office, 2007).
14. James Asker, "US, Russian Space Pact Pledges Unprecedented Trade, Joint Flights," *AWST* 136 (June 22, 1992), 24.
15. Ibid.

16. "NASA's Goldin Foresees Cooperation, Not Sales After Visiting Russian Space Facilities," *AWST* 137 (July 27, 1992).

17. These figures were reported in a 2003 article, indicating that there were likely many more researchers employed in Soviet Russia at the time of initial space station deliberations. Andrew Lee, "Technology in Russia: Russian Evolution," *The Engineer* (November 10, 2003), 7–20.

18. Record Group 255, Records of NASA administrator Daniel S. Goldin, Box 44, Folder Russian Cooperation, Folder 073471, NARA (hereafter Goldin Papers).

19. In 1994, approximately 38 scientific production facilities, design bureaus, factories, and experimental design bureaus were transferred to the management responsibility of the RSA. This restructuring of the Russian aerospace sector came about through Russian Governmental Decree No. 866 signed on July 25, 1994: Box 46, Folder Meeting with Yuri Koptev, RSA, at NASA HQ, September 27, 1994, Goldin Papers. By 1999, 350 aviation companies fell under Rasaviakosmos. "Russian Space Agency Gets Tightest Budget Ever in FY 99," *Aerospace Daily*, March 12, 1999.

20. John Logsdon and James Millar, eds., "US-Russian Cooperation in Human Space Flight: Assessing the Impacts," Space Policy Institute and Institute for European, Russian, and Eurasian Studies, Elliott School of International Affairs The George Washington University, February 2001, available at www.gwu.edu/~spi/assets/docs/usrussia.pdf (accessed April 2, 2010).

21. Tom Cremins and Elizabeth Newton, "Changing Structure of the Soviet Space Programme," *Space Policy* 7 (May 1991), 132.

22. "Potential Russian/Ukrainian Entry Into US Launch Markets," report presented to Dan Goldin, July 2, 1992, Record Group 255, Record no. 073471, Folder Russian Cooperation, Goldin Papers.

23. David, "The Rush," 39.

24. Peter Mason, "Missile Factory Serves up New Fare—Buses: Soviet Arms Plant Now Hoping to Make Nice Things for Western Customers," *Washington Post*, October 2, 1991, A26.

25. For better or for worse, Yuzhny still boasts a highly diversified line of products including spacecraft, rocket complexes, satellite systems, tractors, trolleybuses, windmills, agricultural equipment, and meat-processing equipment. http://www.nkau.gov.ua/nsau/catalogNEW.nsf/ByNamesE/62D29224C09779D3C3256BF8004BF966?OpenDocument&Lang=E (accessed December 31, 2009).

26. "Washington News Initiative Special Report," p. 2.

27. Ibid.

28. David Hamilton, "Piecemeal Rescue for Soviet Science," *Science*, New Series 255 (March 27, 1992), 1632–1634.

29. David, "The Rush," 40.

30. "2/11/97 Phone Calls to ISS Heads of Agencies," Box 55, Folder 074410, Goldin Papers.

31. David, "The Rush," 40.

32. "4/4/97 ViTS with Yuri Koptev re: ISS," Box 55, Folder 074432, Goldin Papers.

33. James Asker, "Gore/Quayle Face-off Foreseen as Clinton Offers Space Plan," *AWST* 137 (July 27, 1992), 22.

34. Vincent Kiernan, "Mir Data to Assist in Designing Extended Orbit Spacecraft," *Space News*, December 9, 1991, I5.

35. Debra Rahn, NASA News Press Release, "US and USSR Expand Space Cooperation," July 31, 1991, Record Number 015583, NHRC.

36. Kathy Sawyer, "US-Soviet Space Swap Revived as Summit Nears: Officials See Mutual Benefits for Exchanges of Crews to Operate Aboard Shuttle, Mir," *Washington Post*, June 30, 1991, A9.

37. Ibid.

38. Debra Rahn, NASA News Release, "NASA and Russian Space Agency Sign Space Agreements," October 6, 1992. The offer took place at the June Washington, DC

Summit Meeting between Bush and Yeltsin. See Andrew Lawler, "Rockwell, NPO Energia to Build Docking Device for Shuttle, Mir," *Space News*, September 14–20. For more details, Record no. 015584, NHRC.

39. Information compiled from: Judy Rumerman, *NASA Historical Data Book Volume VII: NASA Launch Systems, Space Transportation, Human Spaceflight, and Space Science 1989–1998* (Washington, DC: NASA History Division Office of External Relations, 2009), NASA SP-2009–4012.

40. This docking mechanism was based on concepts used in ASTP. Ibid., 264.

41. The following is taken primarily from studies found in Record no. 19801, Folder "US/Russian Human Spaceflight Cooperation Study, 1993," NHRC.

42. "Background Briefing by Administration Official," June 17, 1993, White House Office of the Press Secretary. http://clinton6.nara.gov/1993/06/1993–06–17-background-briefing (accessed February 18, 2010).

43. "Updated Plan for Russian-American Cooperative Programs in Earth Science and Environmental Monitoring from Space," Prepared for the Gore-Chernomyrdin Commission by NASA, NOAA, and the RSA, ROSGIDROMET, June 13, 1994, Record no. 18529, Folder Gore-Chernomyrdin Comm. Mtg. 1994 Draft Briefing Book, NHRC.

44. Tom Murray, "Draft Terms of Reference for the Russian ETF," April 15, 1994, Record no. 18529, NHRC.

45. In the 1990s programs such as the Mission to Planet Earth and its Small Explorer Program, hallmarks of NASA administrator Daniel Goldin's policy mantra of faster, better, cheaper carried atmospheric instruments into orbit a few small packages at a time (as opposed to original plans to send up larger earth-orbiting platforms). See Lisa Shaffer, "International Coordination in the Era of Faster, Better, Cheaper," *Space Policy* 14 (1998), 89–94; W. Jones and Nickolus Rasch, "NASA's Small Explorer Program," *Acta Astronautica* 22 (1990), 269–275, in addition to the scholarship of Henry Lambright.

46. Raymond Roberts, Yuri Milov, Yuri Zonov, Rashid Salikhov, and Leslie Charles, "IAN-USA SAGE III/METEOR-3-M Project," *Acta Astronautica* 38, 479–485.

47. NASA Scientific and Technical Information Branch, *Meteor-3 Total Ozone Mapping Spectrometer (TOMS) Data Products User Guide*, Washington, DC: NASA Reference Publication, 1996, 1.

48. Ibid., 29. This decision was explored and made at the 1992 and 1994 Gore-Chernomyrdin talks, which will be elaborated on in the subsequent chapter.

49. Ibid.

50. "Summary of Results of December 15–16 Gore-Chernomyrdin Commission Meeting, 1993," Record no. 18529, Folder Gore-Chernomyrdin Comm. Mtg. 1994, Draft Briefing Book, NHRC.

51 US Congress, Office of Technology Assessment, *US-Russian Cooperation in Space* OTA-ISS-618 (Washington DC: US Government Printing Office, April, 1995), 56. These figures include the initial $400 million agreement for Shuttle-Mir and ISS cooperation, plus cooperation in other fields and increases to the initial contract detailed below. See Table 9.3 "What the Russians Have Added."

52. US Congress, Office of Technology Assessment, *US-Soviet Cooperation in Space* (Washington, DC: US Government Printing Office, 1985), 53.

53. "Transcript: Gore, Chernomyrdin Remarks at GCC Opening March 10, 1998," Record no. 15584, Folder US-CIS Space Cooperation, NHRC.

54. Ibid.

55. "The Station Concept Overview 13 June, 1994," Box 44, Folder 073851, Goldin Papers.

56. Remarks read by Viktor Chernomyrdin. "Gore, Chernomyrdin Remarks at GCC opening March 10," Tenth Session of US-Russian Commission on Economic and Technological Cooperation (1998), Record no. 015584, NHRC.

57. Judy Rumerman, *NASA Historical Data Book Volume VII: NASA Launch Systems, Space Transportation, Human Spaceflight, and Space Science 1989–1998* (Washington,

DC: NASA History Division Office of External Relations, 2009), NASA SP-2009–4012, 294.

58. Ibid., 294–295.
59. Randy Brinkley, Program Manager Space Station Program, "Review of the International Space Station," March 25, 1994, Record no. 18328, Folder Vest Committee Review 2 Letters 3/25 and 26 1994, NHRC.
60. Ibid.
61. Rumerman, *NASA Historical Data Book*, 296.
62. See Howard McCurdy, *The Space Station Decision: Incremental Politics and Technological Choice* (Baltimore, MD: Johns Hopkins University Press, 2007), for details on the complicated acts of coalition-building that were necessary to keep the space station both funded and supported.
63. Daniel Goldin, "Boeing Plant Expose Club 2/25/94" Record no. 32503 Goldin Speeches, NHRC.
64. Brinkley, "Review of the ISS."
65. Ibid.
66. Transcript: US News and World Report Interview with Dan Goldin, NASA Administration, June 17, 1994, Record no. 073852, Folder Interview—US News and World Report, Goldin Papers.67. Background information for your meeting on April 14, Folder 4/14/97 Meeting with Vice President Gore and Sensenbrenner, Record Number 074434, Goldin Papers. See also Folder 13 April 1994 Hearing on Space Station before House Committee on Space-Committee on Science, Space, and Technology, Record Number 073776, Goldin Papers.
67. V. Khorunov/NPO Energia Head of Power, J. Dunning/LeRC, D. McKissock/HQ Code DE, M. Gross/BAH "US/Russian Electrical Power System," Record no. 19801 US/Russian Human Spaceflight Cooperation Study, NHRC.
68. "Background Information for Your Meeting on April 14, "Record no. 074434, Folder 4/14/97 Meeting with Vice President Gore and Sensenbrenner, Goldin Papers.
69. "Hearings Before the Subcommittee on Space of the Committee on Science, Space, and Technology, US House October 6, 1993," Folder US-Russian Cooperation in the Space Station Program, NHRC.
70. Congress, House, Subcommittee on Space of the Committee on Science, Space, and Technology, *United States-Russian Cooperation in the Space Station Program*, 103rd Congress, October 6 and 14, 1993, 2–3.
71. Ibid.
72. Ibid., 24.
73. Ibid., 25.
74. "Draft 6/17/94 US-Russian Joint Commission on Economic and Technological Cooperation: Joint Statement on Space Station Cooperation Draft Briefing Book," Record no. 18529, Folder Gore-Chernomyrdin Commission Meeting 1994, NHRC.
75. "FY 1995 Congressional Budget ISSA Russian Cooperation Contract Content," Record no. 073776, Folder April 13, 1994 Hearing on Space Station before House Committee on Space-Committee on Science, Space, and Technology, Goldin Papers.
76. These sources of Russian-American conflict are addressed throughout Agendas, Memos, Notes, and Briefings, Record no. 73851, Box 43, Goldin Papers. "See also Status of Bringing Russia into the Space Station Program, Heads of Agencies," Record no. 073768 4/5/94 in particular, binder used at the April 5 ISS Status Summary Briefing for Heads of Agencies.
77. Ibid.
78. Ibid.
79. Ibid.
80. "The Station Concept Overview 13 June, 1994," Box 44, Folder 073851, Goldin Papers.
81. Meeting with RSA June 16, 1994, Box 44, Folder RN 73851, Goldin Papers.
82. Ibid.

83. Ibid.
84. "Background Information for Your Meeting on April 14," Record no. 074434, Folder 4/14/97 Meeting with Vice President Gore and Sensenbrenner, Goldin Papers.
85. "NRL Revamping Control Module for NASA Space Station," NRL press release 7–97r, June 1997, available at http://www.globalsecurity.org/space/library/news/1997/7–97r.htm and "NASA Still Counting on Russia to Launch Service Module in July," *Aerospace Daily*, March 28, 2000, clipping in Record no. 17080, Folder ISS-Interim Control Module NHRC.
86. "The Station Concept Overview," June 13, 1994, Box 44, Folder 073851, Goldin Papers.
87. Matthew von Bencke, *The Politics of Space: A History of US-Soviet/Russian Competition and Cooperation in Space* (Westview Press: Boulder, CO, 1997), 188.
88. C. A. Robbins and B. Rosewicz, "Reaching out: US Hopes to Move Moscow into the West through Deeper Ties," *The Wall Street Journal*, December 13, 1993, A1.
89. Ibid.
90. Cremins and Newton, "Changing Structure of the Soviet Space Programme," 132.
91. Emphasis added; "Your Meeting with the Ukrainian Deputy Prime Minister and the Director General of the National Space Agency of Ukraine at 9 AM on Wednesday May 11," 1994 Record no. 073822, Folder Mtg with Ukrainian Deputy PM and Director Genl of National Space Agency of Uk (NSAU) May 11, Goldin Papers.
92. "Your Meeting with Ukrainian Deputy Prime Minister, Valeriy Shmarov—Tuesday, January 4, 1994," Record no. 073696, Folder 1/4/94 Ukrainian Deputy Prime Minister-Shmarov, Goldin Papers.
93. Ibid.; emphasis in the original.
94. Ibid.
95. "Joint Presentation to Yeltsin in Seattle Draft Script," Record no. 073920, Folder Meeting With Yuri Koptev, RSA at NASA HQ, Wash DC 27 Sept, 1994, Goldin Papers.
96. Joseph Anselmo, "Industry Impacts US Space Policy," *AWST* 149, July 6, 1998, 34.
97. Ibid.
98. Roger E. Bilstein, *The American Aerospace Industry: From Workshop to Global Enterprise* (Prentice Hall: London, 1996).
99. Ibid., 196.
100. Ibid., 200.
101. Von Bencke, *Politics of Space*, 162.
102. Liudmila Bzhilianskaya, "Russian Launch Vehicles on the World Market: A Case Study of International Joint Ventures," *Space Policy* 13 (November 1997), 325–326.
103. As John Logsdon put it, "Russia's commercial partnerships with US aerospace companies play a pivotal role in complementing the ISS engagement. If the ISS project provides Russia an opportunity for highly visible international space cooperation and limited financial support, the real flow of hard currency comes from a variety of commercial contracts. They not only keep the space industry afloat but also help fulfill Russia's ISS obligations." John M. Logsdon and James Millar, "US-Russian Cooperation in Human Spaceflight: Assessing the Impacts," *Space Policy* 17 (2001), 171–178.
104. Von Bencke, *Politics of Space*, 255.
105. David E. Hoffman, *The Dead Hand* (New York: Anchor Books, 2009), 403–404.
106. Sharon Squssoni and Marcia S. Smith, "CRS Report for Congress: The Iran Nonproliferation Act and the International Space Station: Issues and Options," Order Code RS22072 Updated August 22, 2005.

9   An Overview of NASA-Japan Relations from Pencil Rockets to the International Space Station

1. For an overview of joint experiments on NASA missions between 1958 and 1984, see *26 Years of NASA's International Programs* (Washington, DC: NASA, undated).

2. For example, John J. Hudiburg and Michael W. Chinworth state that the annual sign-ing rates for NASA-Japanese technology agreements grew from approximately 2 agree-ments per year in 1966 to 8 agreements per year in 1989. Furthermore, this growth trend has continued with 153 agreements signed between 1994 and in 2003. See John J. Hudiburg and Michael W. Chinworth, "Strategic Options for International Participation in Space Exploration: Lessons from U.S. Japan Defense Cooperation," *American Institute of Aeronautics and Astronautics*, Collection of Technical papers, 1st Space Exploration Conference, Vol. 1, 2005.

3. Numerous organizations, institutions, commissions, and bureaus were involved in space activities during the 1960s. For a detailed account of the participation by various bod-ies, see the Japanese government publication *Space in Japan 1969–70* (Tokyo: Science and Technology Agency, 1970); *Space in Japan 1964* (Tokyo: Science and Technology Agency, 1964); *Space in Japan 1968* (Tokyo: Science and Technology Agency, 1968); *Space in Japan 1966–67* (Tokyo: Science and Technology Agency, 1964).

4. Paul F. Langer, *The Japanese Space Program: Political and Social Implications* (Santa Monica, CA: Rand Corp., 1965), 1.

5. For more details about this team, see Joan Johnson Freese, *Over the Pacific: Japanese Space Policy into the Twenty-First Century* (Dubuque, Iowa, Kendall/Hunt Pub), 58.

6. The word "thesis" was particularly mentioned in the US State Department papers. The ability to orbit useful payloads depends not only on size but on guidance and control systems, which the Tokyo University program had not developed. The State Department papers indicate that Itokawa spoke vaguely of a thrust vector control system for the Mu rocket, but this problem was clearly the major shortcoming of this program.

7. This policy was indirectly endorsed by the State Department's U. Alexis Johnson, who suggested in 1967 that "[i]f liquid fuels are required Japan should buy the technol-ogy from the U.S. rather than wasting resources on development in Japan. Mu class rockets," Johnson added, "can handle any scientific utility satellite that Japan needs so why invest a great deal in the National Space Laboratory Program." Confidential memorandum, U. Alexis Johnson to the United States State Department, June 1, 1967, "Space Activities in Japan," RG59, folder SP – Space and Astronautics, Central Foreign Policy Files, 1967–1969, NARA.

8. John K. Emmerson and Edwin O. Reischauer, *Arms, Yen and Power: The Japanese Dilemma* (New York: Dunellen, 1971), 321.

9. Ashi Shinbun, March 21, 1967, in Yasushi Sato, "A Contested Gift of Power: American Assistance to Japan's Space Launch Vehicle Technology, 1965–1975," *Historia Scientarum* 11:2 (November 2001), 180. It should be noted that ISS later came to play a major role in international collaboration. It was reorganized in 1981 into the Institute of Space and Astronautical Science (ISAS). Since the early 1980s ISAS has been involved in many collaborative international projects involving the exchange of data, scientists, and occasionally instruments on spacecraft. Some of the successful cooperative programs include the 1986 encounter with Comet Halley (John M. Logsdon, "Missing Halley's Comet: The Politics of Big Science," *Isis* 80:2 (June 1989), 254–280, and the Japanese Solar-A mission launched in August 1991). A US-supplied soft X-ray telescope was one of its two major instruments. ISAS was also part of the International Solar Terrestrial Physics (ISTP) program.

10. John M. Logsdon, "U.S.-Japanese Space Relations at a Crossroads," *Science* 255 (January 17, 1992), 294–300, at 294, 297.

11. *Space in Japan*, 24.

12. Takemi Chiku, "Japanese Space Policy in the Changing World," MS thesis, Massachusetts Institute of Technology, 1992.

13. There is very little information regarding the historical and technological aspects of the Tokyo Olympics jointly conducted by NASA and the Ministry of Posts and Telegraphs.

14. Confidential memo, American Embassy to Department of State, November 9 1966, "Japanese Space Program—Itokawa's Version," RG59, Central Foreign Policy Files, Box 3141, NARA.

15. John Krige et.al., Interview with Arnold Frutkin, NHRC, 5.

16. Ibid.

17. Memorandum Richard J. H. Barnes to Robert F. Packard, May 7, 1962, "Scientific Research and Development—International Cooperation in Outer Space," RG59, Box 259, Records Relating to Atomic Energy Matters, 1944–63, NARA.

18. United States Information Agency "Foreign Media reaction to Communist China's Nuclear Device," October 18, 1964, folder Nuclear Testing China, vol. 1, National Security Files, Box 31, LBJL.

19. Memo, U. Alexis Johnson to Department of State, "Japan's Space Program –the views of Nakasone," December 15, 1966, RG59, Box 3151, Central Foreign Policy Files, 1964–1966, folder SP—Space and Astronautics, NARA.

20. Major conclusions with respect to Implication of a Chinese Communist Nuclear Detonation and nuclear capability, folder Nuclear Testing China, Vol. 1, National Security Files, Box 31, LBJL.

21. Robert T. Weber (science attaché, American Embassy, Tokyo) to Department of State, April 30, 1966, RG59, Box 3141, Central foreign policy files, 1964–1966, folder SP—Space and Astronautics, NARA.

22. Memorandum J. Owen Zunhellen, Jr (Consular, American Embassy, Tokyo) to The Department of State, November 27, 1964, "Japanese Request for Assistance in Satellite Project," General Records of the Department of State, Record Group 59, Box 3141, Central foreign policy files, 1964–1966, folder SP—Space and Astronautics, NARA. The US ambassador to Tokyo from 1961 to 1966, Edwin Reischauer, was fine-tuned to Japanese cultural sensitivities. His best-known book *The Japanese* became a bible for understanding Japan during the early 1960s and was coveted by East Asian policymakers; see Edwin O. Reischauer, *The Japanese* (Cambridge, MA: Belknap Press, 1977).

23. Memorandum from James Webb to Assistant to the President William Moyers, September 17, 1965, in John M. Logsdon, *Learning from the Leader: The Early Years of U.S. Japanese Cooperation in Space* (undated, unpublished paper), Space Policy Institute, George Washington University, Washington, DC, 4.

24. We cannot be certain that Humphrey actually used these words. This text is from a Background Paper, Vice President's Visit to Japan, "Cooperative Effort in Outer Space Exploration," December 1965, RG 59, Box 3141, General Records of the Department of State, Central Foreign Policy Files, 1964–1966, Folder SP—Space and Astronautics, NARA.

25. Confidential memorandum, March 11, 1964, "Comments by Profs. Itokawa and Saito on Japanese SpaceActivities," RG59, Box 3141, Central foreign Policy Files, 1964–1966, folder SP—Space and Astronautics, NARA.

26. Confidential memorandum of conversation, May 4, 1966, "Problems in the Japanese Space Program," RG59, Box 3141, Central foreign Policy Files, 1964–1966, folder SP—Space and Astronautics, NARA.

27. Confidential memorandum, Department of State, "Japanese Space Activities" April 7, 1966, Record Group 59, NARA.

28. Carl Tolman (Science Attaché, American Embassy, Tokyo) to Secretary of State, March 9, 1964, Memorandum of Conversation—"Japanese Space Program," RG 59, Box 3141, NARA.

29. At the beginning of 2001, the merger of Science and Technology Agency (STA) with Ministry of Education, Culture and Sports (MoE) created a very large ministry (Ministry of Education, Culture, Sports, Science and Technology or MEXT) in charge of two principle space agencies, the National Space Development Agency (NASDA) and the Institute of Space and Aeronautical Science (ISAS). The two were merged to create a single agency for aerospace JAXA. For more on the merging and emerging policy aspects, see Kazuto Suzuki, "Administrative Reforms and the Policy Logics of Japanese Space Policy," *Space Policy* 21:1 (2005),11–19; "A Brand New Space Policy or Just Papering over a Political Glitch? Japan's New Space Law in the Making," *Space Policy* 24:4 (2008), 171–174; "Transforming Japan's Space Policy-Making," *Space*

Policy 23:2 (2007), 73–80. For more information on JAXA organization, see http://www.jaxa.jp/about/org/pdf/org_e.pdf (accessed September 22, 2009).

30. Michael Schaller, *Altered States: The United States and Japan since the Occupation* (New York: Oxford University Press, 1997).

31. Nishida's remarks made as chairman of SAC were in the SAC monthly report for May 1970. Translation by Kaori Sasaki, International Space University/NASDA, in Logsdon, *Learning from the Leader*, 22.

32. Memorandum from assistant administrator for international affairs to administrator "Japanese post-Apollo visit," July 1, 1971, in Logsdon, *Learning from the Leader*, 23.

33. Logsdon, *Learning from the Leader*, 24.

34. Space Activities Commission, Special Committee on Post-Apollo Programs, *Final Report*, May 27, 1974, translation by Kaori Sasaki, International Space University/NASDA, in Logsdon, *Learning from the Leader*.

35. What follows summarizes Logsdon, *Learning from the Leader*, 40–43.

36. Letter from Acting Secretary of State Robert Ingersoll to NASA administrator James Fletcher, December 18, 1975, in Logsdon, *Learning from the Leader*, 44

37. Memorandum for the deputy administrator from assistant administrator for international affairs, "Report of the Panel on Review of US-Japan Science and Technology cooperation," January 12, 1976, Logsdon, *Learning from the Leader*, 41.

38. Ibid., 42.

39. Memorandum for administrator and deputy administrator from director of international affairs, "Possible Visit to Japan in July," April 13, 1978, in Logsdon, *Learning from the Leader*.

40. Logsdon, *Learning from the Leader*, 43.

41. For more details about the technical aspects of the Japanese contribution to the Space Station, see K. Ida, "The Japanese Experiment Module for Space Station Freedom," *Acta Astronautica* 28 (1992), 43–47. For policy aspects, see Shigeo Kobayashi, "Overview of Japanese Policy on Space Station," *Acta Astronautica* 14 (1986), 11–18. Christian Brünner and Alexander Soucek, "Regulating Iss—an Interdisciplinary Essay," *Acta Astronautica* 60: 4–7, 594–598; Lynn F. H. Cline and Graham Gibbs, "Re-Negotiation of the International Space Station Agreements 1993–1997," *Acta Astronautica* 53:11 (2003), 917–925; Lawrence J. DeLucas, "International Space Station," *Acta Astronautica* 38: 4–8, 613–619; Yoshiya Fukuda, Youko Tanaka, Yoshiaki Iwata, Tomohiro Kusunose, Yuta Kitagawa, and Kazumi Koide, "Trial Production of Iss/Jem Glossary of Terms," *Acta Astronautica* 50:2 (2002), 131–134; Shigeo Kobayashi, "Overview of Japanese Policy on Space Station," *Acta Astronautica* 14 (1986), 11–18; John M. Logsdon, "International Cooperation in the Space Station Programme: Assessing the Experience to Date," *Space Policy* 7:11 (1991), 35–45; "International Involvement in the US Space Station Program," *Space Policy* 1:1 (1985), 12–25; Y. Morishita, N. Saito, and M. Saito, "Jem Present Project Status," *Acta Astronautica* 15:99 (1987), 615–620; M. Saito, K. Higuchi, and K. Shiraki, "Japanese Experiment Module (Jem) Preliminary Design Status," *Acta Astronautica* 16 (1987), 47–53.

42. JEM is a Japanese element that provides laboratory facilities for Japanese material processing and life science research. It also contains an external platform airlock, and robotic manipulator for in-space ("exposed") experiments and a separate logistics module to transport JEM experiments. JEM has a cluster of systems: JEM pressurized module (JEM-PM) is a laboratory for experimental research in areas such as space medicine, life sciences, material processing, and biotechnologies. It contains an airlock to transfer experiments. JEM Exposed Facility (JEM-EF) is an unpressurized pallet structure exposed to the environments of space to support user payloads for the purpose of experimental research in areas such as communications, space science, engineering, materials processing, and earth observation. The pressurized section (ELM-PS) and Exposed Section (ELM-ES) serve as a pressurized and exposed passive storage, respectively.

43. Carl E. Behrens, *The International Space Station and the Space Shuttle* (Washington, DC: Congressional Research Service Report for Congress 7–5700, March 18, 2009).

44. John M. Logsdon, *Together in Orbit: The Origins of International Participation in the Space Station*, NASA History Division, Office of Policy and Plans, NASA Headquarters, Washington, DC 20546, Monographs in Aerospace History #11 November 1998, 3. Perhaps because of its concerns with a rather unstable domestic space program, SAC did not focus on post-Apollo participation. A memo from Tokyo reported that "[w]hen queried recently by embassy officers about how their long range launcher plans would be impacted by an operation space shuttle about 1980, we were informed that this is what the post Apollo subcommittee of SAC was studying. Although we are convinced the Japanese are not yet up to such sophisticated participation," the memo contiued, "we should continue to count them in on our international post-Apollo affairs, because of our overall interest in promoting their program for peaceful applications in space." See Memo to Department of State from American Embassy Tokyo, Japanese Space Program: State of the Art. February 9, 1973, Signed by Ingersoll, Logsdon, *Learning from the Leader*, 63.

45. Brian Harvey, *The Japanese and Indian Space Programs: Two Roads into Space* (London: Springer 2000), 89.

46. "It is very certain that the ISS would bring about few meaningful results if they stick to the traditional scientific and experimental projects. It is recommended that Japan should stop its activities on th ISS project immediately. A quick and silent retreat is the best choice for Japan to avoid wasting money which could be better used on other priorities." Minoru Suzuki, "Alternative International Cooperation in Space Development for Japan—Need for more Cost-Effective Space Applications Projects," *Acta Astronautica* 58 (2006), 430–437.

47. Logsdon, *Learning from the Leader*, 64.

10 NASA and the Politics of Delta Launch Vehicle Technology Transfer to Japan

1. http://www.jaxa.jp/library/space_law/chapter_1/1–2–2–8_e.html; see also http://untreaty.un.org/unts/1_60000/21/5/00040220.pdf.

2. For a nuts and bolts history of the development of Japanese H series rocket, see T. Godai, "H-II: A New Launch Vehicle in the 1990's," *Acta Astronautica* 14 (1986), 143–157; I. Hiraki and Y. Takenaka, "Development of Launch Vehicles for Application Purposes," *Acta Astronautica* 7:8–9, 967–977; A. Konno, M. Endo, Y. Koyari, and Y. Yamada, "Development Status of H-II Rocket Cryogenic Propulsion Systems," *Acta Astronautica* 28 (1992), 127–134; K. Noda and M. Endo, "H-IIA rocket program," *Acta Astronautica* 45:10 (1999), 639–645; R. Sekita, A. Watanabe, K. Hirata, and T. Imoto, "Lessons Learned from H-2 Failure and Enhancement of H-2A Project," *Acta Astronautica* 48:5–12, 431–438; K. Tomioka and Y. Kohsetsu, "H-II Launch Vehicle Development Status in Terms of Vibration, Shock and Acoustic," *Acta Astronautica* 22 (1990), 43–48.

3. Confidential memorandum Department of State to American Embassy, Tokyo, 12/12/66, "Proliferation of Solid Fuel Technology," folder SP-13, Japan, RG 59, box 3141, NARA II.

4. Arms Control and Disarmament Agency, "Space Cooperation with Japan: Arms Control Considerations," September 6, 1966, in John M. Logsdon, *Learning from the Leader: The Early Years of U.S. Japanese Cooperation in Space* (undated, unpublished paper), Space Policy Institute, George Washington University, Washington, DC, 7.

5. Memorandum, Department of State to American Embassy, "Space Cooperation with Japan," December 1967, box 3141, NARA II.

6. Department of State, visit of Prime Minister Eisaku Sato of Japan, November 14–15, 1967, Scientific and Technological Cooperation, undated, U. Alexis Johnson Papers, LBJ Library, cited in John Logsdon, *Learning from the Leader*, 11.

7. John M. Logsdon, ed., *Exploring the Unknown: Selected Documents in the History of the U.S. Civil Space Program*, Vol. 2, *External Relationships* (Washington, DC: NASA SP-4407, 1996), 46.

8. Confidential memo, William P. Bundy to the secretary of state (undated), "U.S./ Japan Space Agreement," RG 59, Box 3006, Central Foreign Policy Files, 1967–69, NARA.
9. U. Alexis Johnson with Jef Olivarius McAllister, *The Right Hand Of Power* (Englewood Cliffs, NJ: Prentice-Hall, 1984), 487.
10. Confidential memo, William P. Bundy to the secretary of state (undated).
11. Confidential memorandum, U. Alexis Johnson to Russel E. Train, August 19, 1969, "U.S. Space Cooperation Agreement with Japan," Box 3006, folder SP- Japan-US, Central Foreign Policy Files 1967–1969, NARA.
12. Johnson, *Right Hand Of Power,* 443–444.
13. Based on interviews conducted by Rebecca Wright and of John Krige, Ashok Maharaj, and Angelina Long, with Arnold W. Frutkin, NHRC. On a side note: some of Frutkin's associates believed that his experience as a navy Office in the Pacific theater during World War II made him less than eager to work closely with Japan, Logsdon, *Learning from the Leader,* 41.
14. Logsdon, *Exploring the Unknown,* Vol. 2, 46.
15. Frutkin interview John Krige, Ashok Maharaj and Angelina Long, op cit.
16. U. Alexis Johnson to Robert C. Seamans, Jr (Secretary of the Air-Force, Pentagon), April 19, 1969, RG 59, Box 3006, Central Foreign Policy Files, 1967–69, NARA.
17. Logsdon's interview with U. Alexis Johnson, 1966 in Logsdon, *Learning from the Leader,* 8.
18. Johnson, *Right Hand of Power,* 87
19. Confidential memorandum, U. Alexis Johnson to David Packard (Deputy Secretary of Defense), June 30, 1969, RG 59, Box 3006, NARA.
20. Oral History Interview, U. Alexis Johnson with Robert McKinzie, Washington, DC, June 19, 1975, Harry Truman Presidential Library, Independence, MO.
21. "List of Pending Cases for the Japanese Space Program," June 30, 1969, RG 59, Box 3006, NARA.
22. http://www.jaxa.jp/library/space_law/chapter_1/1-2-2-8_e.html; see also http:// untreaty.un.org/unts/1_60000/21/5/00040220.pdf.
23. "List of Pending Cases for the Japanese Space Program," June 30, 1969, RG 59, Box 3006, NARA.
24. The OMC administered the International Traffic in Arms Regulations, ITAR. The technologies to be controlled under ITAR were contained in a Munitions Control List—almost, if not all, launch related technologies were on that list. In making its determination of whether or not to issue an export license for a particular technical transaction, the Munitions Control office relied on various Technical Advisory Groups (TAG) with representatives from interested government agencies.
25. Maurice Mountain (director, Strategic Trade and Disclosure, International security affairs, Department of Defense) to John W. Sipes (director, Office of Munitions Control, Department of State), July 9, 1970, in Logsdon, *Learning from the Leader,* 32.
26. Vincent L. Johnson to Maurice J. Mountain, Record no. 14690, folder US-Japan Thor-Delta, NHRC.
27. Vincent Johnson to John W. Sipes, October 30, 1970, RG 59, Box 2962, NARA.
28. Memorandum, Donald Morris to deputy administrator, March 27, 1974, "US Assistance to the Japanese N vehicle," NHRC.
29. From Donald Morris (deputy assistant administrator for International Affairs, NASA) to George M. Low (deputy administrator, NASA), March 27, 1974, "US Assistance to the Japanese "N" Vehicle" Record no. 14690, LEK7/10/5, folder US-Japan Thor-Delta, NHRC.
30. Department of State to secretary of state, September 1974, "Space Cooperation—Visit of Dr. James S. Fletcher," Logsdon, *Learning from the Leader,* 31.
31. Memo Office of International Affairs, NASA to Department of State, November 1, 1974, "Thiokol Castor II Licensed Production in Japan," in Logsdon, *Learning from the Leader,* 31.

32. Memorandum, George M. Low to Medium Launch Vehicle Program manager, April 8, 1974, "More on the Japanese Thor Delta," NHRC.

33. Memorandum, Arnold Frutkin to George M. Lowe, June 18, 1974, "Termination of Advisory Group for Japanese Thor-Delta," George M. Low papers, Rensselaer Polytechnic Institute, Troy, New York.

34. Arnold W. Frutkin to George M. Low, April 25, 1974, "Japanese Launch Vehicle," in Sato, "Contested Gift of Power," 201.

35. Memo from AD/deputy administrator, George M. Low, to I/assistant administrator for international affairs, April 1, 1974, Record no. 14690, "U.S.-Japan Thor-Delta, dates covered 1970 – 1977," NHRC.

36. Memo, deputy administrator to medium launch vehicle program manager, "More on the Japanese Thor Delta," April 8, 1974, Record no. 14690, NHRC.

37. Department of State Cable 1250, September 1974, and Memorandum from Associate administrator for Tracking and Data acquisition to Associate Administrator, NASA support to Japanese synchronous satellite launcher, September 27, 1974.

38. Secret memo, from assistant secretary, Marshall Green, to Ronald I. Spiers, "U.S.-Japan space Cooperation Agreement," November 16, 1971, RG59, Box 2692, folder SP-Japan, Subject Numeric files 1970–1973, NARA.

39. Secret memo, from assistant secretary, Marshall Green, to Ronald I. Spiers, "U.S.-Japan space Cooperation Agreement, November 16, 1971, 78: "The East Asia Bureau believes that it is clearly within U.S. interests to furnish unclassified space technology to Japan that would meet present Japanese space program objectives, including that of placing satellites up to 500 kilograms in to Geostationary orbit."

40. Memorandum from assistant administrator for international affairs to deputy administrator, October 1, 1974. Also Sato, "A Contested Gift of Power," 201.

41. Yasushi Sato, "Local Engineering in the Early American and Japanese Space Programs: Human Qualities in Grand System Building," PhD dissertation, University of Pennsylvania, 2005, 334.

42. Steven J. Isakowitz, et al., *International Reference Guide to Space Launch Systems* (Reston, VA: American Institute of Aeronautics and Astronautics), 2004, 185.

43. Ibid., 186.

44. U. Alexis Johnson to secretary of state, May 1, 1967, "Space Activities: INTELSAT and Japanese Rocket Development," RG59 Box 3006, folder SP-Space and Astronautics, NARA.

45. "Launch Assistance for Space Satellite Projects" (National Security Decision Memorandum—NSDM—187, 1972), *Weekly Compilation of Presidential Documents* 8:42 (October 16, 1972), in Aaron Karp, "Ballistic Missiles in the Third World," *International Security* (Winter 1984/85), 178. This was cited in Leonard S. Spector, *Undeclared Bomb* (Cambridge, MA: Ballinger Publishing Company, 1988), 38.

46. Memorandum secretary of state Washington, DC, to American Embassy, Tokyo, March 10, 1975, "Provision of Geostationary Orbit Insertion," Central Foreign Policy Files, 1973–1976, "1975STATE053438" RG 59 (retrieved from the Access to Archival Databases at www.archives.gov, July 30, 2009]

47. American Embassy, Tokyo, to secretary of state, Washington, DC, September 19, 1974, "Space Cooperation visit of Dr. James S. Fletcher," RG59, Central Foreign Policy Files, 1973–1976, General Records of the Department of State, "1974TOKYO12150" (retrieved from the Access to Archival Databases at www.archives.gov, July 30, 2009).]

48. Gerald M. Truszynski, associate, administrator for tracking and data acquisition, NASA, September 27, 1974, Record no. 14690, folder US-Japan Thor-Delta, NHRC.

49. Georg M. Low to James C. Fletcher, "Synchronous Orbit Transfer or Japanese Satellites," October 4, 1974, Record no. 14690, LEK7/10/5, folder US-Japan Thor-Delta, NHRC.

50. Memorandum secretary of state Washington, DC, to American Embassy, Tokyo, March 10, 1975, "Provision of Geostationary Orbit Insertion," RG59, Central Foreign

Policy Files, 1973–1976, "1975STATE053438" (retrieved from the Access to Archival Databases at www.archives.gov, July 30, 2009)

51. Ka Zeng, *Trade Threats, Trade Wars. Bargaining, Retaliation, and American Coercive Diplomacy* (Ann Arbor: University of Michigan Press, 2004), 152.
52. Ibid.
53. Ronald A. Cass, "Velvet Fist in an Iron Glove. The Omnibus Trade and Competitiveness Act of 1988," *Regulation* (Winter 1991), 50–56, at 52.
54. Ka Zeng, *Trade Threats, Trade Wars,* 157.

11   An Overview of NASA-India Relations

1. For other studies on India, see Angathevar Baskaran, "Competence Building in Complex Systems in the Developing Countries: The Case of Satellite Building in India," *Technovation* 21:2 (2001), 109–121; "Technology Accumulation in the Ground Systems of India's Space Program: The Contribution of Foreign and Indigenous Inputs," *Technology in Society* 23:2 (2001), 195–216; "From Science to Commerce: The Evolution of Space Development Policy and Technology Accumulation in India," *Technology in Society* 27:2 (2005), 155–179; Gopal Raj, *Reach for the Stars: The Evolution of India's Rocket Programme* (New Delhi: Viking, 2000); Amrita Shah, *Vikram Sarabhai, a Life* (New Delhi: Penguin, 2007); U. Sankar, *The Economics of India's Space Programme: An Exploratory Analysis* (New Delhi: Oxford University Press, 2007); Raman Srinivasan, "No Free Launch: Designing the Indian National Satellite," in Andrew J. Butrica (ed.), *Beyond the Ionosphere: Fifty Years of Satellite Communication* (Washington, DC: NASA SP-4217, 1997). Recent work by Asif Siddiqi has indicated new ways of studying the evolution of space programs in emerging space powers. His attempts at integrating the corpus of postcolonial studies—pioneering work by Arjun Appadurai, Warwick Anderson, Gyan Prakash, Itty Abraham, and others—has offered new insights to delineate an alternative framework for understanding postcolonial technoscience in "developing" countries. See Asif Siddiqi, "Competing Technologies, National(ist) Narratives, and Universal Claims: Toward a Global History of Space Exploration," *Technology and Culture* (April 2010), 425–443.
2. Dennis Kux, *Estranged Democracies: India and the United States, 1941–1991* (New Delhi: Thousand Oaks: Sage Publications, 1994); Gary K. Bertsch, Seema Gahlaut, and Anupam Srivastava, *Engaging India: US Strategic Relations with the World's Largest Democracy* (New York: Routledge, 1999); H. W. Brands, *India and the United States: The Cold Peace* (Boston: Twayne Publishers, c1990); Andrew J. Rotter, *Comrades at Odds: The United States and India, 1947–1964* (Ithaca, NY: Cornell University Press, 2000); Robert J. McMahon, *The Cold War on the Periphery: The United States, India, and Pakistan* (New York: Columbia University Press, 1994).
3. Kux, *Estranged Democracies,* 447.
4. For a detailed analysis of modernization efforts by the United States in developing countries, see George Rosen, *Western Economists and Eastern Societies: Agents of Change in South Asia, 1950–1970* (Baltimore: Johns Hopkins University Press, 1985); Michael E. Latham, *Modernization as Ideology: American Social Science and "Nation Building" in the Kennedy Era* (Chapel Hill: University of North Carolina Press, 2000); Nils Gilman, *Mandarins of the Future: Modernization Theory in Cold War America* (Baltimore: Johns Hopkins University Press, 2004); David C. Engerman, "West Meets East: The Center for International Studies and Indian Economic Development," in David C. Engerman, Nils Gilman, Mark H. Haefele, and Michael E. Latham (eds.), *Staging Growth: Modernization, Development, and the Global Cold War* (Amherst: University of Massachusetts Press, 2003), 199–223; Nicole Sackley, "Passage to Modernity: American Social Scientists, India, and the Pursuit of Development, 1945–1961," PhD dissertation, Princeton University, 2004.

5. The availability of recently declassified State Department papers has offered an excellent opportunity to scrutinize the motivations and justifications for initiating a satellite project with India and its legacy.

6. For a recent biographical work on Homi J. Bhabha, see Indira Chowdhury and Ananya Dasgupta, *A Masterful Spirit: Homi J. Bhabha, 1909–1966* (New Delhi: Penguin Books India, 2010).

7. See, for instance, Homi J. Bhabha and W. Heitler, "The Passage of Fast Electrons and Theory of Cosmic Ray Showers," *Proceedings of the Royal Society* 159 A (1937); Homi J. Bhabha, "On Penetrating Component of Cosmic Radiation," *Proceedings of the Royal Society* 164 A (1938). Vikram Sarabhai worked in the field of cosmic ray variations and set up a group, which was undoubtedly the best in this field and which achieved recognition in international science. He was for some years secretary of the internationally instituted subcommittee on cosmic ray intensity variations and was also a member of the cosmic ray commission of the International Union of Pure and Applied Physics. While Tata Institute of Fundamental Research was the cradle of the Indian atomic energy program, Vikram Sarabhai made the PRL the cradle of the Indian space program. His first scientific contribution, "Time Distribution of Cosmic Rays," was published in the *Proceedings of the Indian Academy of Science* in 1942. During this period at Cambridge he also carried out an accurate measurement of the cross-section for the photo fission of 238 U by 6.2 mev r-rays obtained from the 19F (p, r) reaction. This work also formed a part of his PhD thesis. See S. P. Pandya, "The Physicist," in Padmanabh K. Joshi (ed.), *Vikram Sarabhai: The Man and the Vision* (New Delhi: Mapin, 1992), 52–57.

8. For a recent scholarly treatment on his biography, see Shah, *Vikram Sarabhai*.

9. For a detailed account of Bhabha's work on cosmic rays in Bangalore, see Jahnavi Phalkey, "Science, State-Formation and Development: The Organization of Nuclear Research in India, 1938–1959," PhD dissertation, Georgia Institute of Technology, 2007, 157–161.

10. Gyan Prakash, *Another Reason: Science and the Imagination of Modern India* (Princeton: Princeton University Press, 1999), 196; emphasis added.

11. For more on the role of science and technology in national identity, see Carol E. Harrison and Ann Johnson, "Introduction: Science and National Identity," *Osiris* 24 (2009), 1–14.

12. At the recommendation of Vikram Sarabhai, the laboratory was founded following an agreement between the Ahmedabad Education Society and the Karmakshetra Educational Foundation in November 1947. See Government of India, *25 Years of PRL* (Ahmedabad: Physical Research Laboratory, undated), 4.

13. He established the Ahmedabad textile Industry's Research Association (ATIRA), started the Physical Research Laboratory (PRL) at Ahmedabad, took over the management of Sarabhai Chemicals in 1950, established Suhrid Geigy Limited in 1955, assumed the management of Swastik Oil Mills Limited, founded the Ahmadabad Management Association in 1957, and set up Sarabhai Merck Limited in 1958, also took over Standard Pharmaceuticals in Calcutta, established Sarabhai Research Centre in Baroda, and the operations Research Group (ORG) in 1960, was also the prime mover behind establishing the Indian Institute of Management (IIM) at Ahmedabad. See Raj, *Reach for the Stars*, 6–7; Shah, *Vikram Sarabhai*.

14. Kamla Choudhary, *Vikram Sarabhai: Science Policy and National Development* (Delhi: Macmillan 1974), 24.

15. For a closer historical sociology on the establishment of tracking stations in India, see Teasel Muir Harmony, "Tracking Diplomacy: The International Geophysical Year and American Scientific and Technical Exchange with East Asia," in Roger D. Launius, James R. Fleming, and David H. DeVorkin (eds.), *Globalizing Polar Science: Reconsidering the International Polar and Geophysical Years* (New York, NY: Palgrave Macmillan, 2010). Due to a recent division of Uttar Pradesh in the year 2000 Nainital is now brought under the state of Uttarakhand. The data collected from Nainital was

analyzed by SAO and it provided the world science community with precise knowledge about the configuration of the earth and of its gravitational field. Because of the critical location Nainital was connected to the Smithsonian Standard Ellipsoid system—a world system—along with the other 14 stations. The geodetic Cartesian coordinates of all these stations with respect to the same SAO ellipsoid center were made available with a positional standard deviation of 10–15 meters. In order to give an impetus to applications of satellite to geodesy in India, the ISRO set up a satellite geodesy unit. This ISRO unit worked in collaboration with other Indian interests in geodesy such as the Geodetic Branch of Survey of India. For more information on this, see Vikram Sarabhai, P. D. Bhavsar, E. V. Chitnis, and P. R. Pisharoty, *Application of Space Technology to Development*, a study prepared for the United Nations (Unpublished), December 1970, 1.73.

16. For detailed study of the global distribution of tracking stations, see Sunny Tsiao, *"Read you Loud and Clear." The Story of NASA's Spaceflight Tracking and Data Network* (Washington, DC: NASA SP- 2007–4232).

17. Address, Milton C. Rewinkel—consul general for the United States of America, January 31, 1963, RG 59, General Records of the Department of State, Central Foreign Policy file 1963, Box 4184, folder SP 15—Space Vehicle Tracking, NARA.

18. K. R. Ramanathan was an expert on atmospheric ozone and a former president of the International Union of Geodesy and Geophysics and directed the laboratory since its inception in the late 1940s. Both Sarabhai and Ramanathan have led important programs of ground-based research programs, which expanded under the stimulus of the International Geophysical Year (IGY) and the International Years of the Quiet Sun (IYQS). Using counters and monitors at several sites in India, and at Chacaltaya in Bolivia, Sarabhai and his colleagues R. P. Kane, N. W. Nerurkar, G. L. Pai, S. P. Pandya, U. R. Rao, and others studied cosmic rays as they are influenced by the sun's magnetic field.

19. In the 1940s, the Indian Institute of Science (IISC) in Bangalore, the Bose Institute in Calcutta, and the Muslim University at Aligarh were effectively conducting cosmic ray research. Teams of scientists—e.g., Max Milliken and his colleagues, a group headed by Homi Bhabha at IISC, and another team under A. P. Thattee at TIFR—conducted cosmic ray experiments using rubber balloons. Vikram Sarabhai and K. R. Ramanathan started research into space sciences, which led to the establishment of PRL in 1947. By the mid-1950s Physical Research Laboratory (PRL) had become an international center for cosmic ray research. Meanwhile, TIFR established a basic infrastructure, including radio interferometers, large radio telescopes, and a facility for making plastic balloons. From 1959, Indian and foreign scientists within groups from PRL, and the US Air Force used the balloon facility for experiments. In 1961, a real-time satellite telemetry station was established at PRL in collaboration with NASA. For more on the development of scientific institutions in India, see R. S. Anderson, *Building Scientific Institutions in India: Saha and Bhabha*, Occasional Paper Series, No. 11 (Montreal: McGill University Press, 1975), 31; Joshi, ed., *Vikram Sarabhai*, 112; Baskaran, "From Science to Commerce," 155–179.

20. Memorandum Wilmot L. Averill, chief, cooperative programs, office of international programs. The participants were Vikram Sarabhai, Upendra Desai, Arnold Frutkin, and Averill, June 1, 1961, Special Assistant to the Secretary for Energy and Outer Space, Records Relating to Atomic Energy Matters 1955–1963, Box 252, NARA. Launching of an existing experiment developed by Cahill at the University of New Hampshire to study the equatorial electrojet by flying a magnetometer to an altitude of approximately 200 kilometers.

21. According to the agreement NASA agreed to provide a motor-driven dual Yagi antenna, 108 Mc preamplifier, 108 Mc micro lock receiver, FR-100 tape recorder, two FM discriminators, look angles and time for satellite passes, magnetic tapes, and also technical consultants to assist in the installation of the equipments. Memorandum Hugh Dryden

(deputy administrator NASA) to Homi Bhabha, August 11, 1961, RG 59, Central decimal file 1960–63, Box 3112, NARA.

22. Arnold Frutkin to Vikram Sarabhai, July 24, 1961, with an attached memo from Walter W. Stuart, counselor of political affairs, American Embassy, New Delhi, to S. Gupta, joint secretary Western Division, Ministry of External Affairs, New Delhi, RG 59, Central Decimal File 1960–63, Box 3112, NARA.

23. Memorandum, Arnold Frutkin to Homi Bhabha, December 12, 1961, RG 59, Box 250, Folder—Cooperative Space Program, NARA.

24. Memorandum R. Shroff, deputy secretary, to Arnold W. Frutkin, Office of International Program, National Aeronautics and Space Administration, Washington, DC. Record Group 59, Box No. 259, special assistant to the secretary for energy and outer space, Records Relating to Atomic Energy Matters 1955–1963, NARA. The scientists are K. S. Krishnan, A. P. Mitra, K. R. Ramanthan, S. K. Mitra, Vikram Sarabhai, Vainu Bappu, and S. L. Malurkar.

25. A supplementary memorandum of understanding was signed between DAE and NASA on July 1, 1965, for the collaboration in scientific investigations of the upper atmosphere and equatorial electrojet and in the measurement of electron and ion densities and Lyman alpha and X-ray flux in the D region. See government of India, *DAE Annual Report, 1965–66*.

26. *News Release*, National Aeronautics and Space Administration, Release no 63–5, January 14, 1963.

27. The equatorial electrojet is an electrical current at an altitude of 90–100 kilometers in the ionosphere and slowing along the magnetic equator in the sunlit portion of the earth from west to east. It measures about 100–200 kilometers in width, centered on the geomagnetic equator. The electrojet was studied by means of a rocket magnetometer designed to determine the geographic location, height, and intensity of the electrojet during the period when the 11-year cycle of solar activity was near its minimum. The magnetometer permitted the observation of changes in the vertical structure of the electrojet at different times during the day, at different times during the lunar cycle, and during various degrees of magnetic disturbance, particularly at the sudden commencement and during the initial phase of a magnetic storm.

28. Exploration of the upper atmosphere winds was conducted by optical observation of sodium vapor released from a rocket payload. The object of the sodium vapor experiment was to measure upper atmospheric winds by photographing, from the ground, a cloud of sodium vapor released from the rocket and illuminated by the sun.

29. T. Eliot Weil to Wreatham E. Gathright, October 5, 1965, "Proposed Visit of Dr. Homi J. Bhabha," Folder—Cooperative space program, Box 250, NARA.

30. Memorandum Arnold Frutkin to Homi J. Bhabha, secretary Department of Atomic Energy, undated, "Space Cooperation with India" (undated), RG 59, Box 250, special assistant to the secretary for Energy and Outer Space, Records Relating to Atomic Energy Matters 1944–63, NARA.

31. E. V. Chitnis, the then secretary of INCOSPAR along with R. D. John of the Atomic Energy Establishment proceeded to Thumba on January 14, 1963, for site selection. See R. D. John, "Some Reminiscences on Space Construction Programme," *Forerunner* (June 1989), 1–6.

32. Memorandum Richard Barnes to Robert Packard, June 29, 1962, "NASA/ INCOSPAR Cooperative Program," RG 59, Box 250, NARA. For more on Duffy and Cahill's activities in site selection, see Memorandum, Sidney Sober, American consul, to secretary of state, August 14, 1962, "Proposed Collaboration between India and NASA in launching of Sounding Rockets near Geomagnetic Equator," RG 59, Box 3112, NARA.

33. Actually, the magnetic equator passes through Quilon, 32 kilometers away from Thumba, but owning to the safety issues and the number of people to be cleared Thumba was eventually chosen.

34. Arnold W. Frutkin, Progress in International Cooperation in Space Research, *News Release*, May 23, 1963, 5.

35. Arnold W. Frutkin, "The United States Space Program and its International Significance," *The Annals of the American Academy of Political and Social Science* 366 (July 1966), 89–98.

36. Ibid.

37. Kerala was one of the most pro-Soviet states in India during that time period. One could easily spot shops in busy city streets stacked with books published by Soviet press. And students hailing from Kerala always won top prices in Russian-language contests. For more on Kerala's communist sentiments, see Vladimir Gubarev, *Aryabhata The Space Temple* (New Delhi: Sterling Publishers, 1976), 13–14.

38. The Department of Atomic Energy (DAE) under guidance from professor Jacques Blamont of CNES of France supplied sodium vapor release payloads, appropriate photographic equipment, the launching site and supporting facilities, personnel and supporting instrumentation were provided by India. Along with DAE, PRL scientists were involved both in building up ground-based facilities at Thumba and also in making instrumented payloads for use in rockets. See *PRL 25 years*.

39. The spectacle in the skies became a potent symbol. "Egypt wanted NASA to provide rockets for a popular scientific sounding rocket project that involved lofting sodium vapor into the upper atmosphere as a marker to follow wind patterns. Egypt, however, was unwilling to enter into a joint working group, to permit NASA visitation to the premises, or otherwise to satisfy NASA that essential scientific and engineering standards would be met. Frutkin felt that it was especially important to have NASA expertise in this case because the Egyptians wanted to use two rockets in combination that had never been launched together before. When Frutkin declined to have NASA cooperate on this project, Egypt appealed to the State Department. Officials there then pushed for Egypt to buy the sounding rockets commercially in the US. This happened and the State Department assisted in flying the sounding rockets to Egypt, but the Egyptians then excluded the vendor from the premises. The potential NASA-Egyptian cooperation was marked by another problem: heavy handed political motives from the Egyptians. In the early 1960s Egypt was part of the United Arab Republic, as was Syria and the Egyptians wanted to conduct the sodium vapor experiment on the anniversary of UAR's founding in both Syria and Egypt. While the sodium vapor itself would be benign, having pink clouds appear on two sides of Israel with this timing was intended to send a clear political signal to Israel. But Frutkin turned down the project based on Egyptians' unwillingness to cooperate through joint working groups and the like because otherwise, would have been a very sticky political dilemma for his office." See Steve Garber, telephone conversation with Arnold Frutkin regarding international cooperation with "nontraditional Partners," 10/4–5/01, NHRC.

40. Memorandum, Sidney Sober, American consul, to secretary of state, August 14, 1962, "Proposed Collaboration between India and NASA in launching of Sounding Rockets near Geomagnetic Equator," RG 59, Box 3112, NARA.

41. See Arnold Frutkin, *International Cooperation in Space* (Englewood Cliffs, NJ: Prentice-Hall, 1965), 62.

42. The wind patterns of the monsoons in the Indian Ocean merit close study, not only because they give striking evidence of interactions between the ocean and the atmosphere but also because of the possible influence of the sun's particulate radiation on the earth's weather as noted in Victor K. McElheny, "India's Nascent Space Program," *Science* 149 (September 1965), 1487–1489.

43. Memo Vikram Sarabhai to Frutkin, July 12, 1962, Record Group 59, Box 250, special assistant to the secretary for Energy and Outer Space, Records Relating to Atomic Energy Matters 1944–63, NARA.

44. Memorandum, R. Shroff, deputy secretary, Department of Atomic Energy, government of India, to Arnold W. Frutkin, July 28, 1962, RG 59, Box 250, special assistant

to the secretary for Energy and Outer Space, Records Relating to Atomic Energy Matters 1944–63, NARA.

45. Vikram Sarabhai et al., *Application of Space Technology to Development* (a study prepared for the UN, 1970), 1.42.

46. Confidential memorandum, Frutkin to Philip Farley, Department of State, January 18, 1962, "Arcas for India," RG 59, Box 250, folder Cooperative Space Program, NARA. A boosted Arcas was flown from the Thumba Range for scientific studies.

47. Vikram Sarahbai, *Science Policy and National Development*, edited by Kamla Chowdhry (Delhi: Macmillan, 1974), 25.

48. For more on the importance of Vienna Congress for developing countries, see Vladimir Gubarev, *Aryabhata The Space Temple* (New Delhi: Sterling Publishers, 1976).

49. Address given as scientific chairman of the United Nations Conference on the Exploration and Peaceful Uses of Outer Space, Vienna, August 1968. See *Sarabhai*, 28–37.

50. See Carl Q. Christol, "Space Joint Ventures: The United States and Developing Nations," *University of Akron Law Reviews* 8 (1975), 398–415.

51. For a critical view of the "profile," see Ashok Parthasarathi, *Technology at the Core Science and Technology with Indira Gandhi* (Addison-Wesley Professional, 2008).

52. George Joseph, *Fundamentals of Remote Sensing* (Hyderabad: Universities Press, 2005), 16.

53. For a historical account of the origins and development of Landsat, see Pamela Mack, *Viewing the Earth: The Social Construction of the Landsat Satellite System* (Cambridge, MA: MIT Press, 1990).

54. It was used to identify the lost courses of the Saraswati river in the great Indian desert. See Bimal Ghose, Amal Kar, and Zahid Husain, "The Lost Courses of the Saraswati River in the Great Indian Desert: New Evidence from Landsat Imagery," *The Geographical Journal* 145:3 (November 1979), 446–451.

55. For a brief historical overview of remote sensing in India, see Shubhada Savant and Santhosh Seelan, "India's Remote Sensing Programme," *Spaceflight* 48:8 (August 2006), 308–314.

56. P. D. Bhavsar, "Remote Sensing Program In India," *Current Science* 15:2 (September 1985), 15–35.

57. Initially remote sensing was not part of the space program but Sarabhai initiated remote sensing after this UN conference where he heard presentations from American scientists on how they were using remote sensing techniques to detect camouflaged positions in Vietnam. See Baskaran, "From Science to Commerce," 155–179.

58. For an excellent study on the origins and evolution of remote sensing in India, see A. R. Dasgupta and S. Chandrashekar, *Indigenous Innovation and IT-enabled Exports: A Case Study of the Development of Data Processing Software for Indian Remote Sensing Satellites*, a study for the UPIASI Research Project on the Context of Innovation in India: The Case of the Information Technology Industry, Submitted to University of Pennsylvania, Institute for the Advanced Study of India, New Delhi, September 30, 2000.

59. The various departments that attended are: Department of Atomic Energy, Survey of India, India Meteorological Department, Geological Survey Department, Physical Research Laboratory, Indian Institute of Science, Department of Aeronautical Engineering, National Geophysical Research Institute, Forest Research Institute, Indian Institute of Technology, National Physical Laboratory, Airborne Mineral Survey, Indian Agricultural Research Institute, National Institute of Oceanography, and Indian Space Research Organization.

60. C. Dakshinamurti et al., "Remote Sensing for Coconut Wilt," *Proceedings of the International Symposium of Remote Sensing Environment* (1971), Vol. 1.

61. The memo indicated that "because of military connotations of many items of rocket equipment and technology, restrictions have been imposed on the export of many classes of items. Applications for export of rocket equipment or technology must be submitted to the U.S. Department of State for approval...with regard to propellant

technology, I should advise you that the U.S. government is reluctant to permit the export of knowledge of manufacturing techniques," Memo Wreatham E. Gathright, chief, Outer Space Matters, Office of the Special Assistant to the Secretary for Atomic Energy and Outer Space, to Arnold Frutkin, "Space Cooperation with India," October 19, 1961, RG 59, Box 250, special assistant to the secretary for Energy and Outer Space, Records Relating to Atomic Energy Matters 1944–63, NARA.

62. Ibid.
63. Confidential memorandum, Arnold Frutkin to J. Wallace Joyce, "Support of Birla Institute, India," August 25, 1965, RG 59, Box 3140, NARA.
64. Ibid.
65. Letter, Arnold Frutkin to Homi Bhabha, March 10, 1965, RG 59, Box 10, Folder SP Space and Astronautics, Bureau of International Scientific and Technological Affairs, NARA.
66. Ibid.
67. A launch base in the east coast was first hinted at here. By the late 1960s Sarabhai established Sriharikota launch range on the eastern coast close to Madras, now Chennai.
68. Memorandum, Frutkin to Robert F. Packard, officer in charge, Outer space affairs, Office of International Scientific Affairs, Department of State, February 5, 1965, Box 10, Folder Space and Astronautics, NARA.
69. Ibid.
70. Memorandum, Frutkin to Robert Packard.
71. Chowdhry, *Vikram Sarabhai: Science and Development*, 36–37.
72. The Sriharikota range became operational with the firing of RH-125 sounding rocket on October 9, 1971.
73. Secret Memorandum, American Embassy, India to secretary of state, April 30, 1970, "Indian Reaction to Chicom Satellite," RG 59, Box 2962, NARA.
74. Secret Memorandum, American Embassy to secretary of state, April 30, 1970, "US Assistance to Indian Satellite Program," RG 59, Box 2962, NARA.
75. Confidential memo from American Embassy New Delhi to secretary of state, Washington, DC, August 23, 1970, RG 59, Box 2962, Subject Numeric files, 1970–73, Folder Science, NARA.
76. Ibid.
77. Meeting between Frutkin and Sarabhai is mentioned in the memorandum from Anthony C. E. Quainton, senior political officer for India with the State Department, to Harmon E. Kirby, American Embassy, New Delhi, September 24, 1970.
78. Robert A. Clark, jr to Joseph T. Kendrick, December 29, 1970, "U.S.-India Space Cooperation," RG 59, Box 2962, NARA.
79. Confidential memo David T. Schneider to ambassador Meyer, April 5, 1973, "Your appointment with Werner Von Braun," RG 59, Box 25, Records Relating to India 1966–75, Folder: Science, NARA.
80. Rockwell International (formerly North American Rockwell) had asked for authorization for its UK licensee, Rolls Royce, to transfer rocket engine technology and experience; Honeywell Inc. wanted to export certain quantities of two different types of gyros and Kearfott, Division of Singer Co. applied for two licenses, one that would permit its UK licensee, Ferranti, to sell certain gyros and the other that would allow the export of US-made gyros and accelerometers.
81. Confidential Memo John W. Sipes to Joseph J. Sisco, June 27, 1973, "US Posture Towards Indian Space Program," RG 59, Box 2962, NARA.
82. The Soviet Union did not condemn India's test and also was very neutral toward India not signing the NPT.
83. There is a real dearth of materials on Soviet-India space relations. Whatever material that is available is very superficial, Brian Harvey, "Russia: The Indian Connection," *JBIS* 54 (2001), 47–54; Jerome M. Conley, *Indo-Russian Military and Nuclear Cooperation: Lessons and Options for U.S. Policy in South Asia* (New York: Lexington

Books, 2001); U. R. Rao and K. Kasturirangan, *The Aryabhata Project* (Bangalore: Indian Academy of Sciences, 1979); Mikhail Barabanov, "Russian-Indian Cooperation in Space," *Moscow Defense Brief* 1:3 (2005), 27–31.

84. Oral History Transcript, Arnold W. Frutkin, interview by John Krige et al., Washington, DC, August 19, 2007, NHRC.

85. Gopal Raj, *Reach for the Stars*, 57.

86. Ibid., 50. In addition to this, it is important to add that Pramod Kale is well known among NASA officials. He spent three years at Goddard Space Flight Center.

87. Government of India, *Profile for the Decade*, 28.

88. Gary Milhollin, "India's Missiles—With a Little Help from Our Friends," *Bulletin of the Atomic Scientists* (November 1989), 311–315. Most of the scholarship on India's missile and space program often mention Milhollin's piece to state that SLV-3 was built using Scout blueprints. In testimony before the House Committee on Science on June 25, 1998, Milhollin stated that "in 1965, The Indian Government asked NASA for design information about the Scout...NASA obligingly supplied the information. Kalam then proceeded to build India's first rocket, the SLV-3, which was an exact copy of the Scout." See http://www.globalsecurity.org/space/library/congress/1998_h/980625-milhollin.htm (accessed December 12, 2009).

89. The first sounding rocket RH-75 fueled by a Cordite mixture (a mixture of nitroglycerine and nitrocellulose) was launched on November 20, 1967, following which a series of indigenously developed sounding rockets were developed RH-100, RH-125, RH-125s, Menaka I and II, RH-300, RH-300 MK-II, RH-560 (RH denotes Rohini), all using solid fuels. For more on ISRO's sounding rockets, see V. Sudhakar, *Sounding Rockets of Isro* (Bangalore: Indian Space Research Organisation, 1976).

90. A. P. J. Abdul Kalam and Arun Tiwari, *Wings of Fire: An Autobiography* (Hyderabad: Universities Press, 1999), 38. Kalam has been credited for the introduction of project management techniques in the Indian space program and worked on a number of projects before being assigned as the project manager for the SLV-3 project by Sarabhai.

91. The first experimental launch of SLV-3-E1 was conducted in August 1979. It was a failure. The second, designated as SLV-E2, was a success.

92. http://www.america.gov/st/washfileenglish/2004/June/20040622155335ESnamfuaK0.3571588.html (accessed March 17, 2009).

93. http://www.nasa.gov/home/hqnews/2008/feb/HQ_08033_India-agreement.html (accessed March 17, 2009).

94. J. N. Goswami and M. Annadurai, "Chandrayan-1: India's First Planetary Science Mission to the Moon," *Current Science* 96:4 (February 25, 2009): 486–491.

12   Satellite Broadcasting in Rural India: The SITE Project

1. K. Kasturirangan, "Share and Care for a Better World—The Engine for Future of Space," *Acta Astronautica* 54:11–12 (June 2004), 867. The author was the former director of ISRO.

2. Yash Pal, "A Visitor to the Village," *Bulletin of the Atomic Scientists* 33:1 (January 1977), 55. Arthur C. Clark was the consultant to the SITE project since its inception.

3. "Report on the SITE Winter School," *UN Document No. A/AC/105/177*, United Nations Committee on the Peaceful Uses of Outer Space (December 2, 1976), 13.

4. "World Wide Space Activities," report prepared for the Subcommittee on Space Science and Applications of the Committee on Science and Technology U.S. House of Representatives, Ninety –Fifth Congress, First Session (September 1977), 404.

5. Kiran Karnik, "Societal Benefits of Space Technology," *Acta Astronautica* 19:9 (September 1989), 771–777.

6. Romesh Chander and Kiran Karnik, *Planning for Satellite Broadcasting: The Indian Instructional Television Experiment*, reports and papers on mass communication, 78

(Paris: Unesco, 1976); Srinivas R. Melkote, Peter Shields, and Binod C. Agrawal, *International Satellite Broadcasting in South Asia: Political, Economic, and Cultural Implications* (Lanham, MD: University Press of America, 1998); Binod C. Agrawal, *Social Impact of SITE on Adults* (Ahmedabad: Indian Space Research Organization, 1977); *The SITE Experience* (Paris: Unesco, 1983); *Satellite Instructional Television Experiment: SITE Winter School, January 16–28, 1976* (Ahmedabad: Space Applications Centre, Indian Space Research Organization, 1976).

7. Arthur C. Clarke, "Extra-Terrestrial Relays: Can Rocket Stations Give World-wide Radio Coverage," *Wireless World* (October1945), 305–308.

8. For a detailed account of early communication satellites, see David J. Whalen, *The Origins of Satellite Communications, 1945–1965* (Washington, DC: Smithsonian Institution Press, 2002).

9. Hugh R. Slotten, "Satellite Communications, Globalization and the Cold War," *Technology and Culture* 43 (April 2002), 315–350.

10. Dennis Merrill, *Bread and the Ballot: The United States and India's Economic Development, 1947–1963* (Chapel Hill: UNC Press, 1990), 5.

11. William Burr and Jeffrey T. Richelson, "Whether to 'Strangle the Baby in the Cradle': The United States and the Chinese Nuclear Program, 1960–64," *International Security* 25:3 (Winter 2000/2001), 54–99.

12. Secret memorandum J. Wallace Joyce to Mr. Hare, October 10, 1966, "Scientific and Technical Cooperation with India," RG59, Central Foreign Policy files, 1964–1966, Box 3106, NARA.

13. Ibid.

14. Ibid.

15. Vikram Sarabhai, *Science Policy and National Development* (New Delhi: Macmillan, 1974), xiv.

16. Memo, Department of State, October 17, 1966, RG 59, Central Foreign Policy File 1964–66, Box 3106, NARA.

17. Confidential letter from James E Webb to Mr. U. Alexis Johnson, May 19, 1966, deputy undersecretary for Political Affairs, Department of State, RG 59, Box 3140, Folder SP- Space & Astronautics 1/1/64, NARA.

18. Sarabhai, *Science Policy and National Development*, xii.

19. W. W. Rostow, *Stages of Economic Growth: A Non-Communist Manifesto* (Cambridge: Cambridge University Press, 1960). For a closer analysis of Rostow's work and on modernization, see David C. Engerman, Nils Gilman, Mark H. Haefele, and Michael E. Latham, eds., *Staging Growth. Modernization, Development and the Global Cold War* (Amherst: University of Massachusetts Press, 2003); Michael E. Latham, *Modernization as Ideology. American Social Science and "Nation Building" in the Kennedy Era* (Chapel Hill: University of North Carolina Press: 2000); Nils Gilman, *Mandarins of the Future. Modernization Theory in Cold War America* (Baltimore: Johns Hopkins University Press, 2007).

20. John V. Vilanilam, *Mass Communication in India: A Sociological Perspective* (New Delhi: Sage Publications, 2005), 154.

21. Nick Cullather, "Miracles of Modernization: The Green Revolution and the Apotheosis of Technology," *Diplomatic History* 28:2 (April 2004), 231.

22. *Next Ten Years in Space, 1959–1969*, Staff Report of the Select Committee on Astronautics and Space Exploration, 85th Congress, 2nd session, 32.

23. Joyce to Hare, October 10, 1966, "Scientific and Technical Cooperation with India."

24. United States, Congress, House, Foreign Affairs, *Foreign Policy Implications of Satellite Communications, Hearings Before the Subcommittee on National Security Policy and Scientific Developments…91–2, April 23, 28, and 30* (US Govt. Print. Off, 1970).

25. Arnold Frutkin, The India-U.S. Satellite Broadcasting Experiment, folder SITE, NHRC.

26. Bella Mody, "Contextual Analysis of the Adoption of a communications technology: the case of satellites in India," *Telematics and Informatics* 4:2 (1987), 151–158.

27. Ibid.

28. The Ford Foundation paid $65,000 to the government of India to facilitate scientists at DAE and ISRO under the leadership of Sarabhai to conduct studies at MIT—Lincoln labs, on the design configuration of the INSAT, Confidential memorandum, Kenneth Bernard Keating, American Embassy, New Delhi, to secretary of state, Washington, DC, July 21, 1970, "GOI Satellite ITV Program," RG 59, Subject Numeric Files, 1970–73, Box 2962, NARA.

29. For more on the ATS satellite, see Arnold Frutkin, "Direct/Community Broadcast Projects Using Space Satellite," *Journal of Space Law* 3:1 & 2 (January 1975), 17–24; K. Narayanan, "Special Features of ATS-6 Satellite" in *SITE* Winter School, *Satellite Instructional Television Experiment*, 43–52.

30. The 30-foot deployable antenna helps to concentrate the radio frequency power into a narrow beam. At 860 megahertz, the frequency at which SITE operated, this beam is only 2.8 degrees wide and illuminates a circle of about 1,800-kilometer diameter. Since the beam was very narrow it had to be directed to a desired point. Even a small error in pointing will mean that a large area intended to be covered will not receive the signal power. Hence the second objective of 0.1 degree pointing capability.

31. For the SITE broadcasts the satellite received the signals from earth at 6 gigahertz and sent them back down at 860 megahertz. The satellite transponder (which received the transmitted the signals) handled one video picture and two soundtracks simultaneously. See Clifford Block et al., "A Case Study of India's Satellite Instructional Television Project," *AID Report* (January 1977), 4.

32. Ibid., 5.

33. Chander and Karnik, *Planning for Satellite Broadcasting*, 20.

34. Using ATS-6 for the American region was not originally intended. The plan was implemented when the committee voiced protest as to why an advanced technology should not have practical benefits for the American region. See US, Congress, House Committee on Foreign Affairs, Subcommittee on National Security Policy and Scientific Developments, *Hearings, Foreign Policy implications of Satellite Communications*, 91st Cong., 1st sess., 1969.

35. NASA News Release, Release No. 76-157, September 21, 1976, Record no. 005652, ATS 6 Satellite, LEK 3/7/1, NHRC.

36. This amount of new programming exceeds that produced by each US commercial television network for its evening schedule, and a US network can rely on diverse program production sources and a reservoir of experience built up over several decades. Thus meeting the demanding production schedule can be considered a major achievement by Doordarshan. See Block et al., "A Case Study," 7.

37. Ibid.

38. Raman Srinivasan, "*Goods and Gods, Being a Narrative Disquisition on the Poetics of Technology in Post-Traditional India,*" PhD dissertation, University of Pennsylvania, 1994, 118.

39. For production of programs, clusters were set up at the base production centers at Hyderabad, which produces programs for Andhra Pradesh and Karnataka (in languages Telegu and Kannada), in Cuttack, which produces programs in Oriya language, and in Delhi, which produces programs in Hindi for Rajasthan, Madhya Pradesh, and Bihar. These are small studios, which capsule the program in advance and keep up a regular flow of tapes to Ahmedabad for direct telecast. A. Shroff, *SITE: Software Aspect in Satellite Instructional Television Experiment*, SITE Winter School (January 16–28, 1976), Space Applications Centre, Indian Space Research Organization, Ahmedabad, India, 87.

40. Block et al., "A Case Study," 3.

41. Oral History Transcript, Arnold W. Frutkin, interviewed by Rebecca Wright, January 11, 2002, 24, NHRC.

42. US, Congress, House, Committee on Foreign Affairs, Subcommittee on National Security Policy and Scientific Developments, *Hearings, Foreign Policy implications of Satellite Communications*, 91st Cong., 1st sess., 1969, 38.

43. Delbert D. Smith, *Teleservices via Satellite : Experiments and Future Perspectives via Satellite* (Boston: Sijthoff & Noordhoff, 1978), 120.

44. Memo to Clement J. Zablocki by Robert F. Allnutt, assistant administrator for Legislative Affairs. April 24, 1969, "The India-United States Television Satellite Experiment."

45. Frutkin, "Direct/Community Broadcast Projects," 20.

46. Smith, *Teleservices via Satellite*, 121.

47. Dipak C. Talapatra, "The Indian Space Program," *IEEE Aerospace and Electronic Systems Magazine* 8:2 (February 1993), 14.

48. Arnold W Frutkin, statement before the Subcommittee on National Security Policy and Scientific Development, Committee on Foreign Affairs, House of Representatives, April 30, 1970. Record no. 005652, LEK 3/7/1, NHRC.

49. For an anthropological analysis of the Palapa satellite in Indonesia, see Joshua Barker, "Engineers and Political Dreams: Indonesia in the Satellite Age," *Current Anthropology* 46:5 (December 2005), 703–727.

50. The first appearance of this rhetoric was uttered by Vikram Sarabhai at the COPUOS in 1968: "The question has often been asked: 'Can one afford to undertake space research?' But I'm sure there are many here like myself, who will ask: can anyone afford to ignore the applications of space research?" See *Practical Benefits of space exploration: A digest of papers presented at the UN conference on the exploration and peaceful uses of outer space, Vienna, 1968*, United Nations, New York, 1969.

51. See Daniel R. Headrick, *The Tools of Empire: Technology and European Imperialism in the Nineteenth Century* (New York: Oxford University Press, 1981).

52. This becomes more evident when one gives careful attention to the language—unlimited civilizing, modernization, development, and so on—in some documents that were prepared during the planning of SITE. Of particular importance, see the undated document by Arnold W. Frutkin, *The India-U.S. Satellite Broadcasting Experiment*, folder SITE, NHRC.

13 Space Collaboration Today: The ISS

1. John W. Garver, *China and Iran. Ancient Partners in a Post-Imperial World* (Seattle: University of Washington Press, 2006).

2. The narrative line here is due mostly to Roger D. Launius, *Space Stations. Base Camps to the Stars* (Washington, DC: Smithsonian Institution, 2003); John M. Logsdon, *Together in Orbit. The Origins of International Participation in the Space Station. Monographs in Aerospace History #11* (Washington, DC: NASA History Division, 1998); Howard E. McCurdy, *The Space Station Decision. Incremental Politics and Technical Choice* (Baltimore: Johns Hopkins University Press, 1990).

3. McCurdy, *The Space Station Decision*, 199–200.

4. Howard E. McCurdy, "The Decision to Build the Space Station. Too Weak a Commitment?" *Space Policy* 4 (February 1988), 297–306.

5. M. Mitchell Waldrop, "The Selling of the Space Station," *Science* 223:4638 (February 24, 1984), 793–794.

6. Launius, *Space Stations*, 119.

7. McCurdy, *The Space Station Decision*, 200.

8. Logsdon, *Together in Orbit*, 20.

9. Interview, John Krige with Peggy Finarelli, April 20, 2010, NHRC.

10. W. Henry Lambright, "Leadership and Large-Scale Technology: The Case of the International Space Station," *Space Policy* 21 (2005), 195–2003, at 197. See also Logsdon, *Together in Orbit*, 20.

11. Logsdon, *Together in Orbit*, 1.

12. Ibid., 20.
13. Interview, John Krige with Peggy Finarelli, April 20, 2010, NHRC.
14. Logsdon, *Together in Orbit*, 26.
15. Letter Beggs to Griffen, April 12, 1984, cited by Eligar Sadeh, "Technical, Organizational and Political Dynamics of the International Space Station Program," *Space Policy* 20 (2004), 171–188, at 174.
16. Logsdon, *Together in Orbit*, 13.
17. Ibid., 22.
18. These figures were suggested in Memo, Kenneth S. Pedersen, director of international affairs to John Hodge, director, Space Station Task Force, *Strategy for International Cooperation in Space Station Planning*, undated, but about August 1982, reproduced in John M. Logsdon, ed., *Exploring the Unknown. Selected Documents in the History of the U.S. Civil Space Program. Vol. II. External Relationships* (Washington, DC: NASA SP-4407, 1996), Document I-31, 90–100, at 99–100.
19. John Logsdon, "International Involvement in the US Space Station Programme," *Space Policy* 1 (February 1985), 12–25, surveys the many trade-offs that such collaboration involves.
20. Kenneth S. Pedersen, "The Changing Face of International Space Cooperation. One View of NASA," *Space Policy* 2 (May 1986), 120–135, at 131; emphasis in the original.
21. Memo Pedersen to Hodge, *Strategy for International Cooperation in Space Station Planning*, in Logsdon, *Exploring the Unknown*, Vol. II, 92.
22. Ibid. Pedersen also warned against repeating some management and legal arrangements. There should be no teaming between American and foreign companies in the study phase: it could reduce NASA's flexibility when it later came to choose contractors either at home or abroad. It should also not commit itself, as it had in the Spacelab memorandum of understanding (MoU), to purchasing a flight unit produced abroad. And it should be prepared for foreign partners who contributed a piece of hardware to want preferential or free access to the whole station.
23. Ibid., 94.
24. Ibid., 97.
25. Ibid., 99. In August 1972 new guidelines for space cooperation and for dealing with technology transfer were circulated. They differed little from those in place in 1972. The general sense of the guidelines is discussed in the section on ITAR (chapter 14). For the guidelines, see *NASA Fact Sheet. "Space Assistance and Cooperation Policy,"* August 6, 1982, and Attachment, reproduced in Logsdon et al., *Exploring the Unknown*, Document I-32, 100–104.
26. McCurdy, *The Space Station Decision*, 103.
27. Logsdon, *Together in Orbit*, 28–29.
28. Niklas Reinke, *The History of German Space Policy. Ideas, Influences, and Interdependence, 1923–2002* (Paris: Beauchesne, 2007), 233.
29. R. D. Andresen and W. Nellesen, "The Eureca Concept and its Importance in Preparing for the Columbus Programme," *ESA Bulletin* 52 (1987), 57–67; W. Nellesen, "The Eureca Project—From Concept to Launch," *ESA Bulletin* 70 (1992), 17–25. See also John Krige, Arturo Russo, and Lorenza Sebesta, *A History of the European Space Agency, 1958–1987 Vol. II. The Story of ESA, 1973–1987* (Noordwijk: ESA SP-1235, 2000), 62.
30. Reinke, *The History of German Space Policy*, 233.
31. Krige, Russo, and Sebesta, *A History of the European Space Agency, 1958–1987 Vol. II*, 72; Logsdon, *Together in Orbit*, 35.
32. Krige, Russo, and Sebesta, *A History of the European Space Agency, 1958–1987 Vol. II*, 73, 72.
33. F. Longhurst, "The Columbus System. Baseline and Interfaces," *ESA Bulletin* 50 (1987), 88–97, 88–89.

34. Reinhard Loosch, "The International Space Station—The Legal Framework," *Proceedings of an International Colloquium on the Manned Space Station—Legal Issues, Paris, 7–8 November 1989* (Noordwijk: ESA SP-305, 1989), 55–58. Available at http://articles.adsabs.harvard.edu//full/1990ESASP.305...55L/0000055.000. html; Sadeh, "Technical, Organizational and Political Dynamics," 175.

35. J. L. Cendral and G. G. Reibaldi, "The ESA Polar Platform," *ESA Bulletin* 71 (1992), 27–38.

36. Krige, Russo, and Sebesta, *A History of the European Space Agency, 1958–1987 Vol. II,* 645, 650.

37. Reinke, *The History of German Space Policy,* 274. See also F. Engström, J.-J. Dordain, R. Barbera, G. Giampalmo, and H. Arend, "The Columbus Development Programme," *ESA Bulletin* 56 (1988), 10–18, for more details on the elements. Also J. Collett, "The Columbus Free-Flying Laboratory—A Stepping Stone Towards European Autonomy," *ESA Bulletin* 64 (1990), 29–32.

38. McCurdy, *The Space Station Decision,* 202.

39. Interview Krige with Finarelli, April 20, 2010.

40. Reinke, *The History of German Space Policy,* 275.

41. "Resolution on Participation in the Space Station Programme," *ESA Bulletin* 53 (1988), 29–30.

42. Loosch, "The International Space Station," 58.

43. Kevin Madders, *A New Force at a New Frontier* (Cambridge: Cambridge University Press, 1997), 462.

44. Interview Krige with Finarelli, April 20, 2010.

45. For the station cost data, see John J. Madison and Howard E. McCurdy, "Spending without Results: Lessons from the Space Station Program," *Space Policy* 15 (1999), 213–221.

46. For an indication of Shuttle costs, see John Krige, "The Commercial Challenge to Arianespace: The TCI Affair," *Space Policy* 15 (1999), 87–94.

47. For the State of the Union address, see http://history.nasa.gov/reagan84.htm.

48. Interview Krige with Finarelli, April 20, 2010.

49. Logsdon, *Together in Orbit,* 41. This whole section on Canada is based on Logsdon's account.

50. For ESA and Canada, see Lydia Dotto, *Canada and the European Space Agency. Three Decades of Cooperation* (Noordwijk: ESA HSR-25, 2002).

51. This brief summary is thanks to Logsdon, *Together in Orbit,* 36–40.

52. Steven Berner, *Japan's Space program. A Fork in the Road?* (Santa Monica: RAND Corporation, 2005), available at http://www.rand.org/pubs/technical_reports/2005/RAND_TR184.pdf.

53. Joan Johnson-Freese, *Changing Patterns of International Collaboration in Space* (Malabar: Orbit, 1990), 89.

54. Reinke, *The History of German Space Policy,* 277. See also Sadeh, "Technical, Organizational and Political Dynamics," 173–175.

55. Sadeh, "Technical, Organizational and Political Dynamics," 173–175; W. Henry Lambright and Agnes Gereben Schaeffer, "The Political Context of Technology Transfer. NASA and the International Space Station," *Comparative Technology Transfer and Society* 2:1 (2004), 1–30, at 7–8.

56. Interview Krige with Finarelli, April 20, 2010.

57. Ibid.

58. Madders, *A New Force at a New Frontier,* 463. This clause in Article 15 of the IGA engages not only NASA but the US government.

59. Madison and McCurdy, "Spending without Results," 213–221, at 213.

60. Quoted by Sadeh, "Technical, Organizational and Political Dynamics," 176.

61. Quoted in ibid.

62. Launius, *Space Stations,* 150–152.

63. Chapter 8, this volume; Sadeh, "Technical, Organizational and Political Dynamics," 185–186. John M. Logsdon and James R. Millar, "US-Russian Cooperation in Human Spaceflight: Assessing its Impacts," *Space Policy* 17 (2001), 171–178, explore the extent to which these "non-pragrammatic" goals might have been achieved.

64. Launius, *Space Stations*, 178, 179.

65. W. Henry Lambright, "Leadership and Large-Scale Technology: The Case of the International Space Station," *Space Policy* 21 (2005), 195–203, at 198.

66. Michael Riordan, "The Demise of the Superconducting Super Collider," *Physics in Perspective* 2 (2000), 411–425. For the rivalry with Europe, see John Krige, "Distrust and Discovery. The Case of the Heavy Bosons at CERN," *Isis* 92:3 (2001), 517–540.

67. Lambright, "Leadership and Large-Scale Technology," 198.

68. Launius, *Space Stations*, 133.

69. Ibid., 186.

70. Lambright, "Leadership and Large-Scale Technology," 199.

71. Interview John Krige with Lynn Cline, March 30, 2009, NHRC.

72. The section that follows relies extensively on the excellent summary by Reinke, *The History of German Space Policy*, 417–432, and necessarily has more information from German sources than any other. Germany was of course the major European contributor to the ISS.

73. R. D. Andresen and R. Domesle, "The Euromir Missions," *ESA Bulletin* 88 (1996), 6–12.

74. Interview, Krige with Cline, March 30, 2009. Cline pointed out that matters were not as smooth when it came to "figure out how to divvy up the operations and utilization of the Station," since here Russia wanted to be treated as an equal with the United States, and not just be seen as one of America's partners, like the previous participants.

75. Reinke, *The History of German Space Policy*, 418.

76. For more on Columbus and ESA's contributions to the ISS, see J. Feustel-Büechl, "The International Space Station is Real!" *ESA Bulletin* 107 (August 2001), 11–20; A. Thirkettle, B. Patti, P. Mitschdoerfer, R. Klezdik, E. Gargioli, and D. Brondolo, *ESA Bulletin* 109 (February 2002), 27–33; Bernardo Patti, Robert Chesson, Martin Zell, and Alan Thirkettle, "Columbus: Ready for the International Space Station," *ESA Bulletin* 121 (February 2005), 47–51; Martin Zell and Jon Weems, "ESA's 'Real Estate' in Space. Columbus in Orbit," *ESA Bulletin* 136 (November 2008), 33–43.

77. P. Amadieu and J. Y. Heloret, "The Automated Transfer Vehicle," *ESA Bulletin* 96 (November 1998), 14–20.

78. Agreement among the Government of Canada, the Governments of the European Space Agency, the Government of Japan, the Government of the Russian Federation and the Government of the United States of America Concerning Cooperation on the Civil International Space Station, January 29, 1998.

79. Sadeh, "Technical, Organizational and Political Dynamics," 182, 183.

80. Reinke, *The History of German Space Policy*, 428.

81. Ibid., 422.

82. In the case of Europe the critical path items are the data-management system in the Russian module, Nodes 2 and 3, the Cupola, and the ATV, while in the case of Japan they are the centrifuge module (subsequently deleted) and the HTV, private communication, Ian Pryke, October 30, 2008.

83. Logsdon and Millar, "US-Russian Cooperation in Human Spaceflight: Assessing its Impacts."

14 The Impact of the International Traffic in Arms Regulations

1. John Hall (Director) and Paula Geisz of NASA's Export Control and Interagency Liaison Division have provided invaluable help with this chapter. Any errors or inaccuracies are the sole responsibility of the author.

2. For a clear statement of the following, see Margaret G. Finarelli and Joseph K. Alexander (Rapporteurs), *Space Science and the International Traffic in Arms Regulations. Summary of a Workshop* (Washington, DC: The National Academies Press, 2008), 4–5, available at http://www.nap.edu/catalog/12093.html.

3. Alvin S. Bass, *Dissemination of Technical Information Abroad*, June 18, 1970, 7–8, attached to Letter Richard McCurdy, NASA, to Peter G. Petersen, assistant to the president for International Economic Affairs, WHCF [White House Central Files], Subject Files, Folder FG164, NASA 1/1/71- (2 of 2), Nixon Presidential Materials Project.

4. Finarelli and Alexander, *Space Science and the International Traffic*, 4.

5. Ibid., 5.

6. Bass, *Dissemination of Technical Information Abroad*, 9. An annotation on the text by McCurdy remarked that the Office of Munitions Control construed the exemption as regards prime contractors "more strictly than the language appears to intend."

7. Memo Flanigan to David, January 3, 1972, WHCF [White House Central Files], Subject Files, Folder FG164, NASA 1/1/71- (2 of 2), Nixon Presidential Materials Project.

8. Memo Flanigan to David and Rice, December 13, 1971, WHCF [White House Central Files], Subject Files, Folder FG164, NASA 1/1/71- (2 of 2), Nixon Presidential Materials Project.

9. More precisely, all the detailed design, development, and manufacturing production data for the space station is covered by the ITAR/USML. The EAR/CCL covers all hardware specifically designed or developed for the ISS, and the technical data for its operations and use.

10. This section relies on Ryan Zelnio, "A Short History of Export Control Policy," *The Space Review* 9 January, 2006, available at http://www.thespacereview.com/article/528/1.

11. *Public Law 105-261, 17 October, 1998 Strom-Thurmond National Defense Authorization Act for Fiscal Year 1999*, available at http://www.dod.mil/dodgc/olc/docs/1999NDAA.pdf.

12. Joseph Cirincione, "Cox Report and the Threat from China," presentation to the CATO Institute, June 7, 1999, available at http://www.carnegieendowment.org/publications/index.cfm.

13. *United States Congress. House Report 105–851. Report of the Select Committee on U.S. National Security and Military/Commercial Concerns with the People's Republic of China. Submitted by Mr. Cox of California, Chairman*, available at http://www.access.gpo.gov/congress/house/hr105851/index.html.

14. The account that follows is based on *House Report 105–851* (the Cox Report), especially Chapters 5–8. See also Lewis R. Franklin, "A Critique of the Cox Report Allegations of PRC Acquisition of Sensitive U.S. Missile and Space Technology," in Michael M. May (ed., with Alastair Iain Johnston, W. K. H. Panofsky, Marco Di Capua, and Lewis R. Franklin), *The Cox Committee Report: An Assessment* (Stanford University: Center for International Security and Cooperation (CISAC), December 1999), 81–99, at section 3.2.1–3, available at http://iis-db.stanford.edu/pubs/10331/cox.pdf. See also Joan Johnson-Freese, "Alice in Licenseland: US Satellite Export Controls Since 1990," *Space Policy* 16 (2000), 195–204.

15. *Public Law 105-261. Title XV. Matters Relating to Arms Control, Export Controls, and Counterproliferation. Subtitle B. Satellite Export Controls.*

16. *Public Law 105-261. Sections 1511(5), 1513(a)*; emphasis in the original. For the situation in August 2012, see Departments of Defense and State, *Final Report to Congress*, Section 1248 of the National Defense Authorization Act for Fiscal Year 2010 (Public Law 111–84), *Risk Assessment of United States Export Control Policy*.

17. Jeff Gerth, "2 Companies Pay Penalties for Improving China Rockets," *New York Times*, March 6, 2003, available at http://nytimes/com/2003/03/06/world/2-companies-pay-penalties-for-improving-china-rockets.html.

18. Editorial, "Nuclear Pickpocket," *Washington Post*, May 26, 1999, available at http://www.taiwandc.org/wp-9911-htm.

19. Quoted by Cirincione, "Cox Report and the Threat from China."
20. *House Report 105–851 of the Select Committee on U.S. National Security and Military/ Commercial Concerns with the People's Republic of China, Submitted by Mr. Cox of California, Chairman* (the Cox Report), available at http://www.access.gpo.gov/congress/house/hr105851/index.html.
21. The quotations in this paragraph and the next are all from the overview that accompanied all three volumes of *House Report 105–851*, the Cox Report.
22. *House Report 105–851*, Vol. II, 170.
23. Ibid., 172.
24. Ibid.
25. Cirincione, "Cox Report and the Threat from China," at 1.
26. *The Cox Committee Report: An Assessment* (Stanford University: Center for International Security and Cooperation (CISAC), at 38
27. Nicholas Rostow, "The 'Panofsky' Critique and the Cox Committee Report: 50 Factual Errors in the Four Essays," available at www.people.fas.harvard.edu/~johnston/rostow.pdf, A reply by Johnston to the critique of his chapter is also at this URL.
28. Anon., *Balancing Scientific Openness and National Security Controls at the Nuclear Weapons Laboratories* (Washington, DC: National Academies Press, 1999),
29. Cirincione, "Cox Report and the Threat from China," 5.
30. Anon., *Balancing Scientific Openness and National Security Controls at the Nuclear Weapons Laboratories*, at 11.
31. Public Law 106–391, H.R.1654, National Aeronautics and Space Administration Authorization Act of 2000, available online at http://thomas.loc.gov/cgi-bin/query/F?c106:1:./temp/~c106ewB7Za:e942, THOMAS, Library of Congress (accessed on September 26, 2010).
32. Interview John Krige with Robert Mitchell, June 19, 2009, NHRC.
33. Interview John Krige with Charles Elachi, June 9, 2009, NHRC.
34. Interview John Krige with David Southwood, ESA HQ, Paris, July 16, 2007, NHRC.
35. For this clause, see http://www.fas.org/spp/starwars/offdocs/itar/p.126.htm.
36. US Department of Commerce, Bureau of Industry and Security, Update 2012 Conference, "Remarks of Eric L. Hirschhorn," Under Secretary for Industry and Security, July 17, 2012, at http://www.bis.doc/news/2012/hirschhorn_update_2012.html.

15 Conclusion

1. Asif A. Siddiqi, "Competing Technologies, National(ist) Narratives, and Universal Claims. Toward a Global History of Space Exploration," *Technology and Culture* 51:2 (April 2010), 425–443, at 425, 439.
2. The phrase "contact zone" is borrowed from Marie Louise Pratt, *Imperial Eyes. Travel Writing and Transculturation* (London: Routledge, 1992). Pratt defines a "contact zone" as invoking "the spatial and temporal copresence of subjects previously separated by geographic and historical disjunctures, and whose trajectories now intersect." It treats the relationship between them "not in terms of separateness or apartheid, but in terms of copresence, interaction, interlocking understandings and practices, often within radically asymmetrical relations of power," at 7.
3. On the importance of studying knowledge flows, see Siddiqi, "Competing Technologies."
4. Fareed Zakaria, *The Post-American World* (New York: W.W. Norton and Co, 2008), 1.

Index

Printed in the United States of America